Cosmochemistry and the Origin of Life

NATO ADVANCED STUDY INSTITUTES SERIES

Proceedings of the Advanced Study Institute Programme, which aims at the dissemination of advanced knowledge and the formation of contacts among scientists from different countries

The series is published by an international board of publishers in conjunction with NATO Scientific Affairs Division

A	Life Sciences	Plenum Publishing Corporation
B	Physics	London and New York
C	Mathematical and Physical Sciences	D. Reidel Publishing Company Dordrecht, Boston and London
D	Behavioural and Social Sciences	Martinus Nijhoff Publishers
E	Engineering and Materials Sciences	The Hague, London and Boston
F	Computer and Systems Sciences	Springer Verlag
G	Ecological Sciences	Heidelberg

Series C – Mathematical and Physical Sciences

Volume 101 – Cosmochemistry and the Origin of Life

Cosmochemistry and the Origin of Life

Proceedings of the NATO Advanced Study Institute held at Maratea, Italy, June 1-12, 1981

edited by

CYRIL PONNAMPERUMA
Department of Chemistry, University of Maryland, College Park, U.S.A.

D. Reidel Publishing Company
Dordrecht : Holland / Boston : U.S.A. / London : England

Published in cooperation with NATO Scientific Affairs Division

Library of Congress Cataloging in Publication Data

CIP

NATO Advanced Study Institute (1981 : Maratea, Italy)
Cosmochemistry and the origin of life.

(NATO advanced study institutes series. Series C, Mathematical and physical sciences ; v. 101)
"Published in cooperation with NATO Scientific Affairs Division."
Includes bibliographical references and index.
1. Cosmochemistry–Congresses. 2. Life–Origin–Congresses.
I. Ponnamperuma, Cyril, 1923- . II. North Atlantic Treaty Organization. Scientific Affairs Division. III. Title. IV. Series.
QB450.N37 1981 577 82-24099
ISBN 90-277-1544-0

Published by D. Reidel Publishing Company
P.O. Box 17, 3300 AA Dordrecht, Holland

Sold and distributed in the U.S.A. and Canada
by Kluwer Boston Inc.,
190 Old Derby Street, Hingham, MA 02043, U.S.A.

In all other countries, sold and distributed
by Kluwer Academic Publishers Group,
P.O. Box 322, 3300 AH Dordrecht, Holland

D. Reidel Publishing Company is a member of the Kluwer Group

Printed in The Netherlands

CONTENTS

PREFACE

For the first time in human history, developments in many branches of science provide us with an opportunity of formulating a comprehensive picture of the universe from its beginning to the present time. It is an awesome reflection that the carbon in our bodies is the very carbon which was generated during the birth of a star. There is a perceptible continuum through the billions of years which can be revealed by the study of chemistry.

Studies in nucleosynthesis have related the origin of the elements to the life history of the stars. The chemical elements we find on earth, Hydrogen, Carbon, Oxygen, and Nitrogen, were created in astronomical processes that took place in the past, and these elements are not spread throughout space in the form of stars and galaxies.

Radioastronomers have discovered a vast array of organic molecules in the interstellar medium which have a bearing on prebiological chemical processes. Many of the molecules found so far contain the four elements, C, N, O, H. Except for the chemically unreactive He, these four elements are the most abundant in the galaxy. The origin of polyatomic interstellar molecules is an unresolved problem. While we can explain the formation of some diatomic molecules as due to two atom collisions, it is much more difficult to form polyatomic molecules by collisions between diatomic molecules and atoms. There may be other production mechanisms at work such as reactions taking place on the surface of interstellar dust grains.

Organic molecules which may be considered precursors to life have also been found in carbonaceous chondrites. The discovery and the dating of the earth's oldest sediments at around 3.8×10^9 years has helped us to get a glimpse of the face of our planet at the very dawn of terrestrial time.

All these discoveries were examined in the light of the single theme--Cosmochemistry. Within the pages of this volume are

C. Ponnamperuma (ed.), Cosmochemistry and the Origin of Life, vii–viii.

assembled the authoritative statements of the principal lecturers of the NATO Workshop which was held in Maratea in June, 1981. There was ample opportunity for an exchange of ideas and discussion on this subject. The papers published thus reflect some of the thinking generated during a very enjoyable and intellectually stimulating period in Italy. We are grateful to the speakers, the participants and to NATO for having made such an intellectual workshop possible.

Cyril Ponnamperuma

College Park, Maryland
June 1, 1982

COSMOCHEMISTRY AND THE ORIGIN OF LIFE

Cyril Ponnamperuma

Laboratory of Chemical Evolution, Department of Chemistry, University of Maryland, College Park, MD 20742, USA

HOW DID LIFE BEGIN?

One of the first to speculate on the conditions necessary for the origin of life on earth was Erasmus Darwin, the grandfather of Charles Darwin. In his Temple of Nature (1) he had written, "All vegetables and animals now existing were originally derived from the smallest microscopic ones formed by spontaneous vitality." Perhaps this idea had influenced Charles Darwin in his own thinking. Several years later, he wrote to his friend Hooker (2) about some "warm little pond" with all sorts of ammonia and phosphoric salts, light, heat, electricity, etc. present, in which he postulated that a protein compound was chemically formed, ready to undergo still more complex changes.

About this time the physicist John Tyndall (3) argued that every portion of a living organism can be reduced to inorganic matter. In his essay on vitality in 1866, he suggested that one could conceive of the reverse change, from the inorganic to the organic, and that the special arrangement of elements in living bodies led to the phenomenon of life. In 1868, Thomas Huxley (4) delivered a lecture in Edinburgh in which he pointed out that protoplasm was substantially the same over the whole range of living things. To him, the existence of life depended on certain molecules such as carbonic acid, water, and nitrogenous compounds. These compounds are lifeless, but when brought together, give rise to protoplasm.

To the Russian biochemist, A. I. Oparin (5), more than to anyone else today, we owe our present ideas on the scientific approach to the question of the origin of life. In clear and scientifically

C. Ponnamperuma (ed.), Cosmochemistry and the Origin of Life, 1–34.

defensible terms, he pointed out that there was no fundamental difference between living organisms and brute matter. The complex combination of manifestations and properties so characteristic of life must have arisen in the process of the evolution of matter. In 1928, the British biologist Haldane (6) expressed his own ideas on the origin of life. He attributed the synthesis of organic compounds to the action of ultraviolet light on the earth's primitive atmosphere. He suggested that the organic compounds accumulated till the primitive oceans had the consistency of a primordial soup. Twenty years after the appearance of Haldane's paper, in the Rationalist Annual, Bernal (7) theorized before the British Physical Society in a lecture entitled, "The Physical Bases of Life:"

> Condensations and dehydrogenations are bound to lead to increasingly unsaturated substances, and ultimately to simple and possibly even to condensed ring structures, almost certainly contain nitrogen, such as the pyrimidines and purines. The appearance of such molecules made possible still further synthesis. The primary difficulty, however, of imagining processes going this far was the extreme dilution of the system, if it is to take place in the free ocean. The concentration of products is an absolute necessity for any further evolution.

THE RAW MATERIAL

The raw material from which the building blocks of life have evolved consists of the chemical elements of the periodic table. Examination of the crust of the earth, the oceans, and the atmosphere provides us with the information about the abundance of these elements on the earth. Data on the elemental composition of matter beyond the earth come from several sources. The spectroscopic analysis of light from stars reveals the nature of the elements in them. The development of the science of radio astronomy has provided us with the microwave technique of detecting various elements and excited species in intergalactic space. Cosmic ray particles can supply us with samples of extraterrestrial matter. Meteorites and lunar samples have given us valuable knowledge of the composition of our solar system. A reasonable and consistent picture of the abundance of the elements in the universe can thus be obtained.

Hydrogen is by far the most abundant element. It makes up 93% of the total number of atoms in the universe, and 76% of the universe by weight. When atomic abundances are plotted against atomic weights, there is in general an inverse correlation except for a sharp interruption in the case of the elements of the iron group. The relatively high abundance of these elements may be related to the fact that nuclear reactions involving them must absorb energy rather than release it. Beyond helium, all elements combined add up only to a little more than 1% by weight of the entire universe. The theories of Gamow and Hoyle have attempted to outline the processes by which the elements were built up in a stepwise manner. Neutron capture, hydrogen fusion, and helium burning have been invoked to explain the process of nucleosynthesis. There is a parallel between the gradual formation of the elements of the periodic table and the evolutionary development of the stars. The atomic abundance of some of the elements found in the sun are represented in the accompanying Table I. The composition of the sun may be considered to represent the average composition of the solar system. Hydrogen, helium, carbon, nitrogen, and oxygen are the most abundant elements. With the exception of helium, these are the very elements which constitute 99% of living matter.

TABLE I: COMPOSITION OF THE SUN

Hydrogen	87.0 %
Helium	12.9
Oxygen	0.025
Nitrogen	0.02
Carbon	0.01
Magnesium	0.003
Silicon	0.002
Iron	0.001
Sulphur	0.001
Others	0.038

THE NATURE OF THE PRIMITIVE ATMOSPHERE

A true understanding of the nature of the earth's primitive atmosphere is a logical starting point for any discussion of the problem of the chemical origin of life. While it is difficult to answer by direct observation the questions that arise, evidence available from a number of sources indicate that a reducing atmosphere gradually gave way to the oxidized atmosphere of today (8). Information gleaned from astronomy, astrophysics, chemistry, geology, meteoritic studies, and biochemistry can be used to elucidate this problem. The present rarity of the noble gases in the earth's atmosphere in comparison to their distribution in the universe indicates that the primitive atmosphere of the earth was lost, and that the atmosphere that is generally described as primitive was secondary in origin. This atmosphere must have resulted from the outgassing of the interior of the earth during planetary accretion. The secondary atmosphere must have been very similar to the first, and it is this atmosphere which would be considered in our discussion of the primordial atmosphere of the earth. Russell's discovery that hydrogen is the single most abundant element in the universe indicates that the universe as a whole is reducing in nature. A main problem concerning the nature of the terrestrial atmosphere is, therefore, how to explain why an early atmosphere with such a large amount of hydrogen could have given rise to an atmosphere containing free oxygen.

Most of the smaller planets were unable to retain their early atmospheres, while the larger planets such as Jupiter and Saturn contained large amounts of hydrogen, methane, ammonia and water. Analysis of meteorites which have been dated at 4.5 billion years are true examples of the primordial matter from which our planetary system was formed. In these meteorites, the metals are generally found in their reduced form. The study of chemical equilibrium sheds some light on the nature of these reactions which should have taken place in a primordial reducing atmosphere. The thermodynamic properties of carbon, carbon dioxide, methane, ammonia, and water are very well known (see Table II).

TABLE II: EQUILIBRIUM CONSTANTS

	$K_{25°C}$
$C + 2H_2 \rightarrow CH_4$	$8 \cdot 10^{8}$
$N_2 + 3H_2 \rightarrow 2NH_3$	$7 \cdot 10^{5}$
$H_2 + \frac{1}{2}O_2 \rightarrow H_2O$	$4 \cdot 10^{41}$
$CO_2 + 4H_2 \rightarrow CH_4 + 2H_2O$	$7 \cdot 10^{21}$
$S + H_2 \rightarrow H_2S$	$6 \cdot 10^{15}$

Composition of equilibrium mixtures of these materials has been satisfactorily determined. The equilibrium constants for the formation of methane from carbon and hydrogen, or water from oxygen and hydrogen, and ammonia from nitrogen are all of considerable magnitude. It is reasonable to suppose, then, that in a large excess of hydrogen, carbon, nitrogen and oxygen existed in their reduced form.

THE TRANSITION FROM A PRIMITIVE ATMOSPHERE TO ONE CONTAINING OXYGEN

Free oxygen is unique in our planetary system. The oxygen in the present atmosphere appears to have arisen from two sources, from the photodissociation of water in the upper atmosphere by short wave length ultraviolet light, and from plant photosynthesis. Photosynthesis probably evolved when the ozone layer developed in the upper atmosphere and shielded the earth from the ultraviolet light emitted by the sun, thus preventing the further photochemical synthesis of organic compounds in the primitive environment. Heterotrophs which thrived on photochemically produced materials available around them were replaced by autotrophs, which were able to photosynthesize their own food when the supply of biogenic organic matter was depleted. This change may be diagramatically represented by an hourglass. Only those organisms that could incorporate molecules such as the porphyrins and make use of longer wavelengths were able to pass through the bottleneck (Fig. 1). There was, therefore, a wholesale massacre at this junction. The organisms that survived and developed further were able to evolve into the wide variety we see of life on earth today.

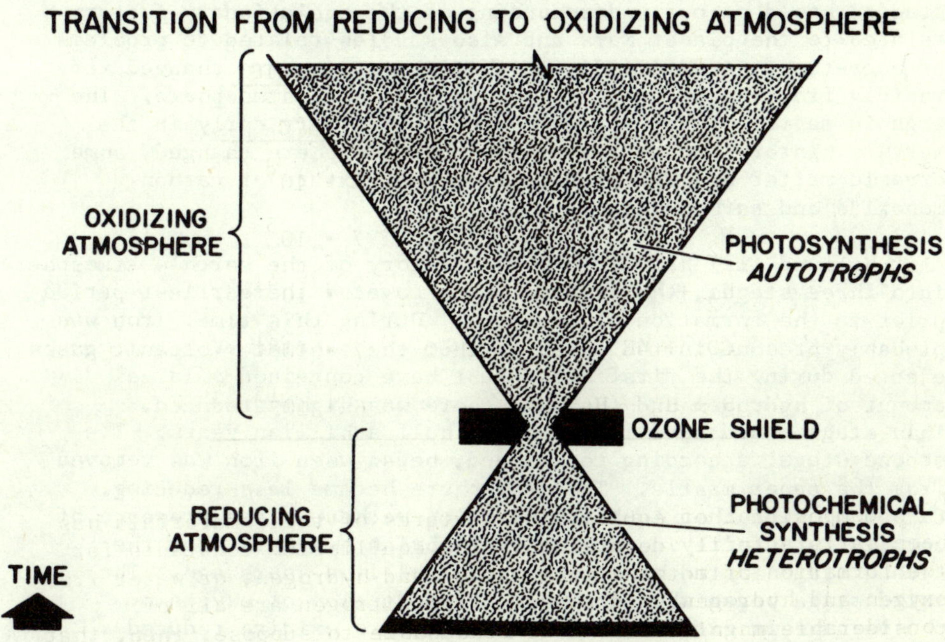

Figure 1. Transition from reducing to oxidizing atmosphere.

In biochemical reactions taking place in contemporary living organisms, carbon has to be reduced before it can be incorporated into amino acids, purines, pyrimidines, etc. This is a universal fact of the biosphere. It is a credit to the great insight of Oparin (9) that even before the reducing nature of the universe was discovered, he recognized that such conditions must have prevailed for at least a billion years.

There are others who disagree with this view. Rubey (10) had suggested that the methane and ammonia atmosphere may not have lasted for more than 10^5 to 10^8 years. Since carbon dioxide is a gas which is primarily evolved from volcanoes today, Rubey maintained that if the early earth's atmosphere was formed by an outgassing from the earth's interior, carbon dioxide must have been the form in which carbon appeared. Using similar arguments, Abelson (11) has postulated that the earth's primitive atmosphere consisted very much of carbon monoxide, carbon dioxide, nitrogen, and some hydrogen. In addition, he has demonstrated that ammonia must have been dissociated by ultraviolet light, and could not have lasted for more than 20,000 years. A further argument suggested by Abelson is that the earliest rocks should have contained an unusual amount of organic matter. More recent studies based on our understanding of the evolutionary factors related to the planet Mars and also studies related to problems of accretion, seem to indicate that the atmosphere changed very rapidly from the reducing to a carbon dioxide atmosphere. The organic matter, then, may have been formed very early in the earth's history and gradually, as the atmosphere changed, some organic matter was synthesized from the mixture of carbon monoxide and carbon dioxide.

Holland (12) had divided the history of the earth's atmosphere into three stages, the first of which covered the earliest period prior to the formation of the core. During this time, iron was probably present in the upper part of the mantle. Volcanic gases ejected during the first stage must have contained a large amount of hydrogen and the atmosphere was highly reduced. This stage could have lasted about half a billion years. The second stage, according to Holland, began when iron was removed from the upper mantle. The atmosphere became less reducing. Carbon dioxide then took the place of methane. The presence of detrityl uraninite in the mid-Precambrian suggests that the second stage lasted at least until 2 billion years ago. The third stage began when the rate of production of oxygen by photosynthesis exceeded the rate needed to oxidize reduced volcanic gases. Berkner and Marshall have developed convincing arguments to show that oxygen accumulated as a result of photosynthesis (13). The authors maintain that the sudden proliferation of life in the Cambrian era was due to the attainment of the Pasteur point, namely concentration of about 1% of the present oxygen pressure

As we saw earlier, Oparin had postulated the need of reduced carbon for organic synthesis. Most organisms have a mechanism to reduce nitrogen before it is first incorporated into living molecules. They also, without exception, have a process devised to eliminate nitrogen in the form of ammonia and urea. This is perhaps a result of the evolution of biochemical pathways in an early anoxygenic atmosphere. Until the cell had achieved a highly organized and protected state, free atmospheric oxygen could not have been allowed, since its presence would have quickly oxidized the early precursors of life and absolutely prevented their evolution or even their existence. The deleterious effect of oxygen on contemporary cell nuclei is also indicative that anoxygenic conditions may have prevailed during the early evolution of the cell. The existence of obligate anaerobes and the universality of anaerobic glycolytic pathways suggests that life arose in non-oxidized environments. As the environment changed from a reducing atmosphere to one containing free oxygen, there may have been a selective advantage in the development of mechanisms for the removal of peroxides as a defense against these lethal components.

ENERGY SOURCES

The energies available for the synthesis of organic compounds under primitive earth conditions are ultraviolet light from the sun, electrical discharges, ionizing radiation, and heat (14). It is evident that sunlight is the principle source of energy. Solar radiation is emitted in all regions of the electromagnetic spectrum. A great deal of this radiation, especially in the ultraviolet region, is shielded today by the ozone layer in the upper regions of the atmosphere. Since reaching its main sequence, the sun has been very stable. The temperature of the sun's surface four and one half billion years ago was almost as great as it is today. With this information, the solar flux at different wavelengths has been calculated. These figures clearly demonstrate the dominant role that ultraviolet light must have played among primitive energy sources. The use of ultraviolet light should be considered as far more important than any other sources of energy that have been used in simulation experiments in the laboratory. However, on account of the intrinsic difficulties that have been involved in using short wavelength ultraviolet light, not many experiments have been performed with this energy source. The absorption of methane extends to 1450 angstroms, water to 1650, and ammonia to 2200. The quantum yield for the photolysis of ammonia is approximately 0.5, for water, 0.3, and for methane, about 0.5. The dissociation of these molecules takes place within the shorter wavelength region. However, the dissociation products may absorb at higher wavelengths.

Next in importance as a source of energy are electrical discharges such as lightning and corona discharges from pointed objects. These occur close to the Earth's surface, and therefore could have effectively transferred the reaction products to the primitive oceans. Since lightning can be easily simulated in the laboratory, many experiments have been performed using this form of energy.

The principle radioactive sources of ionizing radiation on the earth are potassium-40, uranium-38, uranium-235, and thorium-232. Potassium-40 seems to be quantitatively more important than the other three sources at present. Presumably, therefore, it must have been even more important in the past. The energy of potassium-40 is in the form of penetrating beta and gamma rays. In contrast, almost 90% of the energy from uranium-238, thorium-232 is evident as alpha particles, which may not be penetrating enough to have a significant effect. Calculations show that the decay of potassium-40 in the earth's crust today gives rise to 3×10^{19} calories per year. Two billion years ago this would have been 12×10^{19} calories. By comparison, this is only 1/30 the amount of energy available from short wavelength ultraviolet light.

The distribution of energy is a matter of importance. Most radiation from radioactive sources is absorbed in solid matter. Since the earth's crust is about 30 km thick, this type of radiation could not have made a contribution to the synthesis of organic matter in the primitive oceans. However, local regions of high activity might have existed in the past. There may have been micro-environments where the catalytic action of the metals present favored condensation reactions of the first molecules synthesized by ionizing radiation.

Heat from volcanoes was another form of energy which may have been effective. It is reasonable to expect that volcanic activity was more prevalent on the primitive Earth than today. The figure in Table III has been calculated on the assumption of one cubic km of lava emission per year at 1000°C. Besides, heat would also have been available from hot springs around boiling mud pots.

Chemosynthesis by meteorite impact on planetary atmospheres has been suggested as a possible pathway for primordial synthesis of organic matter (15). The reaction is probably a result of the intense heat generated momentarily in the wake of the shock wave following the impact.

EXPERIMENTAL WORK IN THE LABORATORY

The student of chemical evolution has tried to recreate

TABLE III: ENERGY SOURCES

Radiation from the sun (all wavelengths)	260,000 cal cm^2 yr
Ultraviolet light λ < 2500 Å	570 "
λ < 2000 Å	85 "
λ < 1500 Å	3.5 "
Electric discharges	4 "
Cosmic rays	0.0015 "
Radioactivity (to 1.0 km depth)	2.8 "
Heat from volcanoes	0.13 "
Meteorite impact	0.1 "

Darwin's "warm little pond" in the laboratory. In other words, most of these different energy sources that have been considered to be possible have been used in the laboratory to see whether when they act upon the earth's primitive atmosphere, organic molecules are formed.

AMINO ACIDS

The greatest amount of work in the field of prebiological chemistry concerns the origin of amino acids. This may be due to the fact that amino acids are readily formed. A second reason may be that the emphasis on nucleic acid chemistry is only of comparatively recent origin. The protein molecule formerly occupied a pre-eminent position in biochemistry, and it is not surprising that efforts were made to look for its constituents. The development of chromatographic techniques facilitated the direction of minute traces of amino acids in complex mixtures.

Amino acids have been synthesized under simulated primitive earth conditions by the action of ultraviolet light, electric discharges, heat, and ionizing radiation. In some cases, the starting materials were reactive intermediates. It is claimed that 14 of the 20 amino acids found in protein have been synthesized by abiogenic means. In one of the first experiments done simulating the synthesis of organic matter on the primitive earth, Miller exposed a mixture of methane, ammonia, water, and hydrogen to an electric discharge from tesla coils for about a week. Several organic compounds were formed. Among these were the amino acids, glycine, alanine, β-alanine, aspartic acid, and glutamic acid. Of the initial carbon introduced as methane, 2.1% had been converted into glycine.

Little is known about mechanisms of these syntheses. Apart from Miller's experiments, in which the possibility of a Strecker synthesis was suggested, no attempt has been made to find out how amino acids have arisen. Miller suggested two alternative proposals. First, the aldehydes and hydrogen cyanide were synthesized in the gas phase by the spark. These aldehydes and the hydrogen cyanide react in the aqueous phase of the system to give amino and hydroxy-nitriles, which were, in turn, hydrolyzed to amino and hydroxy acids. The second suggestion made was that the amino and hydroxy acids were synthesized according to the first hypothesis, since the rate of production of aldehydes and hydrogen cyanide by the spark, and the rate of hydrolysis of the amino-nitriles were sufficient to account for the total yield of amino acids. Miller's experiments on the mechanism of electric discharge synthesis of amino acids indicate that a special set of conditions is not required to obtain amino acids. A process or combination of processes that yielded both aldehyde and hydrogen cyanide could have contributed to the amount of α-amino acids in the oceans of the primitive earth (16) (Fig. 2).

$$RCHO + NH_3 + HCN \rightleftarrows RCH(NH_2)CN + H_2O$$

$$RCH(NH_2)CN + 2H_2O \rightleftarrows RCH(NH_2)COOH + NH_3$$

$$RCHO + HCN \rightleftarrows RCH(OH)CN$$

$$RCH(OH)CN + 2H_2O \rightleftarrows RCH(OH)COOH + NH_3$$

Figure 2. Strecker Synthesis

The experiments with methane and ammonia in the absence of water to give rise to nitriles has shown that an alternative mechanism may also be valid (17). Thus, if α-aminonitriles are preferentially synthesized by the reaction between methane and ammonia, and these nitriles react with water, hydrolysis can take place to give rise to amino acids. In this case, the amino acids are being formed in the absence of aldehydes.

PURINES AND PYRIMIDINES

The first synthesis of purines under simulated primitive-earth conditions was carried out by Oró, who demonstrated that adenine could by synthesized from a concentrated solution of ammonium cyanide (18). The overall reaction can be depicted as 5 molecules of hydrogen cyanide in the presence of ammonia giving rise to adenine. This synthesis has been confirmed by Lowe et al. Oró's detailed study has shown that 4-aminoimidazole-5-carboximide and formamidine were the probable intermediates in this reaction (Fig. 3). Subsequently, Oró was able to synthesize guanine and xanthine by heating aqueous solutions of amino-imidazole carboximide between 100° and 140°. Yields of 1.5% were obtained for both purines. It is possible that this is a way in which the purines were synthesized on the primitive earth, but the concentrations used by Oró were drastically high for prebiotic conditions. If genuine prebiotic conditions can be satisfied, such as the use of lower concentrations, these reactions will be of great value in understanding the origin of purines on the prebiotic earth.

MECHANISM FOR FORMATION OF ADENINE FROM HCN

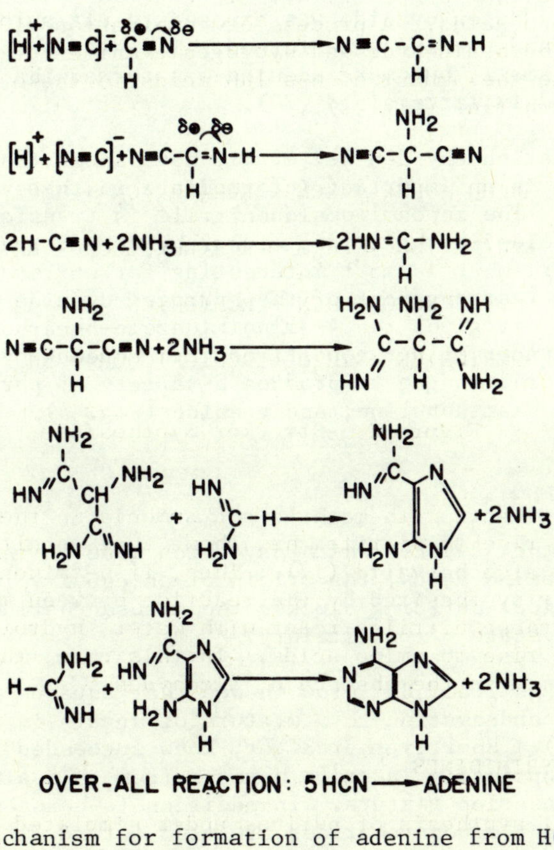

Figure 3. Mechanism for formation of adenine from HCN.

Adenine was also synthesized by Ponnamperuma et al. by the electron irradiation of methane, ammonia, and water (19). Since hydrogen cyanide has been detected in these reactions, the possible pathway may be the same, namely, the base catalyzed reaction of hydrogen cyanide. In the electron irradiation of methane, ammonia, and water, the yield was about 0.01% of the starting methane. This is considerable and is realistic under primitive-earth conditions. The interesting observation was made that the production of adenine was enhanced by the absence of hydrogen. This is not surprising, since methane carbon must be oxidized in order to appear finally in the purines. In any event, the high concentration of organic matter on the prebiotic earth probably arose when most of the hydrogen had escaped from this atmosphere. Although many efforts have been made, purines and pyrimidines have not been definitely identified in the end products of electric discharge experiments. Since a large amount of hydrogen cyanide is formed, it is difficult to understand why the purines could be absent in such experiments.

In further experiments, it was shown that when a dilute solution of hydrogen cyanide was exposed to ultraviolet light, both adenine and guanine were formed. The yields of adenine and guanine were small, but more adenine was formed than guanine. Urea was found in large yield (20).

Investigations by Sanchez et al. have shown that aminomalononitrile is an important intermediate in the synthesis of adenine (21). The monoaminomalononitrile is transformed into 4-aminoimidazole-5-carboximide, either by direct reaction with formamidine or, what is most interesting for our studies, the photochemical rearrangement of the hydrogen cyanide tetramer. By hydrolysis, it gives the 4-aminoimidazole-5-carboximide, which can condense under milder conditions with aqueous cyanide, formamidine, or cyanogen to produce a variety of purines, adenine, hypoxanthine, diaminopurine, and guanine (Fig. 4).

MONOSACCHARIDES

Despite the advances which have been made in the primordial chemistry of amino acids, the origin of an equally important set of biological building blocks, the monosaccharides, has remained in a state of uncertainty.

As early as 1861, Butlerow showed that aqueous formaldehyde will undergo condensation to a mixture of sugars in the presence of alkali (22). Shortly thereafter, Loew succeeded in isolating a mixture of optically inactive hexoses from this alkaline-induced condensation mixture. In addition to hexoses, the lower molecular-weight monosaccharides are also formed in this

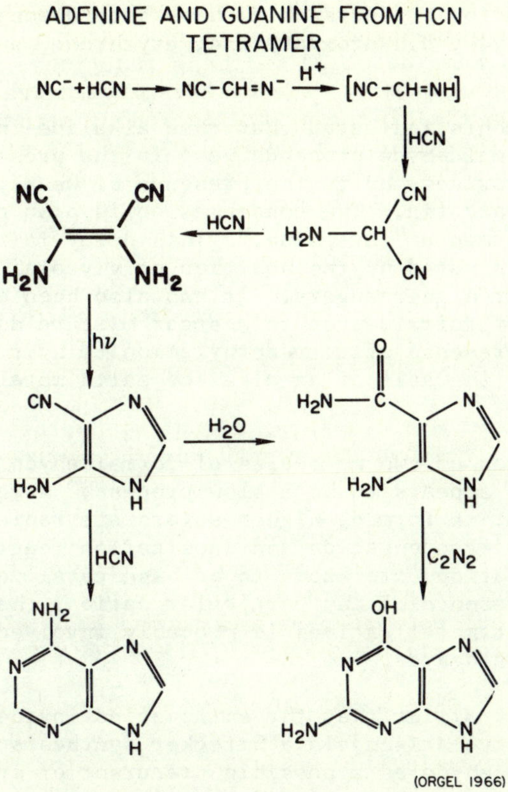

Figure 4. Adenine and guanine from HCN tetramer (After Orgel, 1966).

alkaline induced condensation of formaldehyde. Glycolaldehyde has been isolated and is apparently the product of the condensation of two molecules of formaldehyde (Fig. 5).

$$2\ CH_2O \rightarrow CH_2OH \cdot CHO$$

$$CH_2OH \cdot CHO + CH_2O \rightarrow CH_2OH \cdot CO \cdot CH_2OH$$

$$CH_2OH \cdot CHO + CH_2OH \cdot CO \cdot CH_2OH \rightarrow CH_2OH \cdot CHOH \cdot CHOH \cdot CO \cdot CH_2OH$$

$$2\ CH_2OH \cdot CHO \rightarrow CH_2OH \cdot CHOH \cdot CHOH \cdot CHO$$

$$\text{or } CH_2OH \cdot CO \cdot CHOH \cdot CH_2OH$$

$$2\ C_4 — [C_8]? \rightarrow C_3 + C_5$$

Figure 5. Synthesis of sugars from formaldehyde (Butlerow reaction).

Other condensation products which have been demonstrated include glyceraldehyde, dihydroxyacetone, erythrose, and various pentoses.

Loew's experiments indicated that this alkaline-induced condensation of formaldehyde proceeds best in the presence of alkaline earth hydroxides and in the presence of weakly basic hydroxides of lead and tin. The condensation is also characterized by what has been termed a "lag-phase." This delay in the time of reaction can be eliminated by the addition of glycolaldehyde, dihydroxyacetone, or higher sugars. It has also been observed that condensation of formaldehyde to a sugar mixture does not take place in the presence of tetramethylammonium hydroxide, a strong base, unless the salt of an alkaline earth metal is introduced.

The condensation of two molecules of formaldehyde to form glycolaldehyde thus appears to be a slow process. However, once this primary product is formed, higher sugars are rapidly produced via reversible aldol condensation and enolization reactions. Since aldol condensations are known to be base catalyzed and not dependent on the presence of any particular cation, the catalytic activity of divalent metal cations is probably involved in the formation of glycolaldehyde.

The proposal by Miller that the amino acids in the discharge experiments could have arisen via a Strecker synthesis focused attention on formaldehyde as a possible precursor of sugars in prebiological chemistry. Formaldehyde has been reported in the products obtained when methane, ammonia, and water mixtures were subjected to an electric discharge, β-irradiation or γ-irradiation. Although sugars have not yet been detected in experiments using methane, ammonia, and water as starting materials, the ultraviolet irradiation of formaldehyde can lead to ribose and deoxyribose as well as other sugars (23). Oró and Cox have investigated the formation of 2-deoxyribose from glyceraldehyde and acetaldehyde and from formaldehyde and acetaldehyde. They reported that 2-deoxyribose and its isomer, 2-deoxyxylose, were formed in yields of about 5% from glyceraldehyde and acetaldehyde in aqueous systems. The reaction is catalyzed by divalent metal oxides or ammonia. 2-deoxyribose was also identified as a product when the aqueous starting mixture was composed of formaldehyde, acetaldehyde, and calcium oxide.

Several difficulties are encountered in proposing the use of formaldehyde as the precursor of monosaccharides. Horowitz and Miller have pointed out that the high concentrations of formaldehyde used in some of the experiments were unrealistic in primitive-earth experiments. Objections have also been made to the use of very basic solutions. Abelson maintains that the concentration of free ammonia in the seas and atmosphere was never very large, and that a strongly alkaline ocean never existed. Formaldehyde polymerizes very readily to give paraformaldehyde and may therefore be removed from circulation.

In an effort to circumvent some of these objections, Gabel and Ponnamperuma have used a simulated hydrothermal spring as a reaction medium (24). Aqueous solutions of formaldehyde of various concentrations were refluxed over kaolinite. The products were separated into trioses, tetroses, pentoses and hexoses. At a formaldehyde concentration of 0.5M, only trioses, tetroses and pentoses were formed. When the concentration of formaldehyde was reduced to 10^{-2}M, hexoses were also formed. Ribose was identified in the end products. The simulation of the hydrothermal spring, although possibly occurring only in microenvironments on the primitive earth eliminates the need for postulating a basic primitive ocean, and would have provided an opportunity for the breakdown of the paraformaldehyde. The photochemical synthesis of sugars, however, needs further investigation and may provide another reasonable pathway for sugar synthesis.

NUCLEOSIDES AND NUCLEOTIDES

The satisfactory synthesis of nucleic acid bases and sugars under primitive-earth conditions have led some investigators to explore the abiogenic formation of nucleosides and nucleotides.

When ribose, adenine, and phosphate in dilute aqueous solution 10^{-3} molar was exposed to ultraviolet light, adenosine was formed (25). The phosphate appears to have acted as a catalyst in this reaction. The mechanism has not been investigated. However, it may appear that a sugar-1-phosphate is formed and the phosphate is then replaced by the purine.

It has been suggested that the earth's primordial reducing atmosphere was at least slightly transparent between 2400 Å and 2900 Å, and that the photoactivation of ultraviolet light-absorbing purines and pyrimidines was a possible step in the formation of nucleosides and nucleotides (26). To test this hypothesis, four experiments were performed in which adenine, adenosine, adenosine monophosphate, and adenosine diphosphate, respectively, served as the starting material (27). The conversion of adenine to adenosine, adenosine to adenosine monophosphate, the monophosphate to diphosphate, and the diphosphate to triphosphate, was established. Adenosine was not produced in a detectable amount in the absence of a phosphorous compound. While adenosine was produced in the presence of both phosphoric acid and ethyl metaphosphate, the nucleoside phosphate was detected only with the use of ethyl metaphosphate (28). Although it is unlikely that ethyl metaphosphate could have been a common source of phosphorous on the primitive earth, the result clearly established that the process could occur abiologically. Besides, it is possible that another more abundant phosphorous compound could have substituted for ethyl metaphosphate.

In the synthesis of nucleotides, the heterogeneous reactions which may have taken place on a dried-up ocean bed have also been examined. In simulating these conditions, an intimate mixture of the nucleosides was heated with an organic phosphate. Several phosphates were used in this reaction. Among them were disodium monohydrogen, trisodium, sodium ammonium monohydrogen, ammonium dihydrogen, diammonium monohydrogen, monocalcium, and tricalcium monophosphates and phosphoric acid (29). When the intimate mixture of the nucleoside and the phosphate was heated, phosphorylation took place. The mononucleotides were identified in the end products. The best yields in this reaction were obtained at about 160°. However, a small yield was obtained at temperatures as low as 50°. These experiments simulate an environment with a relative absence of water. However, water is not incompatible with this reaction and does not hinder it unless present in large excess. The conditions under which this reaction proceeds may therefore be described as hypo-hydrous. Amongst the monophosphates that were formed there were equal amounts of the 2', 3', 5', and cyclic monophosphates. It is possible that the presence of certain amino acids might direct the synthesis preferentially to yield a particular isomer.

HYDROCARBONS AND FATTY ACIDS

Although hydrocarbons have been produced in large quantities by the action of various forms of energy on methane, very few detailed studies are available in this field. In experiments performed to compare the effects of different kinds of discharges on methane, very interesting results have been obtained (30). In the high-intensity arc, the product was a clear yellow fluid, which gave chromatograms of well-spaced peaks. On the other hand, the semi-corona yielded a colorless distillate which was poorly resolved by gas chromatography. In the arc discharge, benzene was the most abundant product, and next in order of magnitude was toluene. In contrast, benzene and toluene were virtually absent from the products of the semi-corona discharge. Some of the prominent peaks have been identified by mass spectrometry as 2,2-dimethyl butane, 2-methypentane, 3-methypentane, 2,4-dimethylhexane and 3,4-dimethylhexane.

When the compounds beyond C_9 were examined, it was found that in the semi-corona discharge, the aliphatic fraction consisted of an irresolvable mixture with a maximum around C_{17}. Mass spectrometric and NMR data indicate that these are cyclohexyl derivatives. The use of a molecular sieve indicated that there were no normals or isoprenoid type molecules formed in this system. It is interesting to note that some hydrocarbon deposits have been found which are similar to the spark discharge hydrocarbons (31). A mechanism for obtaining straight-chain hydrocarbons and fatty acids has been suggested by Wilson. This method involves crowding the reaction chains of fatty acids onto a surface so that only the ends are available for reaction, thus preventing branching.

When a mixture of methane and water was exposed to a semi-corona discharge and the products analyzed by gas liquid chromatography and mass spectrometry, several monocarboxylic acids from C_2 to C_{12} were detected. Those above C_6 were characterized as branched carbon chains (32). One area of great interest in the study of chemical evolution is the formation of the normal fatty acids containing 12 or more carbon atoms which are constituent fatty acids of the complex lipids. Although the reaction conditions employed in the semi-corona discharge of methane and water would have favored the production of branched chains owing to the stability of secondary carbon radicals, it is surprising that the formation of only as few as 11 of the possible C_1 to C_{12} isomers were formed. And by contrast to this, highly isomeric mixtures of hydrocarbon are produced when methane alone is subjected to a semi-corona discharge. The present evidence shows that certain C_8 to C_{12} acids which are not excessively branched could have accumulated by the action of a semi-corona discharge passing through methane over an aqueous surface.

SYNTHESIS OF LARGE MOLECULES

In putting two amino acids together to form a dipeptide or two nucleotides together to form a dinucleotide, a dehydration-condensation is necessary. Dehydration could have been achieved by the result of the action of heat on the primordial solution of organic molecules by the evaporation of such a mixture by thermal action. In a celebrated lecture to the British Physical Society entitled, "The Physical Basis of Life," Bernal had suggested that the organic components from the ocean could have been brought from the ocean to the shoreline, absorbed on clay and condensation could have taken place. The Bernalian idea has indeed been tested in the laboratory with encouraging results. Yet another possibility would be to see whether the reaction could take place in water. Since the earth is a very wet planet and if it can be demonstrated that these reactions could take place in water, then a greater opportunity for the formation of polymers and the prebiotic conditions would exist. In living processes, such a system does indeed take place. Enzymes are used to overcome the energy barriers. In the prebiotic system, what would be necessary would be predecessors of such enzymes, or primitive catalysts. A number of experiments have been performed in which both hypotheses have been tested with very satisfactory results.

POLYPEPTIDES

Simulated primordial syntheses of polypeptides have received more attention than the syntheses of other biological polymers. This is probably due to the great emphasis that has been placed on the primordial origin of amino acids. Fox has long been an exponent for the anhydrous or hypohydrous thermal origin of all primordial organic compounds(33). Fox and Harada have shown that in the presence of a proportionally large amount of glutamic acid or aspartic acid, an intimate mixture of all 18 amino acids normally present in proteins can be thermally polymerized at a temperature of 180°-200°C. These polymers have been described as proteinoids. The molecular weights increased from 3600 in a proteinoid made at 160° to 8600 in one made at 190°. The maximum molecular weight which has been obtained is 80,000. The polymers give a positive biuret test, can be hydrolyzed back to amino acids and are attacked by various proteolytic enzymes.

It has been argued that the energy available from volcanic activity is quite small compared to the other available sources of energy. Nevertheless, it cannot be denied that organic reactions, including polymerization of amino acids with the formation of peptide bonds, would occur under conditions simulating a volcano. A glaring weakness in this type of pathway for the origin of polypeptides is the necessity of postulating highly concentrated, intimate mixtures of the amino acids which are to be polymerized.

The synthesis of peptides directly from dilute solutions of amino acids has also been accomplished through the use of such condensing agents as cyanamide and dicyanamide. Both compounds are known to be formed upon ultraviolet irradiation of aqueous solutions of hydrogen cyanide or through the irradiation of a mixture of methane, ammonia and water (34). Ponnamperuma and Peterson have reported that glycylleucine and leucylglycine were formed from a dilute aqueous solution of glycine and leucine subjected to ultraviolet irradiation in the presence of cyanamide at pH 5. Mechanistically, ultraviolet radiation probably increased the concentration of carbodiimide, a tautomer of cyanamide which can act as a condensing agent for amino acids.

$$H_2N - CN \rightarrow HN - C \equiv N \leftrightarrow HN = C = N \rightarrow HN = C = NH$$

Carbodiimides react with carboxyl groups to form an adduct with an extremely good leaving-group on the carboxyl carbon atom. Since amines are much better nucleophiles than water, the amino group of a second amino acid could then attack the carbon atom of the aforementioned adduct forming a tetrahedral intermediate. Disruption of the unstable intermediate would produce the dipeptide and urea.

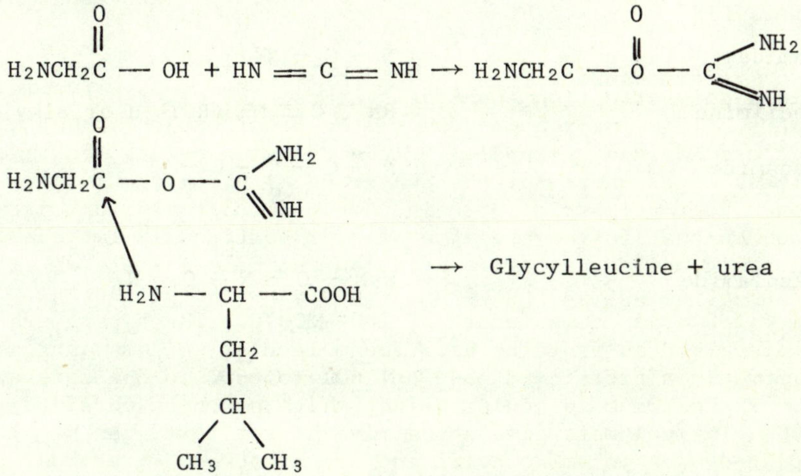

Steinman et al. proposed the formation of an intermediate in the reaction similar to the one in the preceding experiment.

Ponnamperuma and Flores reported the formation of peptides from a methane, ammonia and water mixture that had been subjected to an electric discharge. The peptides appeared to contain three to four amino acid residues (35).

POLYNUCLEOTIDES

Beadle once remarked that primordial polynucleotides could have consisted of an adenine-thymine type of polymer which subsequently developed into the present day polynucleotides. Thermal phosphorylation of nucleosides by inorganic phosphate salts produces some oligonucleotides (36). The temperature required for this reaction is ca. 150° and the yields are quite small. Just as in the synthesis of nucleotises via the same reaction conditions, Rabinowitz et al. have shown that the actual phosphorylating agent is a mixture of polyphosphate salts formed from the various orthophosphate salts used in the experiment.

The condensing agents used for such reactions may be divided into two classes of compounds: the cyanide containing condensing agent and the condensed phosphates. The first class contains cyanide, carboiimides, the hydrogen cyanide tetramer, and in the second we have the linear phosphates and the cyclopolyphosphates (Table IV).

TABLE IV

Cyanamide	$N \equiv C - NH_2$
Carbodiimide	$RN = C = NR$ (R is H or alkyl)
Dicyanamide	$N \equiv C - N(H) - C \equiv N$
Dicyandiamide	$H_2N - C(NH_2) - N(H) - C \equiv N$
Cyanogen	$N \equiv C - C \equiv N$
Hydrogen-cyanide-tetramer	$(NC)(H_2N)C = C(CN)(NH_2)$
Linear polyphosphates	$[O - P(O)(O) - O - P(O)(O) - O - P(O)(O) - O]_n^{-(n+4)}$
Cyclic polyphosphates	$[O - P - O \quad O;\ O \quad P;\ O - P - O \quad O;\ O]_n^{-(n+2)}$

THEORETICAL CONSIDERATIONS

Developments in quantum biochemistry have thrown new light on the origin of biochemical molecules. The work of Bernard and Alberte Pullman of Paris has highlighted some of the important features (37). When a quantum chemist first takes a look at the molecules significant in living organisms, he is struck by the remarkable fact that almost all of the biomolecules which are essential to living processes are conjugated systems, rich in pi electrons. The three fundamental units, the nucleic acids, the proteins, and the energy-rich phosphates exhibit this phenomenon of electronic delocalization.

The most significant constituents of the nucleic acid molecule are the purines and pyrimidines. These are conjugated heterocycles. While the proteins do not at first sight appear to be conjugated, the overall structure implies some degree of delocalization. In the helica structure of the protein, there is a matrix of hydrogen bonding which permits a certain amount of electronic transfer. In the case of the energy rich phosphates, the mobile electrons of one phosphoryl group interact with the electrons of another phosphoryl group or with the pi electrons of an organic radical. Among other conjugated molecules which are important are the porphyrins. The porphyrin is made up of four pyrrole rings joined together to a central metal atom. The outer structure of the porphyrin is an alternating sequence of double bonds, implying a high degree of conjugation and electronic delocalization. The conclusion that one is obliged to draw from even a superficial consideration such as this is that the basic manifestations of life are intimately connected with the existence of highly conjugated compounds. For some reason, these compounds were chosen by nature as a vehicle of life. Electronic delocalization is perhaps the single greatest quantum effect in biochemical evolution.

Let us for a moment consider how this outstanding characteristic of biomolecules can account for their occurence as the principle building blocks of living matter. Two major factors appear to be stability requirements and functional advantages. The major result of delocalization is an increment in stability. Quantitatively, this increment is defined as resonance energy. This thermodynamic stabilization must have played an important part in the selection of early molecules. It may be considered to be a period during which there was a struggle for survival. There was a selection of biomolecules, and those which were finally chosen were the ones which were more stable. This fact is confirmed by the extraordinary unity of biochemistry. The same limited number of compounds perform the same functions over the entire plant and animal kingdom. To take one example, chlorophyll, which is used by plants, is very similar to helium, which is used by animals. Both compounds are synthesized by the same sequence of reactions. Stabilization by electron delocalization may also have played a part in the orientation of small molecules to give the large polymers such as the nucleic acids and proteins.

In summarizing the possible role of electronic delocalization, the following observations may be made. Evolutionary selection used the most stable compounds. On account of electronic delocalizations, these compounds were best adapted for biological purposes. Life did not originate with the appearance of the conjugated compounds but the possibility of life as we know it was made more probable by their appearance.

INTERACTION BETWEEN NUCLEIC ACIDS AND PROTEINS

Perhaps the major problem facing the student of chemical evolution is the elucidation of the origin of living systems through the origin of nucleic acid-directed protein synthesis. There is no experimental evidence at present bearing on the identity of the primitive precursors of the present translation system. It is conceivable that primitive forms of translation could have been rather different from the present one. One might speculate that early translation may have been either imprecisely or ambiguously defined. This could follow from the fact that the accuracy and precision of the present system depends on the precise function of many, if not all, of its highly evolved and sophisticated components, activating enzymes, and ribosomal proteins. Since the accuracy and precision of at least some of these proteins are final products of the evolution of the entire system, it is conceivable that the precursors to the present protein synthetic system could have been ones in which their precise roles would be of less importance, and the role of the nucleic acids corresponding of more importance than seen today. The hypotheses for the origin of the genetic code have then to assign importance to the function of nucleic acid in determining codon assignments and the ordered arrangement of the present code. One of the possibilities is that the structure of the set of codon assignments and of the protein synthetic machinery somehow reflects recognition of some form of catalytic interaction between the nucleotides and amino acids, perhaps only during the primitive stage of the evolutionary history of the cell. To examine this possibility, studies of amino acid oligonucleotide interactions have been undertaken. The purpose of these studies has been to establish what types of interaction are possible between these two classes of compounds, then to learn how to use the information from simple interpretable results to interpret more complex systems, and thirdly to determine the degrees of selectivity obtainable in systems of increasing chemical complexity.

Many studies have shown that a variety of interactions can occur between nucleic acids and amino acids depending upon composition, conformation, state of polymerization, and environment of the reacting species. These studies have led to the conclusion that a degree of specificity does exist, although its origin has not yet been elucidated. When the two reacting species are simple, one cannot expect to observe specificity of this sort implied in the biological use of the term. What one can look for at the simplest level is evidence for selectivity of some sort, cases where the strength of binding of an amino acid to a nucleotide under given conditions is to some extent a function of the composition of both interacting species. While results of this sort from simple protein nucleic acid systems would not be spectacular from a biological point of view, they are nevertheless a first step in any systematic study of a role for amino acid/nucleic acid interactions in the evolution of living systems. Working with monomeric species in aqueous media can permit the effects of individual factors to be assessed, and to provide basic information necessary to interpret the more complicated polynucleotide amino acid interaction. If interactions exist, what are they? Are they selective? Can they lead to nucleic acid directed protein synthesis, or help one to understand the system? Can they be applied to understand a more complex system? Can they lead to the addition of constraints which can achieve genetic coding in a defined prebiotic system?

The use of immobilized nucleic acids on ion-exchange columns, studies with NMR looking at the proton shifts, and the possible role of templates in catalyzing the activated amino acid polymerization have given rise to some possible suggestion that stereochemical factors may have played a role in the origin of the genetic code. Another possible system would be to look at the biological process as we see it today in living organisms. This could then be looked at in its more simple form and examined in such a way as to see whether the most primitive of the nucleic acid components in the present translation apparatus could have been effective in a much more simply organized translation and replication process.

THE ANALYTICAL APPROACH

The laboratory studies have established that the molecules necessary for life can be made in the laboratory. A question of paramount importance is whether this process did indeed take place on the earth or elsewhere in the universe. Can we go back in time to the earliest stages of the earth, of the solar system, and dig up some evidence for the presence of prebiologically synthesized organic molecules? Many attempts have been made to resolve this problem. When did the transition occur from the chemical to the biological systems? As the biological record is pushed farther, the period in which chemical evolution could have taken place is compressed. The age of the earth may be represented in a 12-hour diagram (Fig. 6). In such a system the base of the Cambrian is 600 million years ago. At one time it was thought that no life existed before the Cambrian. Paleontologists had argued that there were no skeletons, and therefore there was no life. However, the work of the micropaleontologists has pushed the presence of life much further back than 600 million years ago. At one billion years ago occurs the Bitter Springs formations in Australia, in which microstructures have been observed. From three to three and a half billion is the Swaziland sequence in South Africa.

There are only three locations on the earth which are indisputably recognized as older than 3.6 billion. There is the Isua region of West Greenland, the Saglefjord region of the Nain Province of Northern Labrador, and the Sighbhum granitic complex of eastern India.

The first evidence for life older than the Cambrian is the discovery of microfossils in the thin sections from the Gunflint chert by Barghoorn and Tyler in 1954. Since then, many deposits from the Proterozoic have been found which yield microfossils. A preservation of fine structural features as permanent comparisons of extant species and ecological patterns has resulted in the unequivocal acceptance of many Proterozoic microfossils as true remains of Precambrian organisms.

Among the earliest microfossils which meet the rigorous criteria are the simple spheroids which have been found in the 3.4 billion year old Swartkoppie chert which immediately overlies the Upper Onverwacht of South Africa (38). The microfossils which have recently been found in the Pilbara Block of Western Australia appear to be of less certain origin. These appear to be poorly defined and resemble some of the microstructures found in the Onverwacht cherts. It is unlikely that they would even be considered as biogenic if they were not associated with stromatolites. While it seems possible and perhaps likely that at least some of the early Archean microstructures are true microfossils of once-living organisms, the evidence is still equivocal. At present, while such techniques as size distributions are applicable to a limited extent, the signs of micropaleontology cannot differentiate between those fossils which are from primitive living organisms and those which may have been abiotic.

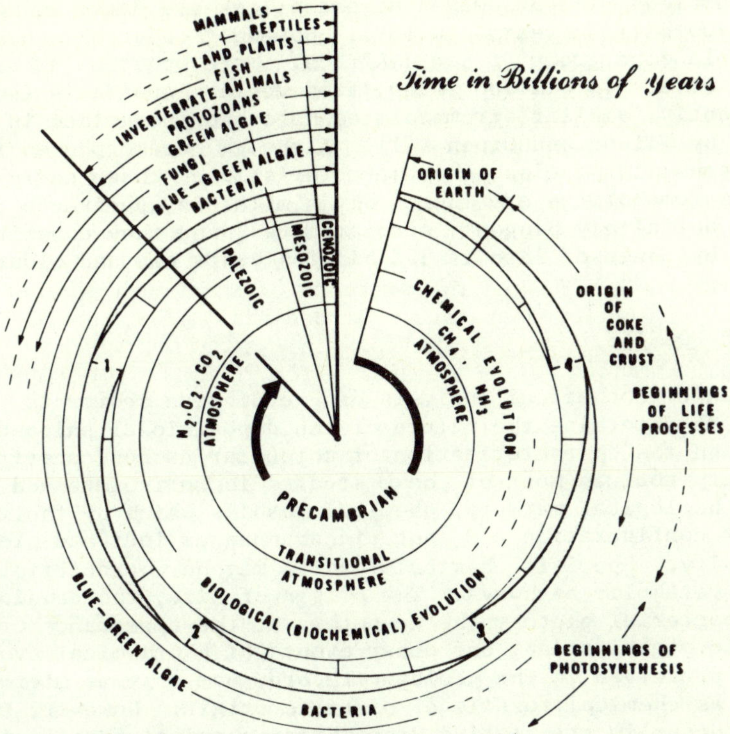

Figure 6. Geologic clock.

MICROPALEONTOLOGY

Stromatolites are macroscopic sediments formed from the trapping and binding of detrital and precipitated minerals by sheaths of algae mats. These mats represent a complete ecology. The trapping of sediments by cyanophites form a compact and opaque layer which forces the community to migrate up. As the process continues, layer upon layer of detritus is deposited between layers of organic debris and trapped microorganisms. While present day stromatolites are limited in size and to harsh environments which exclude Metazoan grazers, stromatolites fluorished in the Precambrian time.

Although some non-biological sedimentary processes can result in similar layered structures, stromatolites may be considered the best indication of life in the Archean. There are several known examples. The Bulowayan stromatolites are dated at about 2,700 million years old (39). Somewhat older are those in the Pangola supergroup of South Africa. More spectacular are those dated at 3.4 billion years, from western Australia (40). More recently, similar stromatolites have been described in Zimbabwe by Wilson and Orpen (41). Although metamorphosed to lower greenschist facies, the characteristic of these early Archean stromatolites are incredibly similar in appearance to younger, definitely biogenic stromatolites, and they constitute the best evidence of life at 3.2 billion years old and older.

MOLECULAR FOSSILS

Organic geochemical analysis of Precambrian sediments attempts to determine the nature of the depositional paleoecosystem through the characterization of molecular carbon constituents. The primary goal in most of these studies is to isolate and identify biological markers, chemical fossils which by their molecular configuration and isotopic abundances indicate biological activity. Specific chemical fossils may be characteristic of specific metabolic pathways. The origin of life, the development of bacterial photosynthetic pathways, the appearance of eukaryotic organization, and other events of biochemical evolution could be preserved in the geologic record, not only as microfossils, but also as chemical fossils of discrete origin. However, two problems occur in translating Precambrian chemical fossils into a representation of a paleoecosystem: the alteration of primary organic compounds and postdepositional contamination. In recent sediments, the extractable components are similar to biochemicals synthesized by contemporary microorganisms. The primary organic compounds, however, are rapidly altered by bacterial degradation and digenetic processes. Biopolymers are rarely hydrolyzed to their components. Functional groups are easily replaced by hydrogen, and molecular configurations are destroyed through rearrangement and racemization. With increased temperature, thermal cracking and condensation of organic compounds, kerogen, an insoluble macromolecule with no defined structure, may be formed. A vast portion of organic carbon in the crust of the earth existed in soluble kerogen. Highly metamorphic temperature altered the kerogen further, until only carbon-carbon bonds are present in the form of graphite. The molecular information is completely lost, and isotopic information may be obscured as well. Organic compounds may have been introduced into Precambrian sediments after deposition. Seemingly permeable chert is sufficiently porous and permeable to groundwater to contribute significant amounts of recent organic contaminants. On the other hand, the insoluble carbonaceous material which may well constitute 98% or more of the total carbon in a Precambrian sediment, is generally considered to be syngenetic, as it is physically too large to be mobile.

Certainly more information can be derived from the soluble fraction than from the kerogen or graphite. Yet it may be virtually impossible to distinguish whether the observed molecules date from the time of deposition or were introduced earlier. In therefore exploring the possibility of using molecular fossils as criteria for the presence of life, one has to be extremely cautious about contamination and digenesis. In the work on the Isua sediments, no attention has been paid to the extractable organics. However, the presence of graphite has been carefully studied and most of the graphite samples seem to give us an isotope fractionation which falls in the biological range. However, there are some samples which may give us anomalous figures. In the absence of further evidence to the contrary, these figures may suggest that life on earth is as old as the oldest rocks. If this is indeed the case, it may not be possible to find on earth evidence of any of the prebiological processes that we have attempted to verify in the laboratory. Evidence for such processes may have to be looked for in other places in the solar system -- the moon, the meteorites, and the surfaces of planets.

ANALYSIS OF LUNAR SAMPLES

The Apollo missions provided a unique opportunity for the analysis of samples brought back from the moon. Samples from the Apollo missions 11-16 were analyzed in the laboratory. These studies included an examination of the lunar material for total carbon, the organic carbon isotope fraction, microfossils, and mineralogy. Sequential treatment of this sample by benzene, methanol, water, and hydrochloric acid provided extracts for examination by chromatographic and spectrometric methods. To minimize contamination, the analyses were carried out in a clean laboratory with filtered air, and the entire sequence of solvent extractions of the lunar dust was accomplished in a single glass vessel. The total carbon determined by measuring the volume of CO_2 evolved when a one gram sample was outgassed at 150° at a pressure of less than one micron, and then burned at 1,150°. The values ranged from 140 to 200 micrograms per gram. The most consistent values were between 140 and 160. The amount of carbon which could be converted into volatile carbon-hydrogen compounds was determined by pyrolizing about 30 mg of the dust in an atmosphere of hydrogen and helium. The resulting involatile compounds were estimated by hydrogen flame ionization detector. The average value obtained was 40 micrograms per gram. Isotope measurements on the total sample gave us a δC^{13} value of +20, relative to the PDB standard. These figures are considerably higher than those reported for intact meteorites (42).

Observations of dust, surfaces of microbreccias in thin sections made by light and electron microscopy yielded no evidence of indigenous biological structures. To detect extractable organic compounds, some of the lunar dust was treated with a mixture of benzene and methanol. Fluorescence excitation and emission spectra of these extracts are suggestive of porphyrins. But similar responses have been obtained in exhaust products from the tests of the lunar descent rocket engines. It is possible that these porphysics were formed from components of the rocket exhausts.

Capillary gas chromatography showed that any single normal alkane from C^{12} to C^{13} was not present at concentrations above 2 x 10^{-5} micrograms/gram. When the sample was treated with 6-HCl, hydrocarbons from C^1 to C^4 were identified, suggesting that carbides were present in the samples. Although the limits of detectability of the techniques employed were in the nanogram range, normal alkanes, isoprenoids, hydrocarbons, aromatic hydrocarbons, fatty acids, amino acids, sugars, and nucleic acid bases were not present at this level of concentration. It is readily understood that the surface samples of the moon were exposed to high intensity ultraviolet light and micrometeorite bombardment and carbon-carbon bonds would have easily broken under such conditions.

CARBONACEOUS CHONDRITES

For over a century, meteorites have been examined for the presence of organic compounds. The Alais meteorite was analyzed by Berzelius, the Kaba by Wohler, and the Orgueil by Berthelot. It is now generally agreed that carbonaceous chondrites do contain polymeric organic matter. However, results obtained in the past have been ambiguous as to the origin of the detected extractable organic matter. Was it truly extraterrestrial in origin, or was it a result of terrestrial contamination? The analysis of the Murchison meteorite provided the first unambiguous and conclusive evidence for the presence of extraterrestrial organic compounds in meteorites. The Murchison Type-2 carbonaceous chondrite fell on September 28, 1969, near Murchison, Victoria, Australia. Several pieces were collected soon after the fall, and later, during the months of February and March, 1970. The stones selected for analysis were those with the fewest cracks, the least exterior contamination and of massive appearance. The samples used in the study contained 2% by weight of carbon and 0.16% weight of nitrogen. In these samples, aliphatic hydrocarbons were identified. The samples were symmetrical in distribution with a maximum retention time of C18. Comparison with hydrocarbons produced from a spark discharge experiment showed a great similarity. Both appeared to be saturated alkanes containing the same dominant analogous series. These similarities suggest that the hydrocarbons in the Murchison meteorite may have originated abiogenically. Studies on the aromatic hydrocarbons reveal a similar result (43). The benzene extracts were analyzed by gas chromatography and gas chromatography combined with mass spectrometry. A thermal origin for the aromatic hydrocarbons in the Murchison meteorite is suggested by the resemblance to the aromatic synthesized during the pyrolysis of methane. In both groups of aromatics, polynuclear compounds containing an even number of carbon atoms predominated. The relatively simple composition of the aromatic hydrocarbon fraction is of interest in view of the more complex nature of the aliphatic hydrocarbons.

In the search for amino acids, interior pieces of the meteorite were pulverized, extracted with boiling water, and analyzed by gas chromatography combined with mass spectrometry. A large number of amino acids were identified in the Murchison meteorite, including glycine, alanine, valine, proline, glutamic and aspartic acid. In addition, there were a number of non-protein amino acids, n-methyl glycine, β-alanine, 2-methyl alanine, α-amino butyric acid, β-amino butyric acid, and γ-amino butyric acid. Of the amino acids with assymetric centers, both d and l enantiomers were detected. The presence of non-protein amino acids and amino acids with equal amounts of d and l enantiomers strongly suggests that these amino acids were prebiotic in origin. These provide us with the first conclusive results concerning extraterrestrial amino acids (44).

The studies similar to those undertaken on the Murchison meteorite have been applied to the Murray, which fell in Kentucky in 1951, and to the Mighei, which fell in Odessa in 1866 (45). These meteorites have been contaminated by exposure and by handling. The techniques of gas chromatography and mass spectrometry enable us to separate the indigenous amino acids and hydrocarbons from terrestrial contamination. Thus far, three examples were found which showed amino acids of extraterrestrial origin.

More recently, the exploration of the Antarctic gave a new impetus to the study of meteorites. The Japanese geologist Kenzo Yanai, on an expedition to the Yamato highlands of the Antarctic, discovered eight pieces of meteorite outside his tent in 1971. When these meteorites were studied, he found they happened to be eight different meteorites. The following year, 300 were identified, in a subsequent year, 600, and now, over the years, 4000 new meteorites have been brought back from the Antarctic. Of these, about 40 fall into the category known as "carbonaceous chondrites." Two of these have been analyzed in the laboratory, the Alan Hills and the Yamato, and they provide supportive evidence of the observations made on the Murchison, the Murray, and the Mighei (46).

A striking feature of the Antarctic meteorites is that they are uncontaminated. The exterior and interior have the same amount of amino acids. Short of going to the asteroid belt, these are the best meteorites that are available to us for analysis.

The search for the heterocyclic bases which occur in the nucleic acids, purines, and pyrimidines has been conducted for meteorites as well. Earlier, adenine and guanine were recorded in the Orgueil meteorite. Since then, there has been some controversy with regard to the presence of these compounds in carbonaceous chondrites. However, more recently, the work of Schwartz and others have clearly demonstrated that the bases do occur in meteorites (47). In other words, not only could the components of the proteins, but also of the nucleic acids, occur in carbonaceous chondrites. The prebiological processes which have been postulated for the earth and simulated in the laboratory appear to have taken place elsewhere in the solar system and verified for us in the carbonaceous chondrites.

JUPITER, SATURN AND TITAN

The study of the giant planets from ground-based observatories has given us a fair knowledge of the composition of these atmospheres. They appear to be highly reduced, containing large amounts of hydrogen. helium, and the hydrocarbons. The Voyager mission, with its number of experiments, especially the infrared observation, has given us a further handle to the composition of these giant planets and their satellites. It is clear that if hydrocarbons such as methane acetylene, ammonia, and phosphate are available, many organic reactions could take place. Simulation studies in the laboratory have shown very clearly that a large number of the organic molecules that were postulated as prebiotic could be synthesized under these conditions (48). As a matter of fact, the pictures that can be produced of their colored material -- the red spot of Jupiter, for example -- could be explained by the formation of organics from methane and ammonia. The presence of organic molecules in the giant planets has, therefore, been established by the observations of the Voyager spacecraft and by the laboratory simulation studies.

Of great interest to us is the possible reactions that may be taking place on Titan, the giant satellite of Saturn. Titan appeared in the pictures sent back to Earth by the Voyager mission as a fuzzy, orange ball. The analytical data point out to an atmosphere consisting mainly of nitrogen, and up to about 2% of methane. Furthermore, for the first time, hydrogen cyanide has been observed in a planetary atmosphere. When the data obtained from the Voyager mission gave us an atmosphere of 98% nitrogen and 2% methane, a large number of experiments were done where this mixture was exposed to a variety of energy sources. A whole range of organic molecules was synthesized (49). These corresponded very largely to those that were formed, or identified in the Voyager mission. The data available to us clearly established the presence of organic molecules in the giant planets and the satellite of Saturn, Titan.

As regards life, the conditions within these may be such that the convection would take the synthesized molecules into the center of the planet to high temperatures and perhaps break them down. The chances of life originating under such conditions may be minimal. In the case of Titan, though it has an atmosphere, the temperature is probably so bitterly cold that the chances of any of the molecules coming together may be small. But the process of prebiotic evolution may be taking place under these conditions.

CONCLUSION

The studies which have started in the laboratory with the Berkeley cyclotron or the electric discharge or Urey and Miller have taken us through our solar system to outer space. We are optimistic that the path of chemical evolution will be outlined in the laboratory. The biochemical knowledge which has been amassed within a few years has given us a deep insight into some of nature's most secret processes. With this understanding to help us, the time needed to solve our problem may not be long. We cannot deny the immensity of the prospect for any man's philosophical position or shrink from its pursuit on account of the difficulty of the task.

REFERENCES

1. Darwin, E.: 1803, "The Temple of Nature, Johnson, London.

2. Darwin, C.: 1959, "Life and Letters 3," Notes and Records of the Royal Society, 14, no. 1, London.

3. Tyndall, J.: 1871, "Fragments of Science for Unscientific People," Longmans, Green and Co., London.

4. Huxley, T.H.: 1968, "On the Physical Basis of Life," in Collected Essays, Vol. 1, Macmillan, London.

5. Oparin, A.I.: 1924, "Proischogdenie Zhizni," Moscovsky Robotchii, Moscow.

6. Haldane, J.B.S.: 1928, Rationalist Annual 148, pp. 3-10.

7. Bernal, J.D.: 1951, "The Physical Basis of Life," Routledge, London.

8. Urey, H.C.: 1952, "The Planets, their Origin and Development," Yale University Press, New Haven.

9. Oparin, A.I.: 1938, "Origin of Life," Dover Publications, New York.

10. Rubey, W.W.: 1955, in "Crust of the Earth" (ed. A. Poldervaart), Geological Society of America, New York, pp. 631-650.

11. Abelson, P.H.: 1966, Proc. Nat. Acad. Sci. U.S. 55, pp. 1365-1372.

12. Holland, H.D.: 1962, in "Petrologic Studies," A volume to honor A.F. Buddington, Geol. Soc. Am., pp. 447-477.

13. Berkner, L.V. and Marshall, L.C.: 1964, in "The Origin and Evolution of Atmospheres and Oceans" (ed. P.J. Brancazio and A.G.W. Cameron), John Wiley and Sons, New York, pp. 86-101.

14. Miller, S.L. and Urey, H.C.: 1959, Science 130, pp. 245-251.

15. Hochstim, A.R.: 1968, Proc. Nat. Acad. Sci. U.S. 59, pp. 22-28.

16. Miller, S.L.: 1957, Biochem. et Biophys. Acta 23, pp. 480-89.

17. Ponnamperuma, C. and Woeller, F.: 1967, Curr. Mod. Biol. 1, pp. 156-58.

18. Orò, J.: 1963, Ann. N.Y. Acad. Sci. 108, pp. 464-81.

19. Ponnamperuma, C., Lemmon, R.M., Mariner, R. and Calvin, M.: 1966, Proc. Nat. Acad. Sci. U.S. 49, pp. 737-40. Also published in "Extraterrestrial Life" (E.A. Shnoeur and E.A. Otteson), NAS-NRC, Washington, D.C.

20. Ponnamperuma, C.: 1965, in "The Origins of Prebiological Systems and of Their Molecular Matrices (ed. S.W. Fox), Academic Press, New York, pp. 221-42.

21. Sanchez, R.A., Ferris, J.P. and Orgel, L.E.: 1968, J. Mol. Evol. 38, pp. 121-28.

22. Butlerow, A.: 1861, Justus Liebigs Annalen Chemie 120, p. 295.

23. Ponnamperuma, C.: 1965, in "The Origins of Prebiological Systems and of Their Molecular Matrices" (ed. S.W. Fox), Academic Press, New York, pp. 221-42.

24. Gabel, N.W. and Ponnamperuma, C.: 1967, Nature 216, pp. 453-55.

25. Ponnamperuma, C., Sagan, C., and Mariner, R.: 1963, Nature, 199, pp. 222-26.

26. Sagan, C.: 1957, Evolution 11, pp. 40-55.

27. Ponnamperuma, C., Sagan, C., and Mariner, R.: 1963, Nature 199, pp. 222-26.

28. Schramm, G.: 1965, in "The Origins of Prebiological Systems and of Their Molecular Matrices" (ed. S.W. Fox), Academic Press, New York.

29. Ponnamperuma, C. and Mack, R.: 1965, Science 148, pp. 1221-23.

30. Ponnamperuma, C. and Woeller, F.: 1964, Nature 203, pp. 272-74.

31. Ponnamperuma, C. and Pering, K.: 1966, Nature 209, pp. 979-82.

32. Allen, W.V. and Ponnamperuma, C.: 1967, Curr. Mod. Biol. 1, pp. 24-28.

33. Fox, S.W. and Dose, K.: 1972, "Molecular Evolution and the Origin of Life," W.H. Freeman, San Francisco.

34. Schimpl, A., Lemmon, R.M. and Calvin, M.: 1965, Science 147, pp. 149-50.

35. Flores, J.J. and Ponnamperuma, C.: 1972, J. Mol Evol. 2, pp. 1-9.

36. Ponnamperuma, C. and Mack, R.: 1965, Science 148, pp. 1221-23.

37. Pullman, B. and Pullman, A.: 1962, Nature 196, pp. 1137-42.

38. Knoll, A.H. and Barghoorn, E.S.: 1977, Science 198, pp. 396-98.

39. Schopf, J.W.: 1972, in "Exobiology" (ed. C. Ponnamperuma), North-Holland, Amsterdam, pp. 16-61.

40. Walter, M.R., Buick, R. and Dunlop, J.S.R.: 1980, Nature 284, pp. 443-445.

41. Orpen, J.L. and Wilson, J.F.: 1981, Nature 291, pp. 218-20.

42. Ponnamperuma, C.: 1970, Atomes 25: 174, pp. 169-74.

43. Pering, K.L. and Ponnamperuma, C.: 1971, Science 173, pp. 237-9.

44. Kvenvolden, K.A., Lawless, J.G., and Ponnamperuma, C.: 1971, Proc. Nat. Acad. Sci. 68, pp. 486-90.

45. Ponnamperuma, C.: 1978, in "Proceedings of the Robert A. Welch Foundation Conferences on Chemical Research. XXI. Cosmochemistry" (ed. W.O. Milligan), Robert A. Welch Foundation, Houston, pp. 137-97.

46. Kotra, R.K., Shimoyama, A., Ponnamperuma, C., and Hare, P.E.: 1979, J. Mol. Evol. 13, pp. 179-83.

47. Van der Velden, W. and Schwartz, A.W.: 1977, Geochim. Cosmochim. Acta 41(7), pp. 961-968.

48. Ponnamperuma, C. and Woeller, F.: 1969, Icarus 10, pp. 386-92.

49. Gupta, S., Ochiai, E. and Ponnamperuma, C.: 1981, Nature 293, p. 5835.

SYNTHESIS OF THE CHEMICAL ELEMENTS

V. E. Viola, Jr.

Department of Chemistry and Cyclotron Facility
Indiana University
Bloomington, IN 47405

ABSTRACT

The origin of the chemical elements in nature is understood in terms of nuclear reactions which occur in a variety of cosmological settings. In order to develop a self-consistent explanation of these processes, extensive information from many diverse fields of science is required. First, one must know the characteristic abundances of the elements and their isotopes in the universe, for it is these values that any successful theory of nucleosynthesis must reproduce. Second, it is necessary to investigate all possible nuclear processes that might give rise to these abundances. Finally, an appropriate cosmological environment must be found which is capable of generating the essential nuclear reactions. The major sources of element production will be reviewed. These include:

1) Cosmological processes associated with the big bang;
2) Nucleosynthesis in stars during stellar evolution from main sequence stars through the supernova stage, and
3) Formation of elements via interactions of galactic cosmic rays with the interstellar medium.

The relationship of recent measurements relevant to light element nucleosynthesis in the big bang and the question of the expanding universe is discussed, as is the possibility of forming super-heavy elements during supernova explosions.

I. INTRODUCTION

During the 15 billion years or so that have elapsed since the

C. Ponnamperuma (ed.), Cosmochemistry and the Origin of Life, 35–69.

big bang - from which we trace the origin of our universe - a complex array of evolutionary processes has occurred. These phenomena, including the emergence of galaxies, stars, planetary bodies, and their constitutent chemical elements, have been vital predecessors to the formation of intelligent life on our planet (or elsewhere). The synthesis of atomic nuclei via nuclear reactions in various astrophysical environments is an especially important precusor for the development of life.

It is currently believed that the primordial universe was composed largely of the simplest constituents of matter, the elementary particles. Until such a time in our history when the element carbon had evolved, there was no possibility for even the simplest organic molecules to form. Similarly, the existence of increasingly complex molecules of biological significance depended upon the existence of other elements, such as nitrogen, oxygen, phosphorus, iron,etc. Of further significance, in order for living systems to sustain themselves, a constant source of energy was essential. As we shall discuss below, this energy is provided by nuclear reactions which synthesize the elements in stars.

Despite what might at first appear to be an exceedingly complicated task, during recent years great progress has been made in understanding the origin of the elements. Important early theories of nucleosynthesis were proposed by Gamow (1); Burbidge, Burbidge, Fowler and Hoyle (2), and Cameron (3). For a detailed review of both historical as well as more recent developments in this field the reader is referred to several more complete treatments (4-6).

The present model for the origin of the elements draws upon many diverse fields of science: astronomy, astrophysics, nuclear science, elementary particle and theoretical physics, atomic and molecular spectroscopy, geochemistry, etc. Each of these has been vital to the development of our present understanding of nucleosynthesis. In Fig. 1 we schematically show the idealized picture that is followed in the present development of the subject. The model contains three basic components, all of which must be self consistent:

(1) The basic principles - First of all a cosmological setting must be proposed that is consistent with the observed behavior of matter in the universe. Into this environment we then introduce the presently-known fundamental particles and the basic forces of nature. Thus, we must first review the salient properties and basic laws which describe the particles and forces.

(2) Interaction processes - Given the system defined by the above principles, we must then ask what reactions will occur. Since in this discussion we are primarily concerned with the formation of atomic nuclei, our main concern here will be nuclear stability and reactions. In

order to understand these we must draw on extensive experimental and theoretical results derived from the study of nuclear science.

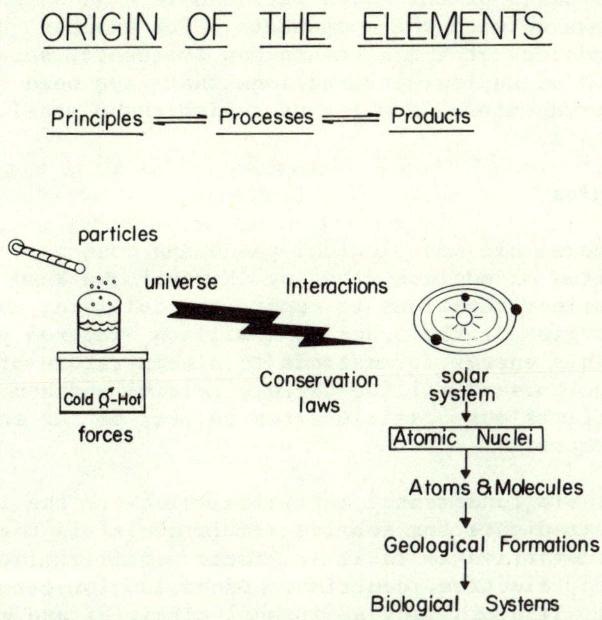

Fig. 1. Recipe for discussing the origin of the elements. Add the fundamental particles to the universe and allow the basic forces to act (principles). The interactions that result, governed by nature's conservation laws (processes) will produce the existing universe (products).

(3) Products - Finally, the products that are predicted from the interactions that occur in the model system must correspond to observed experimental data. Specifically, this means that the abundances of the chemical elements and their isotopic ratios in nature must be reproduced.

The outline for this presentation is as follows. First, the basic components of the model outlined above will be discussed. Then the big bang origin of the universe will be considered, with an emphasis on the element formation in this event. Next we shall describe the synthesis of most of nature's elements during stellar evolution, followed by the formation of Li, Be and B in interactions between galactic cosmic rays and the interstellar medium. This final subject will lead us back to the consideration of the eventual fate of the big bang of our universe; i.e. will our universe expand forever or will it someday collapse again into another hot, dense fireball of fundamental particles?

II. THE MODEL

In practice, there has been considerable interplay between the various parts of the model outlined in Fig. 1, as indicated by the arrows between the components. For example, our knowledge of the abundances of the elements has frequently served as a guide to the types of nuclear interactions that have been responsible for their production. Now let us review these model components in more detail.

A. Principles

In general all cosmological phenomena must be considered as possible sites of nucleosynthesis. However, we know that in order for most nuclear reactions to occur, the colliding nuclei must possess energies of the order of a million electron volts (MeV) or more. This energy corresponds to a temperature of about 10^{10}°K (that is 10 thousand million degrees Kelvin) and hence, this criterion restricts our possible sites to very hot or energetic phases of the universe.

The stable fundamental particles which are the important starting ingredients for forming atomic nuclei in these environments are summarized in Table I. These particles include the proton, neutron, electron, neutrino, and photon (or electromagnetic radiation). Although many additional particles are known, they have lifetimes of 10^{-6} seconds or less. Thus, only those particles that live sufficiently long to survive on a cosmological time scale play a significant role in the structure of our present universe. It should be noted here that the neutron in free space is radioactive and decays into a proton with an average lifetime of about 16 minutes. However, inside atomic nuclei the neutron is stable, i.e. it retains its identity for the lifetime of the nucleus in which it resides.

"FUNDAMENTAL" PARTICLES		mass	charge	spin
Baryons	protons ^1_1H	1.0078 u	+1	1/2
	neutrons ^1_0n	1.0087 u	0	1/2
Leptons	electron $^0_{-1}e$	0.00054 u	-1	1/2
	neutrino $^0_0\nu$	very small	0	1/2
Photon	light, x-ray, γ-ray, etc.	0	0	1

Table I. Particles of concern in nucleosynthesis.

Each of the fundamental particles can be characterized by three primary properties: mass, charge and spin. Other properties also serve to distinguish the particles more completely, but these are not necessary for the present discussion. There also exists for each particle an antiparticle. Antiparticles are essentially the same as the particles in Table I, except that all their properties are reversed. For example, the antielectron (or positron) is an electron with a positive charge (+1). An amount of energy equal to twice the mass of a particle is required to create an antiparticle. When a particle and antiparticle of the same type collide, the two annihilate one another, usually with the creation of two photons, each having the mass-energy equivalent of the particle (thus conserving energy in the creation-annihilation process).

The two particles in which most of the present mass of the universe resides are the proton and the neutron, which belong to a class of particles called baryons. These have nearly identical masses and differ only in that the proton has a charge of +1 and the neutron has a neutral charge (0). Although we have listed the proton and neutron as "fundamental" particles in Table I, it is presently believed that each of these has a substructure consisting of simpler particles such as quarks (7) (hence, the quotes around fundamental). The understanding of neutron/proton structure promises to be one of the exciting areas of physics in the next several years. Fortunately, as far as element synthesis is concerned, it is adequate to treat the neutron and proton as "fundamental" without altering our conclusions here.

The electron and neutrino belong to another class of particles called leptons. The electron has a very small mass relative to the baryons and has a negative charge. Although electrons are critical to the remaining talks in this institute, we shall say very little about them here since under the conditions of nucleosynthesis, the temperatures are so high that all atoms are completely ionized. The primary importance of the electron in this description will occur when we discuss its antiparticle and nuclear β decay, a radioactive decay process whereby a neutron changes into a proton, an electron and an antineutrino. The neutrino is a neutral particle with a very small mass and no charge. The mass of the neutrino plays an important role in our discussion of the big bang and its eventual fate. Attempts to measure this mass are currently being pursued by a number of investigators (8) and the results have important potential consequences for both astrophysics and elementary particle physics. Both electrons and neutrinos have very important roles to play in the evolution of stars (9), but these will be discussed here only superficially.

The final particle is the photon, or electromagnetic radiation. The photon has no mass or charge, and is the basic particle of the electromagnetic field. It is known by several names, depending upon its wavelength; for example, light, ultraviolet, infrared, microwave and γ-radiation are all different forms of the photon. γ rays are very short wavelength radiation which originate in nuclei.

The basic forces - nuclear, electromagnetic and gravity - determine how the fundamental particles will interact with one another - i.e. they represent the glue whereby the fundamental particles are held together. These include:

(1) The nuclear force - The nuclear force is attractive and consists of two components: the strong nuclear force which governs interactions among baryons and the weak force which also includes interactions of leptons. The strong nuclear force is the strongest of the forces and is about 10^{11} times stronger than the weak force. Both are effective only when the particles are very close together ($\sim 10^{-12}$ cm), whereas gravity and electromagnetism exert their influence over much longer distances. The important point is that in order to generate nuclear reactions, it is necessary to bring baryons (or nuclei) very close together.

(2) The electromagnetic force - This force represents a repulsion between charged particles of the same charge and attraction between opposite charges. Thus, nuclei attract electrons to form the atoms. The force between two protons is repulsive and hence, they must be accelerated toward one another in order to bring them together in nuclear reactions. The electromagnetic force between two charged particles is about 1/137 times as strong as the strong nuclear force.

(3) The gravitational force - This force is attractive and depends upon the mass of the constituent particles in a system. Hence, it will primarily affect protons and neutrons, and to a much lesser extent electrons and neutrinos (if they have mass). This force is only about 10^{-39} times as strong as the strong nuclear force. But, for massive astronomical bodies such as the sun, with a mass of 2×10^{33} grams (about 10^{56} atoms), it can become very important. In the past decade major progress has been made in the theoretical interrelationships among the forces. The grand unification theory (10) of Weinberg, Salam and Glashow has successfully described the weak and electromagnetic forces within a single framework. The question as to whether or not the theorists can also incorporate the strong nuclear force - and perhaps even gravity - into a complete theory of unification remains a formidable challenge for them in the future.

Two additional fundamental relationships will also be necessary to understand nucleosynthesis. The first of these is the well-known Einstein equation:

$$E = Mc^2 \qquad (1)$$

That is, energy (E) and mass (M) are interconvertible. In many nuclear reactions that occur in stars, substantial mass is converted into energy. This relationship determines the amount of energy our sun (or any star) emits. The second fact to keep in mind is that for matter

$$T \propto \rho^{1/3}$$

where T is temperature and ρ stands for the density. Under the force of gravity matter contracts, thereby increasing its density. The resultant gravitational pressure causes massive amounts of cosmological matter to heat up, leading to the formation of stars. In this way very high temperatures can be obtained and provide a means of increasing the energies of nuclear particles to the point where they can react with one another, converting mass into energy and providing a long-term energy source for our universe. At the same time new elements can be created.

B. Nuclear Stability and Interactions

Extensive studies of nuclear reaction processes and the systematic behavior of nuclear matter have played a central role in the development of theories of the origin of the elements. As a starting point in generating the observed solar system abundances of the elements, it is valuable to ask the question: Assuming all possible combinations of neutrons and protons may exist, what atomic nuclei can actually be observed? Measured trends in nuclear stability serve as an important guide in answering this question.

In Fig. 2 a plot of neutron number (N) versus proton number (Z) is shown for the nuclei found in nature (about 270 of them). It is apparent that light nuclei prefer to have approximately equal numbers of protons and neutrons, whereas increasingly heavy elements tend to have an excess of neutrons due to the increasing importance of electromagnetic repulsion for high Z elements. In Fig. 3 the energetic behavior of nuclei is summarized in terms of a plot of the average binding energy (the average energy with which the nucleons in a nucleus are held together). In such a plot nuclei that have the highest binding energies are held together most strongly and hence are the most stable. Note that ^{56}Fe is at the peak of this curve, making it the most stable nucleus in nature. This fact has important consequences for the synthesis of heavy elements. Fig. 3 can be used to estimate the energetic trends in nuclear reactions. For example, reactions between light nuclei (<u>fusion</u>)

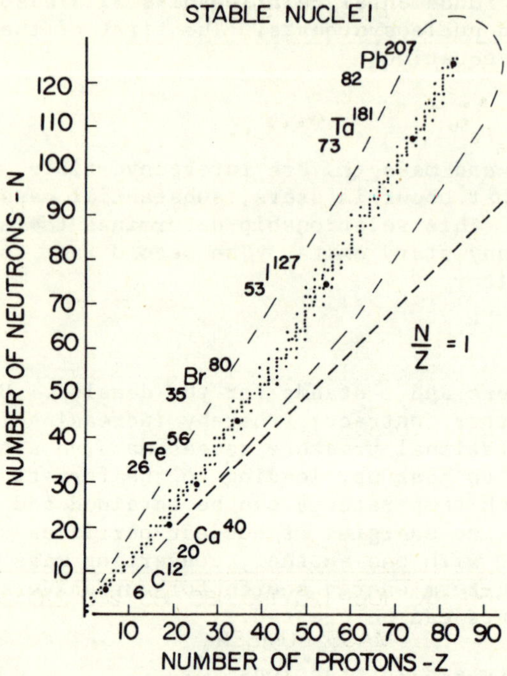

Fig. 2. Plot of the proton number (Z) versus neutron number (N) for the stable nuclei observed in nature (dots). The long-dashed line encompasses most of the nuclei that have been observed in the laboratory and the short-dashed line represents nuclei with equal neutron and proton numbers. Representative stable nuclei are indicated by heavy dots.

will generally lead to more stable nuclei and thus will release energy. Fusion reactions which lead to nuclei heavier than iron have the opposite energetic trend and therefore must absorb energy. On the other hand, by splitting a heavy nucleus into two pieces (fission) to form more stable lighter nuclei, energy will also be released. Examination of the finer details of Fig. 3 shows many other features of nuclear structure which correlate strongly with the observed elemental abundances. For example, for light nuclei the mass numbers A = 4n (^{24}Mg, ^{28}Si, ^{32}S etc.) are unusually stable with respect to their neighbors. In addition nuclei with even mass numbers (even numbers of protons and neutrons) are more stable than those with odd mass numbers. Nuclear shells at N = Z = 2, 8, 20, 28, 50, 82 and 126 are also important in determining nucleosynthetic pathways. When considering whether or not a nuclear process will release or absorb energy, Fig. 3 provides an important predictive tool.

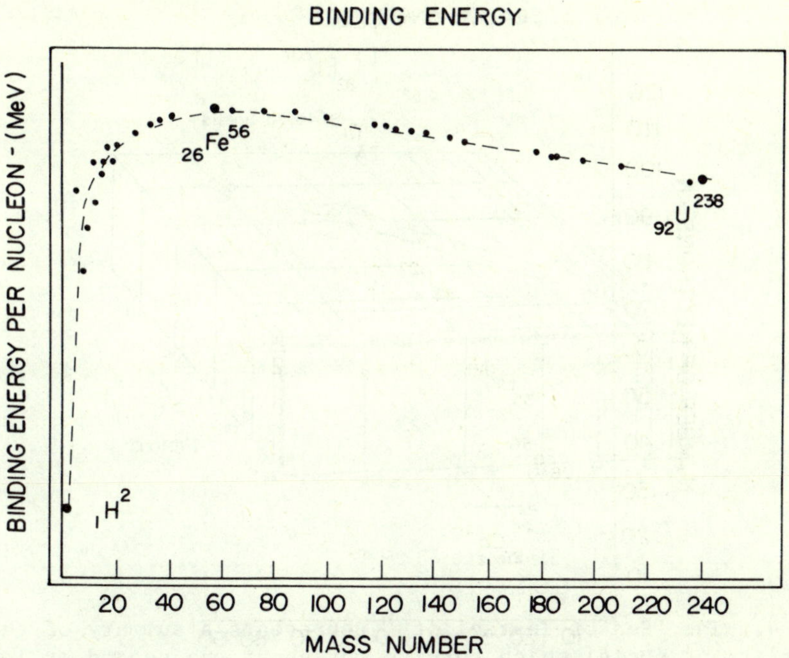

Fig. 3. Plot of the binding energy per nucleon versus mass number for the stable nuclei. The binding energy represents the strength with which the neutrons and protons in a nucleus are held together.

The results of Figs. 2 and 3 are combined in Fig. 4 to present an allegorical summary of nuclear stability trends, known as the "sea of nuclear instability" (11). Here neutron and proton numbers are shown in the horizontal plane and the vertical represents the degree of nuclear stability (or binding energy). Sea level corresponds to nuclei which are sufficiently stable to retain their identity for about one second or more. It is observed that the nuclei stable enough for us to study in the laboratory (~ 1900 of them) form a rather narrow peninsula extending into the sea of instability. All other species are submerged below sea level; i.e., they are very unstable and rapidly disintegrate. However, as new techniques evolve, more and more of these nuclei are being brought to the surface (12), thus expanding our knowledge of nuclear stability. Further topographical structure arises when quantum mechanical effects imposed by closed shells (or magic numbers) are included. These are indicated by the solid lines for the magic proton and neutron numbers.

Uranium, element 92, is the heaviest element observed in nature, whereas elements as high as atomic number 106 have been observed in the laboratory (13). Beyond the known limit of existing nuclei there is a small island of nuclei near Z = 114 and

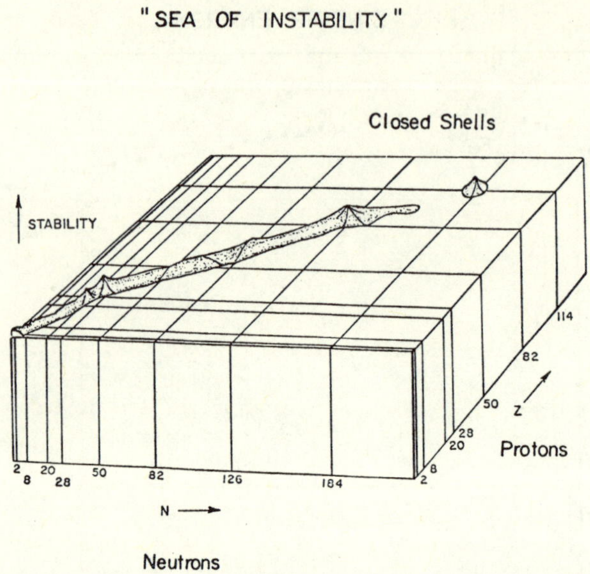

Fig. 4. The "Sea of Instability" represents a summary of the stability of nuclei which survive for about one second or longer. Neutron and proton numbers are plotted in the horizontal plane, whereas stability is indicated by the vertical dimension. Solid lines correspond to closed nuclear shells. The "island of stability" near 114 protons and 184 neutrons represents "superheavy nuclei", which have not yet been observed.

and N = 184. These are the hypothesized superheavy elements which, if they exist, are a consequence of the stability associated with nuclear shells at the corresponding proton and neutron numbers. Thus far, searches for the existence of these species in both nature and in the laboratory have been unsuccessful. (For a thorough review of the more or less current status of this subject the reader is referred to reference 13).

Once it is known whether or not a nucleus is stable, the probability of forming it in nuclear reactions must be considered. Extensive measurements of the probabilities of nuclear reactions, or the reaction rates, have been made in many laboratories throughout the world, in particular at the California Institute of Technology. Based upon these measurements one can evaluate the course of any chain of nuclear events that might lead from the fundamental particles to more complex systems.

In Fig. 5 we show a representative plot of reaction probability (defined in terms of a cross section, σ) as a function of projectile bombarding energy for the reaction between two ^{4}He ions. This reaction is important in ^6Li and ^7Li synthesis. The peaks at low energy demonstrate quantum effects of nuclear structure, whereas the smooth decrease with increasing energy is due to the statistical behavior of nuclear systems.

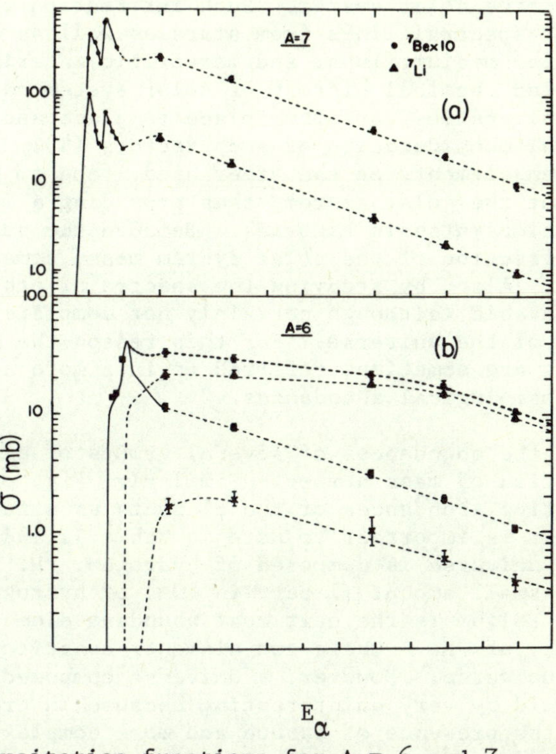

Fig. 5. Excitation functions for A = 6 and 7 production in the $\alpha + \alpha$ reaction: (a) ^7Be and ^7Li (sum of ground state and first-excited state); (b) ^6Li ground state, solid lines based on detailed balance for $(\alpha,d)^6Li$, x-$(\alpha,2p)^6He$ reaction; ▲$(\alpha,pn)^6Li$ reaction; ■-$(\alpha,d)^6Li$ reaction, and ●-total ^6Li cross section.

By combining the known behavior of nuclear energetics and the measured probabilities for the reactions of interest in nucleosynthesis, one can then hope to predict what elements might be formed in cosmological processes and also estimate the relative abundances of these products. In cases where experimental data are lacking, one must then rely upon calculated nuclear properties and reaction rates in order to make these estimates.

C. Solar System Abundances

The critical test of any theory of nucleosynthesis is its ability to reproduce the solar system abundances of the elements and their isotopic ratios. Determination of such abundance ratios is a complex task dependent upon a knowledge of the composition of all components of the solar system. Such information is derived from the study of spectral lines from stars as well as analysis of earth, interstellar medium, lunar and meteoritic material. Because of the varied chemical history of solar system material, elemental abundance ratios vary from place to place and thus require very careful consideration of such factors (4). Isotopic ratios for a given element, on the other hand, tend to be rather uniform throughout the solar system, thus providing a more rigorous test of nucleosynthesis theories. Because the sun constitutes the major fraction of the solar system mass, some simplifications result. Also, by studying the spectra of other stars, one notes considerable (although certainly not complete) uniformity for much of the universe. For this reason the solar system abundances are sometimes referred to in a more adventuresome spirit as cosmological abundances.

In Table II the abundances of several groups of elements are listed as a function of mass number. Similarly, Fig. 6 gives a plot of the relative abundances of the elements as a function of atomic number. It is important to note in Table II and Fig. 6 that most of the universe is composed of hydrogen, ^{1}H. In addition, there is a small amount (1 part in 10^{5}) of hydrogen of mass 2 (deuterium). Helium is the next most abundant element and is composed primarily of ^{4}He. These two elements constitute most of the mass of the universe. However, a universe composed only of this material would be very uninteresting because in order for life to evolve, the presence of carbon and more complex elements is essential.

Another important feature of the solar system abundances is that nuclei with mass numbers A = 5 and 8 do not exist in nature. Both are very unstable nuclei which do not hold together for any appreciable length of time. This fact severely limits the pathways for synthesizing heavier elements from ^{1}H and ^{4}He and has very important consequences for our theory of nucleosynthesis. It is also interesting to note that the elements lithium, beryllium, and boron (Li, Be, B) have very small abundances. This is due to the fact that these nuclei are extremely fragile and are easily broken up in stars. As the elements become gradually heavier, one finds that the elements carbon through neon represent about 1.8 percent of the nuclei in nature and for increasingly heavier elements, the abundances become still smaller. In Fig. 6, it is worth noting that there is a peak in the abundance curve corresponding to iron. This peak is a consequence of the unusual nuclear stability

Hydrogen: ^{1}H (^{2}H)	71%
Helium: ^{4}He (^{3}He)	27%
A = 5, 8	0
Li, $_{4}Be$, $_{5}B$	10^{-6}%
^{12}C - ^{20}Ne	1.8%
Sodium - Titanium (Z = 11 - 22)	0.2%
Iron Group (Z = 23 - 28)	0.02%
63 ≤ A ≤ 100	10^{-5}%
A > 100	10^{-6}%

Table II. Abundances of the Elements in the Solar System (by Mass).

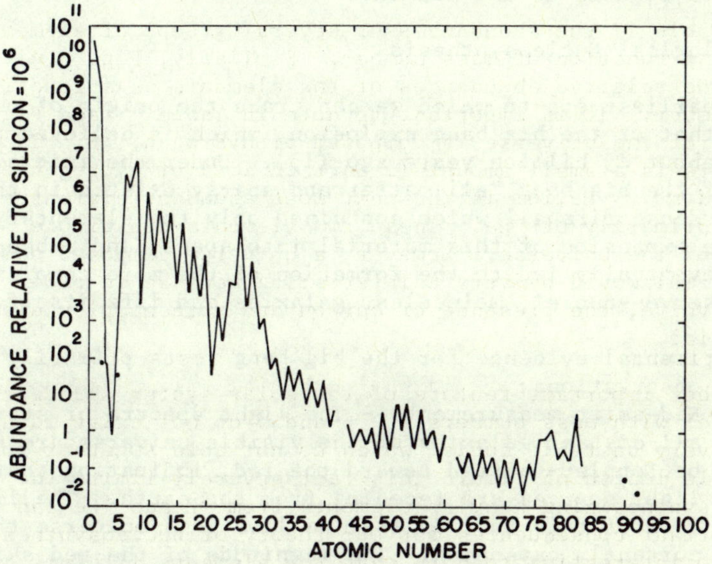

Fig. 6. Solar system abundances of the elements relative to silicon as a function of atomic number.

of ^{56}Fe. It is also apparent that for heavy elements beyond iron, there is some structure in the abundance curves. This is associated with features of nuclear shell structure that become important in the synthesis of heavy elements.

In addition to the abundance ratios of the elements, there is also a large amount of data comparing the ratios of the isotopes of a given element. For example, the ratio of ^{7}Li to ^{6}Li in nature is known to be 12.6. Similar isotopic ratios have been measured with high accuracy for all of the elements and are found to be remarkably uniform throughout the solar system. Observation of anomalies in the isotopic ratios frequently indicates unusual evolutionary features of the sample in question (14).

Thus, our model of nucleosynthesis begins with the basic principles, applies existing knowledge of nuclear stability and interactions, and the final result should then describe the abundances and isotopic ratios of the elements. Research over the past 25 years had indicated that there are three major sources that are responsible for the synthesis of the elements. These include, (1) cosmological nucleosynthesis in the big bang, (2) nucleosynthesis during stellar evolution, and (3) nucleosynthesis in the interstellar medium via galactic cosmic ray interactions. We shall now turn to a discussion of each of these processes in more detail.

III. MAJOR SOURCES OF THE ELEMENTS

A. Cosmological Nucleosynthesis

The earliest era to which we can trace the origin of our universe is that of the big bang explosion, which is believed to have occurred about 15 billion years ago (15). Under the initial conditions of the big bang, all matter and energy existed in the form of a hot, dense fireball which contained only the elementary particles. The expansion of this material into space, and subsequent cooling, eventually led to the formation of the more complex systems we now observe--nuclei, molecules, galaxies and life itself.

Experimental evidence for the big bang rests primarily in two important observations:

(1) Red-shift measurements-- The light spectra of stars in all distant galaxies of the visible universe are known to be Doppler-shifted toward the red, indicating that the light sources are receding from the earth. The implication of this fact is that we live in a universe that is currently expanding. The magnitude of the red shift increases for increasingly distant galaxies, which permits calculation of the expansion rate of the universe. This rate is expressed in terms of the Hubble constant H, where H is usually taken (16) to be 55 km/sec-Mpc (see, however, Science 207, 167 (1980) for possible evidence that H_0 may be twice as large). From this experimental value one can estimate that the initial expansion of the universe began about 15×10^9 years ago (or even less if

the new value of H_0 is accepted).

(2) The universal 2.8°K background radiation-- Radioastronomy measurements have shown that there exists an isotropic background radiation in the universe (17) which corresponds to a black-body source at a temperature of 2.8°K. The isotropy of this background radiation field implies that it is of cosmological rather than galactic origin. On this basis it is presumed to be the remnant of the radiation field associated with the big bang explosion. In addition to providing us with strong evidence for the big bang origin of our universe, the temperature associated with this radiation provides an essential clue as to the big bang environment. For example, from the 2.8°K result it is possible to deduce that the ratio of photons to baryons is about 10^9 in the present universe, and was presumably the same at the time the elements were synthesized.

From these two facts a great deal can be inferred about the primordial condition of the universe. The theory that has evolved to describe the big bang - the standard model - involves the following basic assumptions:

(1) General relativity is the correct theory of gravity;
(2) The universe is homogeneous and isotropic on a macroscopic scale;
(3) Only the presently known fundamental particles were present during nucleosynthesis;
(4) The baryon number of the universe (baryons minus antibaryons) is positive; and
(5) The lepton number (leptons - antileptions) is less than the number of photons.

With these assumptions the principles of theoretical physics permit one to reconstruct a time evolution of the universe - assuming these principles are the same throughout the universe. For a more complete, but very readable, account of the standard model, see Ref. 15.

Since our objective here is to synthesize nature's elements, we shall not concern ourselves with the earliest stages of the big bang. Instead, we shall choose a time 1 second after the beginning of big bang expansion, when the temperature of the universe was about 10^{10}°K. At this stage the universe consisted of a sea of photons, electrons, positrons, neutrinos and antineutrinos, plus a few baryons, all in statistical equilibrium. The fact that any baryons at all existed is an important consequence of the grand unification theory (10). The neutrons and protons existed in equilibrium with one another according to the following equations:

$$^{1}H + e^{-} \rightleftharpoons {}^{1}n + \nu$$

$$^{1}n + e^{+} \rightleftharpoons {}^{1}H + \bar{\nu}$$

The essential first step in the synthesis of more complex nuclei - the combination of a neutron and a proton to form deuterium (^{2}H) - is not possible at such high temperatures because the ^{2}H nuclei instantaneously disintegrate.

However, as the universe continued to expand, the temperature eventually decreased to a point where ^{2}H nuclei could survive for a finite length of time. At this point, approximately three minutes after the initial explosion (15), the temperature had dropped below about 10^{9}°K, where the following series of reactions - all rather well studied in the laboratory (18) - became possible

$$^{1}H + ^{1}n \rightarrow ^{2}H + \gamma$$

$$^{2}H + ^{1}n \rightarrow ^{3}H + \gamma \; ; \; ^{3}H + ^{1}H \rightarrow ^{4}He + \gamma$$

$$^{2}H + ^{1}H \rightarrow ^{3}He + \gamma ; \; ^{3}He + ^{1}n \rightarrow ^{4}He + \gamma$$

$$^{3}He + ^{4}He \rightarrow ^{7}Be + \gamma ; \; ^{7}Be + e^{-} \rightarrow ^{7}Li + \nu$$

The formation of nuclei heavier than ^{7}Li in the big bang is strongly inhibited by the instability of nuclei with mass A = 5 and 8, shown in Table II.

With continued expansion and decreased density, the universe eventually cooled to a point where nuclear reactions could no longer be sustained. The remaining neutrons then decayed to protons as follows:

$$^{1}n \rightarrow ^{1}H + e^{-} + \bar{\nu} \qquad \text{(average lifetime = 16 minutes)}$$

Thus, the unreacted protons and neutrons from the big bang formed the large residual hydrogen abundance in the universe.

The primary nuclear reaction product of the big bang was ^{4}He. One strong argument in favor of this hypothesis is that the abundance of ^{4}He appears to be rather uniform (~ 24-28%) everywhere we look in the universe, thus giving rise to the conclusion that most of the helium must have been made at approximately the same time. Otherwise, there would be much greater differentiation in helium compositions among the stars of different ages. Thus, we believe that cosmological nucleosynthesis in the big bang produced primarily ^{4}He. In addition, trace amounts of deuterium ($D/H \sim 10^{-5}$), ^{3}He ($^{3}He/H \sim 10^{-5}$), and ^{7}Li ($^{7}Li/H \sim 10^{-9}$) were also formed; we shall return to this subject in the final section of this paper.

Beyond this brief instant in the expansion of the universe, nucleosynthesis was no longer possible due to the decreased temperature and density of the expanding medium. Chemical reactions did not become possible for an additional 700,000 years, when the temperature had cooled below about 3000°K, where H, He and Li atoms could form. On a macroscopic scale the universe has continuted to expand and cool to its present size and temperature of 2.8 °K. If it were not for microscopic inhomogeneties in the expanding big bang dust, this would be the state of the universe today. Such an environment, containing only H, He and Li, would hardly be conducive to the origin of life.

B. Stellar Evolution

In the aftermath of the big bang, cosmological dust consisting largely of hydrogen and helium atoms filled the expanding universe. If it were not for the force of gravity and the existence of density flucutations, element synthesis would have ceased at this stage. However, eventually the attractive force of gravity began to act upon regions of higher than average density to produce massive lumps of matter in local regions of space. This process represented the beginning of galaxy and star formation - and at the same time provided us with a new environment for synthesizing elements. In this section we shall focus on the evolution of stars formed from the big bang dust, rather than on the much more poorly understood process of galaxy formation. It is assumed that nucleosynthesis in both environments is similar, other than for the time scale involved.

As localized gas clouds began to condense from the big bang dust, the associated increase in density produced a re-heating of the constituent matter. At sufficiently high temperatures, the hydrogen atoms became ionized into protons and electrons. Under increasing gravitational pressure the density of the nuclei in the core of an embryonic star can eventually reach very high values, corresponding to about 100 g/cm^3, or a temperature of about 15 million °K. Such conditions are certainly extreme when compared with those existing for hydrogen on earth, which is about 300°K and 10^{-4} g/cm^3. On the other hand, this is much less dense than nuclear matter, which has a density of about 10^{14} g/cm^3. It is important to realize that only the core of a star reaches the maximum temperature and density; for example, the temperature at the surface of our sun is only about 5,700°K, while its core is thought to have a temperature of about 14,000,000 °K. The density and temperature profiles characteristic of stars like our sun are illustrated in Fig. 7.

1. <u>Hydrogen Burning - Main Sequence Stars</u>. When the core of a star reaches a temperature of approximately 10 to 20 million °K and densities approaching 100 g/cm^3, the protons in the core acquire sufficient kinetic energy that nuclear reactions again become possible. The process of hydrogen burning (or proton burning) characterizes main sequence stars, of which our sun is an example (19).

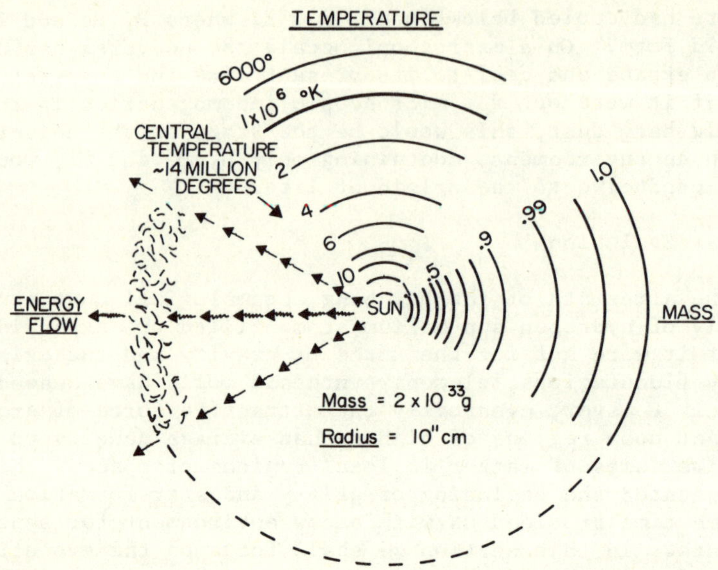

Fig. 7. Schematic drawing of the structure of the sun, showing the temperature and fraction of mass as a function of solar radius.

About 90 percent of the stars in the universe are main sequence stars. Such stars burn protons into helium by means of the following series of nuclear reactions, also illustrated in Fig. 8.

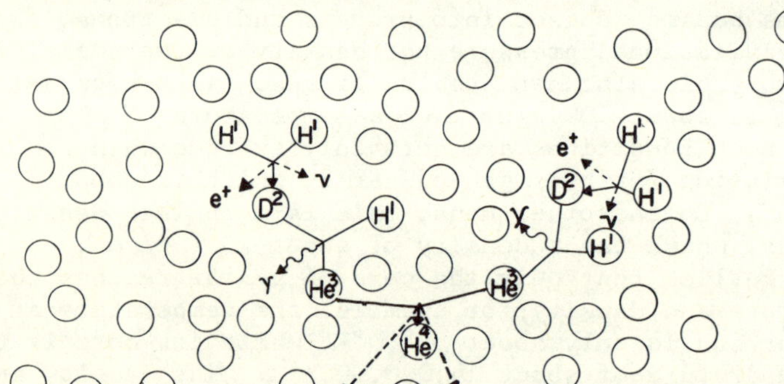

Fig. 8. Hydrogen burning: The fusion of ordinary hydrogen in main sequence stars.

$$^{1}H + {}^{1}H \rightarrow {}^{2}H + \beta^{+} + \nu$$

$$^{2}H + {}^{1}H \rightarrow {}^{3}He + \gamma$$

$$\underline{^{3}He + {}^{3}He \rightarrow {}^{4}He + 2\,{}^{1}H}$$

$$\text{NET: } 4\,{}^{1}H \rightarrow {}^{4}He + 2\beta^{+} + 2\nu + 26.7 \text{ MeV}$$

In this way the element helium is synthesized during hydrogen burning. The above reactions (called the ppI cycle) have been studied in the laboratory and several variations are known to be possible, depending on the core temperature of the star. These include the "pep" reaction ($^{1}H + e^{-} + {}^{1}H \rightarrow {}^{2}H + \nu$); the ppII and ppIII cycles which involve intermediate states of lithium nuclei, and the CNO cycles which uses ^{12}C as a catalyst for burning H to He in later generation stars. The net effect in each case is to convert ^{1}H into ^{4}He. These are examples of fusion reactions and the amount of energy released makes this reaction chain one of the most efficient energy sources known. For example, there are about 1.5×10^{11} calories liberated per gram of hydrogen burned in the above reaction, more than 20 million times the amount of energy liberated in the chemical burning of a gram of carbon. The energy released in hydrogen burning serves to stabilize a condensing star, counteracting the force of gravity; thus the star appears to be a stable body. As long as the nuclear fuel lasts, the star continues to provide a constant source of energy in space.

The mass of a star determines the rate at which it burns nuclear fuel and thus how long it will live. The heavier the star, the faster it burns. In main sequence stars the slowest step is the fusion of two protons to make deuterium. Note that positrons (antielectrons) and neutrinos are produced in the first of the above equations. This is unusual in nuclear reactions and causes the nuclear process to proceed very slowly, accounting for the relatively long lifetime of main sequence stars. From an experimental knowledge of how fast this reaction occurs and how much hydrogen exists in the sun, it is possible to calculate that our sun will continue to shine for another 5 billion years or so. The fact that neutrinos are emitted in this reaction provides the opportunity to observe directly the nuclear reactions that occur in the sun. Since neutrinos interact very weakly with matter, they are able to escape from the core of the sun and constantly bombard the earth. Davis and collaborators have constructed a massive detector for solar neutrinos in the Homestake gold mine in South Dakota and have been measuring these particles for the past decade (19). One of the striking conclusions of these studies is that the observed rate of hydrogen burning in the sun is a factor of three less than calculated. The explanation of this discrepancy constitutes one of the major problems in solar physics today. One possibility, presently the subject of intensive investigation, is that the solar neutrinos undergo transformation to another state during their ten minute trip from the sun to the earth (8).

In summary, hydrogen-burning reactions stabilize a condensing star and generate energy by producing helium from hydrogen. However, the net product of this process only results in a small additional amount of helium, which was already present in the initially formed star. In order to synthesize more complex elements, we must examine more advanced stages of stellar evolution.

2. Helium Burning - Red Giant Stars. As a main sequence star becomes older, it begins to develop into two phases:

(1) A core composed largely of the helium produced during hydrogen burning, and
(2) an outer envelope consisting largely of unburned hydrogen.

Hydrogen burning continues at the interface between the core and the envelope. However, at temperatures of 15 million °K reactions between helium nuclei are inhibited because of the electromagnetic repulsion between the two nuclei. Furthermore, studies of the nuclear stability of the elements lithium, beryllium, and boron (Z = 3, 4 and 5) show that these are extremely fragile nuclei and that at temperatures above about 1 million °K they disintegrate. For this reason Li, Be and B cannot be formed in any appreciable amount in stars (see Table II). Thus element synthesis temporarily subsides in the core.

If the mass of the star is sufficiently large, the force of gravity begins to contract the core once again, leading to still higher temperatures and densities. This causes the envelope of the star to expand greatly and gives rise to a new stage in the evolution of the star, called the red giant stage. Stars which are not heavy enough to sustain more advanced stages of nuclear burning simply exhaust their hydrogen fuel and undergo no further evolution. These are known as white dwarf stars, which represent the stellar graveyard.

During the red giant stage of a star the gravitational force continues to contract the core. When the temperature reaches about 10^8°K, which corresponds to a density of 10^5 g/cm^3, a new type of nuclear reaction becomes possible. Of the several possible reactions that might conceivably lead to the production of heavier elements from hydrogen and helium at such temperatures, laboratory studies have led us to believe that only one is likely. This reaction, called helium burning, is represented by the equation

$$^4He + {}^4He \rightarrow [^8Be]^* + \gamma \qquad \text{(lifetime = } 10^{-16} \text{ seconds)}$$

$$[^8Be]^* + {}^4He \rightarrow {}^{12}C + \gamma$$

To produce a helium-burning reaction, three ^{4}He nuclei must collide almost simultaneously, as shown in Fig. 9. The chances of this occurring are low due to the very short lifetime of the ^{8}Be intermediate, which is about 10^{-16} seconds. As a consequence, the red giant stage of a star can last for millions of years. Thus, the element-synthesis chain skips over the elements lithium, beryllium, and boron to produce carbon. At this stage in the star's development the basic element for the formation of biological compounds has now been synthesized.

Once helium burning begins, the core of the star is again stabilized and a new equilibrium situation results. Under this new condition, gravitational contraction and expansive nuclear burning again offset one another. At the same time it becomes possible to produce oxygen by means of the following reaction

$$^{4}He + {}^{12}C \rightarrow {}^{16}O + \gamma$$

The evolutionary cycle of our schematic star thus far is indicated in Fig. 10. The star has begun by burning hydrogen on the main sequence and converting this into helium nuclei. As the helium concentration increases, the core of the star heats up further to the place where three helium nuclei fuse and form ^{12}C and, depending upon the conditions of the star, ^{16}O. At this stage if the star is of sufficiently low mass, it will burn out and become a white dwarf. On the other hand, if the mass is sufficiently great, much more complicated sets of nuclear reactions can occur. These are described below.

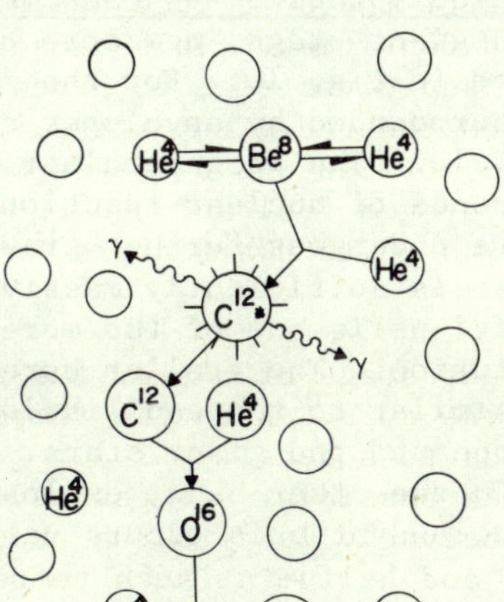

Fig. 9. Helium Burning: The Fusion of Helium in Red Giant Stars.

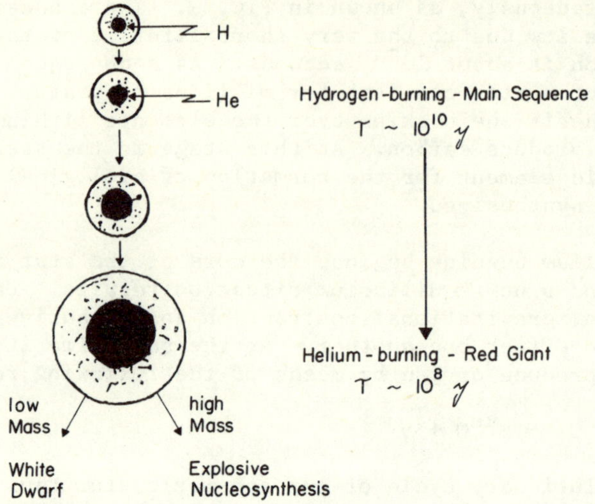

Fig. 10. Schematic diagram of stellar evolution from the main sequence to the red giant phase. Hydrogen-burning core in the main sequence phase evolves into ^{4}He core. This eventually undergoes helium-burning, leading to a greatly expanded envelope in the red giant phase. Low mass stars become white dwarfs whereas heavier stars undergo more advanced stages of nuclear evolution.

3. Explosive Nucleosynthesis -- Massive Stars. As a star passes through the red giant stage, new core conditions eventually develop, as illustrated in Fig. 10. For the most part the core contains ^{12}C and ^{16}O surrounded by envelopes composed of helium and hydrogen, respectively. The large nuclear charge of ^{12}C and ^{16}O prevent the occurrence of nuclear reactions at these temperatures. Hence, the core undergoes further gravitational contraction and heating if the star is sufficiently massive. Its subsequent fate under these conditions is one of the more poorly understood phases of stellar evolution. The stellar core may continue to evolve via processes similar to the equilibrium situations that existed in main sequence and red giant stars, although on a much shorter time scale. On the other hand, explosive conditions may develop under which nucleosynthesis occurs very rapidly.

If the core temperature and density reach about 600 million °K and 5×10^5 g/cm^3, new types of nuclear burning begin. The first type that becomes possible is called carbon burning. This reaction involves the fusion of the ^{12}C and ^{16}O remnants from helium burning which form still heavier nuclei. These reactions are complicated (20) but can be represented by equations of the type

$$^{12}C + ^{12}C \rightarrow ^{20}Ne + ^{4}He$$

$$^{16}O + ^{16}O \rightarrow ^{28}Si + ^{4}He$$

$$^{12}C + ^{16}O \rightarrow ^{24}Mg + ^{4}He$$

Because these reactions can occur relatively rapidly at high temperatures, the evolution of the star proceeds much faster at this stage and a much more varied nuclear composition develops. As the life cycle of a heavy star continues, a new core composed largely of nuclei near ^{28}Si evolves. At temperatures near 1 x 10^9°K and densities about 1 x 10^6 g/cm^3 a process known as silicon burning (or the e-process; e for equilibrium) begins. Because of the large electric charge on nuclei such as silicon, it becomes increasingly difficult for fusion reactions between two ^{28}Si nuclei to proceed. However, because of the high temperatures and the variety of nuclear reactions that can occur in this advanced stage of stellar evolution, reactions involving ejection of an α-particle (^4He) by γ-rays [(γ,α) reactions] and the inverse process of ^4He capture [(α,γ) reactions] begin to occur. This complex reaction chain is summarized as follows:

$$^{28}Si + \gamma \rightleftarrows ^{24}Mg + ^{4}He$$

$$^{4}He + ^{28}Si \rightleftarrows ^{32}S + \gamma$$

$$^{32}S + ^{4}He \rightleftarrows ^{36}Ar + \gamma$$

$$\text{Net:}\quad ^{28}Si \overset{(\alpha,\gamma)}{\rightleftarrows} ^{32}S \overset{(\alpha,\gamma)}{\rightleftarrows} ^{36}Ar \overset{(\alpha,\gamma)}{\rightleftarrows} ^{40}Ca \overset{(\alpha,\gamma)}{\rightleftarrows} \cdots A = 56\ (^{56}Fe)\ \text{etc.}$$

These reactions can go in either direction but the reaction going toward the right always occurs to some extent. This chain of reactions primarily produces nuclei with mass numbers A = 32, 36, 40, 44, 48, 52, and 56, which turn out to be unusually abundant in nature. In addition, because of the much richer composition of nuclear matter that exists in the stellar core during silicon burning, a much more diverse batch of other nuclear reaction products is possible (20). This leads to the synthesis of many other nuclei below ^{56}Fe, but in smaller quantities.

Of particular importance in the silicon-burning process is the fact that it stops near mass number A = 56. Recall that we stated earlier (Fig. 3) that ^{56}Fe is nature's most stable nucleus. Because of this fact, fusion reactions which produce nuclei heavier than ^{56}Fe in the stellar core absorb energy rather than releasing it. Thus, when an iron core develops in a star, the absence of energy-liberating nuclear reactions removes the major source of stellar support for resisting gravitational contraction.

To summarize our account of stellar evolution to this point, we have described a rather complex star, containing most of the elements up to iron in various layers. This is indicated in Fig. 11. Most of the elements needed to sustain life have now been constructed. At each stage of evolution, the processes become less efficient and more diverse, accounting for the steadily decreasing abundances of the elements, observed in Fig. 6 and Table II. The difficulty in producing nuclei beyond iron creates a sink for Fe-like nuclei, producing the peak at iron in Fig. 6, and also accounts for the low abundances of the heavier elements. Further, the unusual stability of nuclei in the iron region also means that nuclear reactions can no longer act as a source of energy to sustain the structure of a star against strong attractive forces of gravity that exist at these very high densities.

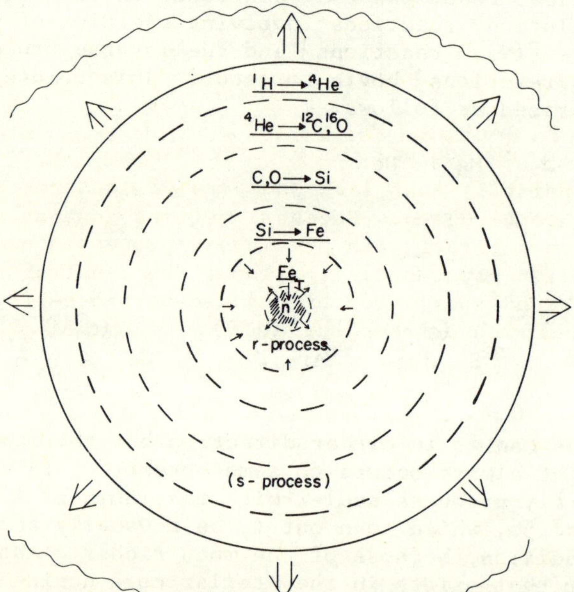

Fig. 11. Schematic diagram of stellar structure at the onset of the supernova stage. Nuclear burning processes are indicated for each layer. The r-process is associated with the disintegration of iron nuclei in the central region of the star; a process which liberates neutrons.

4. Heavy Element Production - The r-Process. The accumulation of the iron group elements in the core of a star leads to catastrophic conditions. Because nuclear reactions can no longer release energy and privide their stabilizing influence, the

gravitational force causes the core to collapse; that is, an implosion of the core upon itself results, indicated in Fig. 11 in the central region. The implosion process occurs on a time scale as short as seconds, during which the density of nuclear matter may reach 10^8 g/cm^3 with a corresponding temperature of 6×10^9 °K in the center of the core. This is followed by a massive shock wave that leads to explosion of the star, and is believed to be associated with supernovae.

There are two important consequences of gravitational collapse and the rapid heating which follows. First, the temperature increase triggers a varied array of nuclear reactions throughout the outer envelopes of the star. This leads to an enrichment of nuclear species for the elements previously formed. Second, the conditions in the very center of the core, where the temperature and density are highest, cause the iron nuclei to break up by means of photodisintegration reactions, leading to the following schematic processes:

$$^{56}\text{Fe} + \gamma \rightarrow 13\ ^{4}\text{He} + 4\ ^{1}\text{n}\ ;\ ^{4}\text{He} + \gamma \rightarrow 2\ ^{1}\text{H} + 2\ ^{1}\text{n}$$

$$^{1}\text{H} + e^{-} \rightarrow {}^{1}\text{n} + \nu$$

The important point is that large numbers of neutrons are produced in the central core region. Because neutrons have no electric charge, they can interact with previously processed nuclear material, including the iron group nuclei, without the constraint of electromagnetic repulsion experienced by charged-particle projectiles. This serves to enrich further the variety of nuclei that can be produced.

This stage of nucleosynthesis is called the r-process (r for rapid) and is assumed to produce most of the elements beyond iron according to the following series of reactions:

$$^{56}_{26}\text{Fe} \xrightarrow{(n,\gamma)} {}^{57}\text{Fe} \xrightarrow{(n,\gamma)} {}^{58}\text{Fe} \xrightarrow{(n,\gamma)} \rightarrow \rightarrow {}^{79}_{26}\text{Fe}$$

etc.

These reactions produce highly neutron-rich nuclei which are well submerged below the sea of instability in Fig. 4. As this process of neutron addition continues, nuclear beta-decay (conversion of a neutron into a proton) becomes increasingly probable, thus producing the next higher element, as shown below:

$$^{79}_{26}\text{Fe} \rightarrow {}^{79}_{27}\text{Co} + e^{-} + \bar{\nu}$$

$$^{79}\text{Co} \xrightarrow{(n,\gamma)} {}^{80}\text{Co} \xrightarrow{(n,\gamma)} {}^{81}\text{Co}$$

$$^{81}\text{Co} \rightarrow {}^{81}\text{Ni} + e^{-} + \bar{\nu}, \text{ etc.}$$

This sequence of neutron-capture and beta decays produces heavier and heavier elements.

It is the r-process that forms the heaviest elements in nature and must account for the possible existence of any "superheavy" elements. There is no other nuclear reaction mechanism known by which we can account for the production of the amounts of uranium and thorium that we find in nature today. A schematic diagram of the r-process is shown in Fig. 12. The upper limit to element synthesis in the r-process is imposed by nuclear fission reactions which become increasingly probable as the nuclear charge increases beyond Z ≈ 90. Fission involves division of the nucleus into two nuclei of roughly the same mass numbers. Thus, the nucleosynthesis process is terminated and nuclear material simply cycles between the buildup of very heavy elements and their fission into intermediate-mass elements. It is not clear at what point fission becomes dominant in the r-process chain of mass buildup, but most

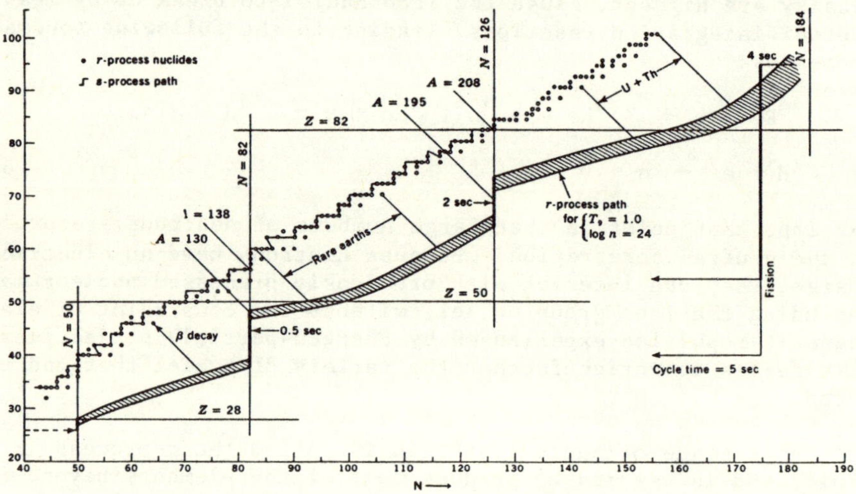

Fig. 12. Diagram of the r-process path on a proton versus neutron number plot. The stable heavy isotopes are indicated by the open circles. The dashed region shows the path followed by the r-process, which is terminated by nuclear fission (heavy vertical line). The s-process path is given by the solid line through the stable heavy isotopes. Fig. is from Ref. 21.

probably (21) it occurs around mass number A ≅ 270, which would mean that the superheavy elements (A ~ 300) are not produced in nature. (Hopefully, this prediction by the author will someday be demolished). At the present time attempts to discover these elements experimentally, both in nature and via nuclear reaction studies in the laboratory, have been unsuccessful (13). However, these searches continue and it may be possible that eventually one will observe the superheavy elements.

It is presently hypothesized that a supernova explosion in the vicinity of our developing solar system may have triggered its condensation 4.6×10^9 years ago. Evidence for this is based on the observation of anomalously high abundances of neutron-rich isotopes in certain meteorites, a result which implies fresh supernova material was incorporated into these bodies at the time of their formation (14).

Although there are other significant processes responsible for element production (to be discussed below), the r-process is thought to conclude the life cycle of a first generation star; that is, a star composed of original big bang material. Following gravitational collapse, the supernova core may form a dense neutron star (density $\sim 10^{14}$ g/cm^3) or perhaps even a black hole (density $\sim 10^{16}$ g/cm^3). Neutron stars are believed to be the explanation for the existence of pulsars, which are small, massive sources of periodic radiowave emission. Black holes are hypothetical astronomical bodies that are so dense, even radiation cannot escape (hence, they are black).

The supernova explosion itself ejects processed nuclear material out into the interstellar medium where the temperatures and densities are much lower. Nearly all of the elements essential for life are now present and it can be estimated that this stage of nucleosynthesis was reached over 10^{10} years ago somewhere in the universe. This material then cools, attracts electrons to form neutral atoms and molecules and the entire cycle begins anew. First, the gravitational force begins to condense matter to form second generation stars, or - in the case of smaller amounts of mass - planets, meteorites, and cosmic dust are formed. In this way succeeding generations of stars, richer in nuclear reaction possibilities evolve. Our sun must be at least a second generation star because we see evidence that it contains heavy elements in addition to its hydrogen and helium content.

The life cycle of a star is depicted in Fig. 13. If one recalls the abundances of the elements shown in Table II and Fig. 6, it is clear that none of the successive stages of element synthesis needs to be very efficient in order to produce nature's elements. Hence, even after the complete evolution of a star, 98% of the material still will be in the form of hydrogen and helium.

At this point, our basic picture for stellar evolution and the production of elements in nature is roughly complete. Many refinements are required to give a thorough description of these processes; in fact, many pieces of experimental information are still lacking or incompletely understood. Nonetheless, the model represents our best current understanding of the origin of the elements.

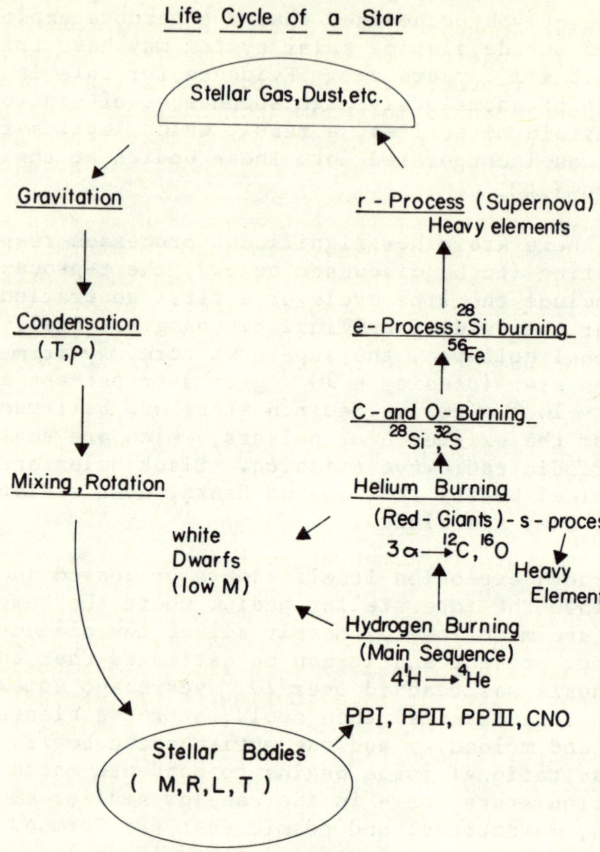

Fig. 13. Life cycle of a star. M, R, L, T Refer to the mass, radius, luminosity and temperature of a star.

5. The s-process. Up to now we have emphasized the production of new elements in the initial cycle of the star's lifetime. In later generation stars the presence of previously processed nuclear material makes it possible to form elements in many new ways. Among the most important of such mechanisms is the s-process (s for slow). This process, like the r-process, involves sequential neutron-capture and beta-decay steps, but it takes place in relatively stable stars where neutrons are produced at a slow, steady rate. For example, in red giant stars neutrons can be produced by means of reactions of the type

$$^{13}C + {}^{4}He \longrightarrow {}^{16}O + {}^{1}n$$

$$^{17}O + {}^{4}He \longrightarrow {}^{20}Ne + {}^{1}n$$

When sizeable amounts of iron group elements are present as seed nuclei, it is possible to build up heavy elements much the same way as in the r-process. However, unlike the r-process--where many neutrons are captured in a single nucleus in a matter of seconds--in the s-process a single nucleus captures a neutron every few thousand years or so (hence, slowly) and therefore beta-decay can occur before another neutron is captured. Since red giants exist for millions of years, the s-process can exert a strong influence on the production of heavy elements in stars. The difference in time scales between the s-process and the r-process results in the formation of different isotopes in the elements. For example, compare the following chain of s-process reactions on ^{56}Fe seed nuclei with that for the r-process on ^{56}Fe in the previous section:

$$^{56}_{26}\mathrm{Fe} \xrightarrow{(n,\gamma)} {}^{57}\mathrm{Fe} \xrightarrow{(n,\gamma)} {}^{58}\mathrm{Fe} \xrightarrow{(n,\gamma)} {}^{59}\mathrm{Fe} \rightarrow {}^{59}\mathrm{Co} + e^- + \bar{\nu}$$

$$^{59}\mathrm{Co} \xrightarrow{(n,\gamma)} {}^{60}\mathrm{Co} \rightarrow {}^{60}\mathrm{Ni} + e^- + \bar{\nu} \text{ etc.}$$

The r-process tends to form the heavier isotopes of a given element, whereas the s-process forms the lighter isotopes. This is illustrated in Fig. 14 where both the s- and the r-process paths are shown for a given range of atomic nuclei. The s-process can be studied extensively in nuclear reactors and is thought to be understood fairly well. It is a continuing process, as evidenced by the fact that we see the atomic spectra of element 43, technetium, in some stars. Technetium is extinct on earth because the longest lifetime of any of its isotopes is 10^6 years.

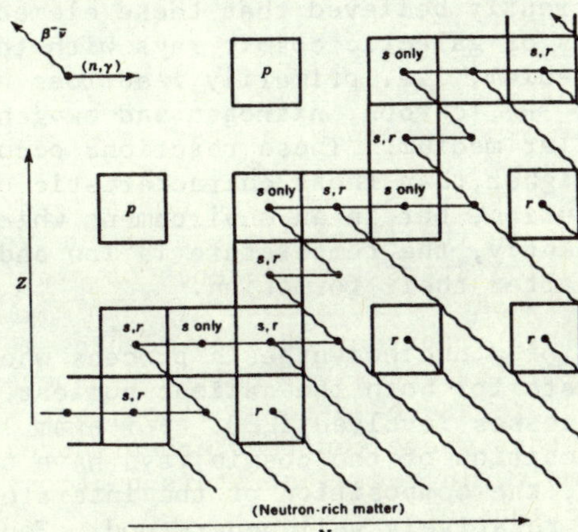

Fig. 14. Plot of N versus Z (as in Fig. 3 and 12) showing production of isotopes of the elements by the r-, s- and p-processes. The squares represent the stable isotopes of an element. Wavy lines indicate the beta-decay path of neutron-excess isotopes produced in the r-process. The solid line through the center of the stable isotopes shows the s-process path of neutron capture.

One other process, the p-process, is responsible for forming the lightest isotope of many heavy elements. In general the abundances of the p-process isotopes are extremely small because they are formed by (p,γ) and (γ,n) reactions on the s- and r-process residues.

In the various processes ranging from hydrogen burning through the s- and r-processes one sees that stellar evolution results in the formation of all of the elements between carbon and uranium, and perhaps heavier. Simultaneously, stellar evolution leads not only to the formation of the elements, but also to the richness of cosmological phenomena that we observe in our universe, ranging from main sequence stars to supernovae.

C. Nucleosynthesis in the Interstellar Medium

From our previous discussions it is seen that cosmological nucleosynthesis in the big bang and element synthesis during stellar evolution can account for nearly all the elements of the periodic table. However, three elements have been omitted in our scenario: the elements lithium, beryllium, and boron (LiBeB). As stated earlier, these elements are known to be extremely fragile and consequently cannot survive during stellar evolution. Some ^{7}Li may remain from the big bang, but the remaining isotopes of these elements, ^{6}Li, ^{9}Be, ^{10}B and ^{11}B, must have been produced by some other mechanism. One compelling clue as to the origin of LiBeB is that these elements are enriched by a factor of about 10^{6} in the galactic cosmic rays relative to solar system abundances, as shown in Fig. 15. Hence, it is currently believed that these elements have their origin in interactions of galactic cosmic rays with the gas and dust of the interstellar medium; i.e. primarily reactions of protons and ^{4}He nuclei with other ^{4}He, carbon, nitrogen and oxygen nuclei present in the interstellar medium. These reactions occur at rather high energies, much higher than those characteristic of the big bang and stellar evolution, but in an environment which has a very low density. Consequently, the temperature is low and the LiBeB products can survive after their formation.

This is one case of a nucleosynthesis process where rather extensive knowledge exists for both the salient nuclear reactions and the astrophysical processes involved (22). For example, the energy spectrum and the composition of the cosmic rays have been widely studied. Furthermore, the composition of the interstellar medium is also thought to be relatively well understood. Hence, measurement of nuclear reaction cross sections for these systems allows one to calculate the elemental abundances observed for lithium, beryllium and boron to test the validity of this proposed mechanism. In Table III theory and experiment are compared. It is found that one is able to reproduce the abundances of ^{6}Li, ^{9}Be, ^{10}B, and ^{11}B quite well with this model. However, the isotope ^{7}Li is greatly underproduced, which further strengthens the belief that ^7Li is synthesized in the big bang.

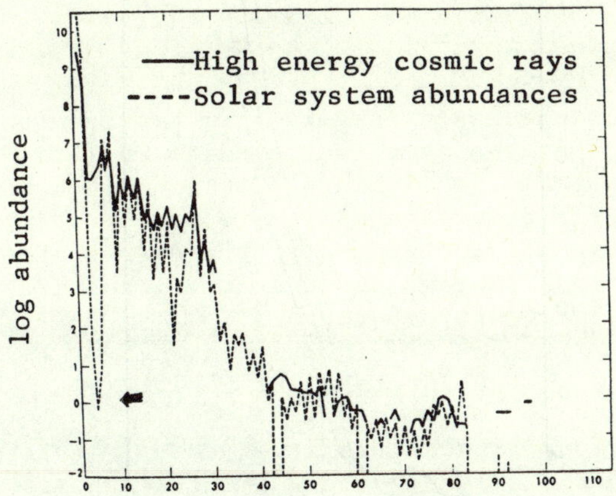

Fig. 15. Relative abundances of elements in the solar system and in the cosmic radiation measured at the top of the earth's atmosphere.

	^6Li	^7Li	^9Be	^{10}B	^{11}B
experiment	2.0(1.4)	25(1.4)	0.38(2)	1.1(3)	4.3(3)
calculated	2.6	3.4	0.43	2.1	5.5

Table III. Total abundances (Si = 10^6) for LiBeB produced in collisions between galactic cosmic rays and the interstellar medium. Numbers in parentheses are error factors. Experimental abundances are from Ref. 24 and the calculations from Ref. 23.

In fact, if one assumes that the additional ^7Li necessary to match its abundance in the solar system comes from the big bang, it is possible to infer basic conditions which characterized the big bang from this information. In Fig. 16 we show a plot of the ratio of the abundance of ^7Li which we attribute to the big bang divided by the abundance of deuterium (which is thought to have been formed only in the big bang) as a function of the matter density of the universe. The solid curves are the predicted $^7Li/^2H$ ratios for the big bang as a function of the density of the universe, based upon calculations involving all possible nuclear reaction cross sections for these species (18). It is observed that the ratio of ^7Li to deuterium corresponds to a present matter density of the universe of about 6×10^{-30} g/cm^3. This density would have been much greater at the time of the big bang, since our universe has been expanding for 15 billion years.

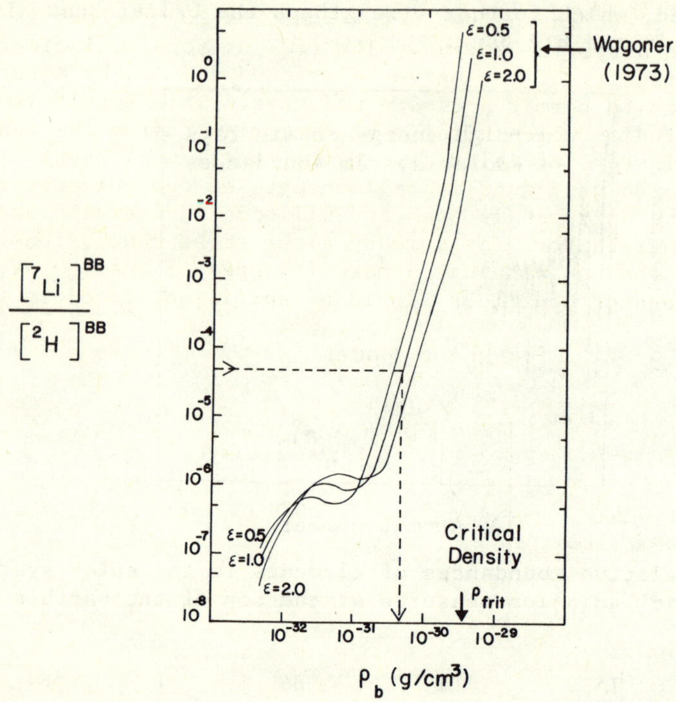

Fig. 16. Solid lines show a plot of the ratio of ^{7}Li to ^{2}H formed in the big bang as a function of the present density of the universe, as calculated by Wagoner (18). The dashed line corresponds to the observed ^{7}Li/^{2}H ratio and the corresponding density. The density required to close the universe, ρ_{crit}, is shown by the heavy arrow. Curves marked ξ = 0.5, 1.0 and 2.0 correspond to different assumptions concerning the parameters involved in the calculation.

An important question related to this density then is whether or not the universe will continue to expand forever. Do we live in an ever-expanding (open) universe, or will it eventually stop expanding under the force of gravity and contract again (a closed universe)? This is illustrated in Fig. 17 which shows the average separation between galaxies as a function of time calculated from Einstein's field equations for a baryon-dominated universe (Friedmann model). Two assumptions are made concerning the critical density of the universe, ρ_c. In the case where $\rho > \rho_c$, a primeval fireball corresponding to the big bang might once again result. If $\rho < \rho_c$, then the universe should expand forever. This is a subject of considerable debate in current astrophysical theory. Since the ^{7}Li to deuterium ratio is seen to be such a critical function of the density of the universe, one can estimate whether or not the universe is open or closed from these data. The estimated density required to close the universe is about 6×10^{-29} g/cc, as indicated by the arrow in Fig. 16. The present results indicate that the matter density of the universe is thus too low by about a factor of ten to permit a closed universe. This result is in agreement with studies which attempt to determine ρ from the mass associated with clusters of galaxies. Hence, these nucleosynthesis data also impinge on very fundamental concepts of our universe and indicate that the universe is open and will continue expanding forever. However, other factors may someday alter these conclusions. For example, if the neutrino has a mass (8) greater than 50 eV, then the total mass of neutrinos would be sufficient to close the universe.

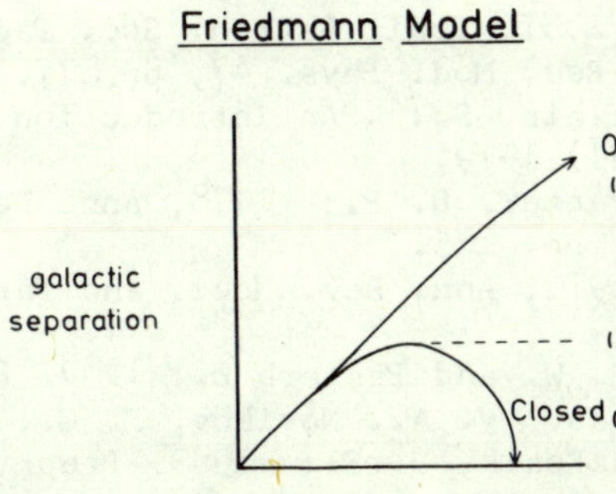

Fig. 17. Average separation between galaxies as a function of the age of the universe for two assumptions concerning the density: $\rho < \rho_c$ (open universe) and $\rho > \rho_c$ (closed universe).

IV. SUMMARY

From our previous discussions we have seen that cosmological nucleosynthesis is primarily responsible for the formation of hydrogen and ^{4}He, with a small amount of deuterium, ^{3}He and ^{7}Li also being contributed to the earth's elements. Subsequently, stellar evolution synthesizes all the nuclei between carbon and uranium (or heavier), and the abundances calculated on the basis of these two models agree rather well with experimental data. Furthermore, the elements lithium, beryllium, and boron can also be understood when one adds interactions of galactic cosmic rays with the interstellar medium. Undoubtedly, cosmic ray interactions also contribute very small amounts to the abundances of other elements, but these represent only a minor perturbation on the major element abundances. Thus, with the above processes we can synthesize the atomic nuclei that make up our universe and provide an energy source for subsequent evolution of our solar system and surrounding galactic phenomena. At this stage, then, the basic materials are present to permit the subsequent evolution of planetary bodies, atoms, molecules and eventually life.

ACKNOWLEDGEMENTS

The author wishes to thank the John Simon Guggenheim Foundation for partial financial support during the preparation of this paper. This research program is supported by grants from the U.S. Department of Energy and National Science Foundation.

References

(1) Gamow, G.: 1949, Mod. Phys. 21, pp. 367.
(2) Burbidge, E. M., Burbidge, G. R., Fowler, W. A. and Hoyle, F.: 1957, Rev. Mod. Phys. 29, pp. 547.
(3) Cameron, A. G. W.: 1957, Publ. Astron. Soc. Pac. 69, pp.201.
(4) Trimble, V.: 1975, Rev. Mod. Phys. 47, pp.877.
(5) Audouze, J. and Vauclair, S.: "An Introduction to Nuclear Astrophysics" (Reidel) 1979.
(6) Rolfs, C. and Trautvetter, H. P.: 1978, Ann. Rev. Nucl. and Particle Science 28, pp. 115.
(7) Greenberg, O. W.: 1978, Ann. Rev. Nucl. and Particle Science, 28, pp. 327.
(8) Reines, F., Sobel, H. W. and Pasierb E.: 1980; Phys. Rev. Lett. 45, pp. 1307; Lyubinsov, V. A., Novikov, E. G., Nozik, V. Z., Tretyakov, E. F. and Kosik, V. S.: 1980, Preprint ITEP-62, ITEP, Moscow; Daum, M., Eaton, G. H., Frosch, R., Hirschmann, H., McCulloch, J., Minehart, R. C. and Steiner, E.: 1979, Phys. Rev. D 20, pp. 2692.
(9) Clayton, D. D.: 1968, "Principles of Stellar Evolution and Nucleosynthesis", (McGraw-Hill).
(10) Weinberg, S.: 1974, Rev. Mod. Phys. 46, pp. 255; Scientific American 50, July, 1974.
(11) Seaborg, G. T.: 1968, Ann. Rev. Nucl. Sci. 18, pp. 53.
(12) Breuer, H., Wolf, K. L., Glagola, B. G., Kwiatkowski, K. K., Mignerey, A. C., Viola, V. E., Wilcke, W. W., Schroder, W. U., Huizenga, J. R., Hilscher, D. and Birkelund, J. R.: 1980, Phys. Rev. C 22, pp. 2454; Guerrau, D., Galin, J., Gatty, B., Tarrago, X., Girard, J., Lucas, R. and Ngo, C.: 1980, Z. Phys. A 295, pp. 105; Symons, T. J. M., Viyogi, Y. P., Westfall, G. D., Doll, P., Greiner, D. E., Farrggi, H., Lindstrom, P., Scott, D. K., Crawford, H. and McParland, C.: 1979, Phys. Rev. Lett. 42, pp. 40.
(13) Lohdi, M. K.: 1978, "Proc. of the Int. Sym. of Super-Heavy Elements", Lubbock, TX, March.
(14) Schramm, D. N. and Clayton, R. N.: 1978, Scientific American, 239, pp. 124.
(15) Weinberg, S.: 1977, "The First Three Minutes", (Basic Books, Inc., NY).
(16) Gott, J. R., Gunn, J.E., Schramm, D.N. and Tinsley, B.M.: 1974, Astrophysical J., 194, pp. 543.
(17) Weiss, R.: 1980, Ann. Rev. Astronomy and Astrophysics, 18, pp. 489.

(18) Schramm, D. N. and Wagoner, R. V.: 1977, Ann. Rev. Nucl. Sci. 27, pp. 37.
(19) Bahcall, J. N. and Davis, R., Jr.: 1976, Science 191, pp. 264.
(20) Woosley, S. E., Arnett, W. D. and Clayton, D. D.: 1974, Astrophysical J. Suppl. 26, pp.231.
(21) Seeger, D. A., Fowler, W. A. and Clayton, D. D.: 1965, Astrophysical J. Suppl 11, pp. 121; Mathews, G. J. and Viola, V. E.: 1976, Nature 261, pp. 382.
(22) Austin, Sam M.: 1980, Prog. Part. Nucl. Phys. 40, pp. 14.
(23) Read, S. M., Viola, V. E., Glagola, B. G. and Mathews, G. J.: 1981, Indiana University Report INC-4007-3.
(24) Reeves, H. and Meyer, J. P.: 1978, Astrophysical J. 226, pp. 613.

THE LARGEST MOLECULES IN SPACE: INTERSTELLAR DUST

J Mayo Greenberg

Laboratory Astrophysics, University of Leiden
P.O. Box 9504, 2300 RA Leiden, The Netherlands

Abstract.

The bulk of complex molecules in the space between the stars is shown to be in the small frozen particles of interstellar dust. Each dust grain typically contains some 10^9 atoms of oxygen, carbon and nitrogen in an amorphous molecular mixture. As a result of chemical processing of the particles by ultraviolet photons over times spanning ~ 10^8-10^9 years a substantial portion of each dust grain is converted into complex organic molecules whose maximum molecular weight is limited only by the size of the grain. Laboratory studies of evolution of analog grain materials shows that molecular weights of the order of 500 are readily created and that there is an excellent probability of much more complex molecules being produced. The organic dust component consitutes about one tenth of a percent of the total mass of the Milky Way and far outweighs any estimates of the total mass of all the planets. A planet like the earth is continually accreting matter from space and there was a high probability that in the first five hundred million years after its crust formed it passed through several dark clouds and accreted from a hundred million to ten thousand million tonnes of the organic material of the interstellar dust during each passage. It is suggested that this rain of material could have provided the molecular templates for the origin of life.

C. Ponnamperuma (ed.), Cosmochemistry and the Origin of Life, 71–112.

I. INTRODUCTION

Although the major obvious constituents of our Milky Way are the stars, the space between the stars is far from empty. Indeed it is from the matter which fills this space that new stars are continually being born. What is this interstellar material and what role can it possibly play in the story of the origin of life? These are the questions which we will attempt to answer in the course of this chapter.

a) Atoms.

Most of the interstellar medium is hydrogen which was created in the earliest stages of our universe. However as we have seen in the chapter by Dr. Viola, the formation of the heavier elements has been an ongoing process ever since the first stars were born. What we see today, on the grand scale, is a distribution of the elements which have been produced by stars and ejected back into space. The relative abundances of the elements, as observed on the average, are called cosmic abundances. A summary of abundances, for the principle elements relative to hydrogen, is shown in Table 1. Following helium -which plays no role in

Table 1. Selected abundances of the elements (From A.G.W. Cameron, 1973, Sp.Sci.Rev. 15, 121)

Element	Relative number of atoms
H	1
He	1.2×10^{-1}
C	3.70×10^{-4}
N	1.17×10^{-4}
O	6.76×10^{-4}
Ne	6.3×10^{-4}
Na	1.3×10^{-6}
Mg	3.4×10^{-5}
Si	3.2×10^{-5}
S	2.8×10^{-5}
Ca	1.6×10^{-6}
Fe	2.6×10^{-5}

the interstellar chemistry- the group of elements comprising oxygen, carbon and nitrogen constitute about one part in a thousand by number relative to hydrogen. The next most abundant group -magnesium, silicon and iron- are down further by a factor of ten. Sulphur is also a relatively abundant element, being in the class of the latter. It is obvious that in space the elements which are required for organic molecules are the most abundant.

b) Molecules and dust.

Since the advent of the discovery of formaldehyde (H_2CO) in the interstellar gas in 1969 (1.1) a wide range of molecular species have been detected. The most abundant of these is carbon monoxide (CO) but a listing of the molecules detected to date (see Table 2) is convincing evidence that there is a very active

Table 2. Molecules and ions identified in the interstellar gas.

CN	HCO^+	C_2H_2	CH_3CHCN
CH	HNC		CH_3C_2H
CH^+	OCS	HC_3N	HC_5N
H_2	H_2O	HCOOH	
OH	SO_2	H_2CNH	CH_3COOH
		H_2NCN	
CO	C_2H		
CS	N_2H^+	H_2C_2O	C_2H_5OH
SiO	HCO	HC_4	$(CH_3)_2O$
SO	HCS^+		CH_3CH_2CN
SiS		CH_3CN	HC_7N
NS	NH_3	$(NH_2)HCO$	
C_2	H_2CO	CH_3OH	HC_9N
NO	H_2CS	CH_3SH	
	HNCO		$HC_{11}N$
H_2O	C_3N	CH_3COH	
HCN	HNCS	CH_3NH_2	

chemical factory in space (1.2). However, as we shall see, these are only the "tip of the iceberg". The most complex and abundant molecules are not in the gas but rather must be in small solid particles called interstellar dust which float about in the gas. What these dust particles are, both chemically and physically, will therefore be our principle theme. However, first let us consider where and how much dust there is, and how this relates to the amount and presence of hydrogen and other molecules.

c) Distribution of gas and dust.

In Figure 1 we see an example of a spiral galaxy (M51). The spiral arms are seen because they contain the brightest stars. However, the spiral arms are perhaps even better delineated by the dark lanes at the inner edges of the arms. These dark lanes are produced by the blocking of the light of the stars by concentrations of the interstellar dust acting just like a smoke screen. In Fig. 1 a superposition of the peak gas contours is seen to follow very much the dark dust lanes. This shows that where the dust is dense so is the gas; i.e. the dust is correlated with the gas. Fig. 2a shows how a spiral galaxy looks when seen edge-on. The dark band separating the two halves of the galaxy is again produced by the light blocking power of the dust which is concentrated in a thin disk within the galaxy of stars.

Figure 1. Photograph of a spiral galaxy (M51) seen face-on. The dark bands at the inner edges of the spiral arms are the clearly defined concentrations of the dust. Superimposed on the photograph are solid lines showing the peak in the continuum radiation at 1415 MHz. These lines appear to coincide well with the dust lanes (From Mathewson et al. 1972, Astron. Astrophys. 17, 468).

It turns out, from observations of the gas, that the gas is also concentrated within a thin disk along with the youngest population of stars. Our own Milky Way, as seen from the location of the sun indicates clearly in Fig. 2b the presence of the dust in the mid plane.

Figure 2a.

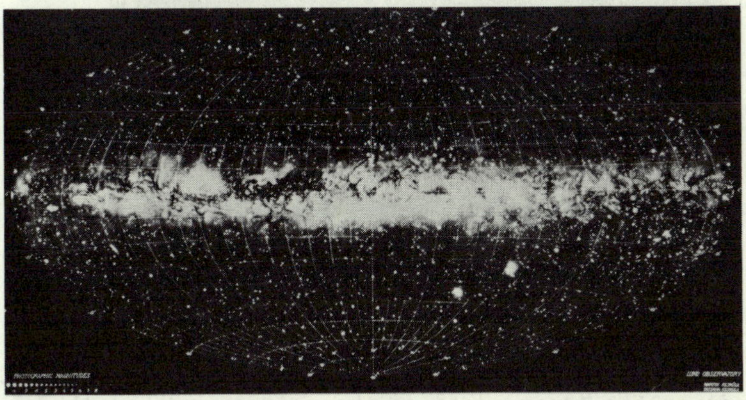

Figure 2b.

Fig. 2.

(a) The distant spiral galaxy NGC 4565. This galaxy seen on edge exhibits a dark band in its central plane, indicating the concentration of extinction by interstellar dust.

(b) Composite photograph of the Milky Way. The dark patches running along the central plane show the obscuration by extensive distributions of dark clouds of interstellar dust (Courtesy Lund Observatory).

On a more localized scale the dust is concentrated in molecular clouds and in regions of new star formation. An excellent example is the Orion nebula (Fig. 3) which is both a site for

Figure 3. The Orion nebula: A region of protostars, young stars, gas, molecules and dust.

active star formation and -not just coincidentally- also molecules. Although the dust may not be apparent in this photograph, it is well observed both by its blocking of star light and by its infrared emission caused by heating near the stars. Not far from Orion is a very dense cloud in the shape (projected) of a horse head (Fig. 4). Here the dust is so concentrated that essentially no stars can be seen through it. Those we do see in the "horse's head" are foreground stars. A further example of how dust can be localized -this time in an isolated region- is shown in Fig. 5. In the case of Orion we are seeing dust involved in the star formation process both before and after. However, there are also clouds of dust which float in space and which are only accidentally seen in the vicinity of stars. Such a case is shown in Fig. 6 where the dust is seen by scattering of the light of the stars in the Pleiades constellation. The relative velocities of the stars and the clouds have

Figure 4. A very dark cloud-the horsehead nebula-in the Orion region. Note the bright rim at its edge (Mount Wilson and Palomar Observatories).

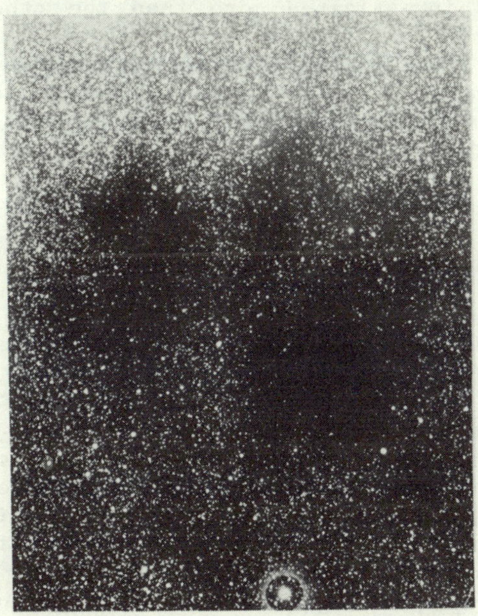

Figure 5. A Bok globule. An isolated cloud of dust and gas.

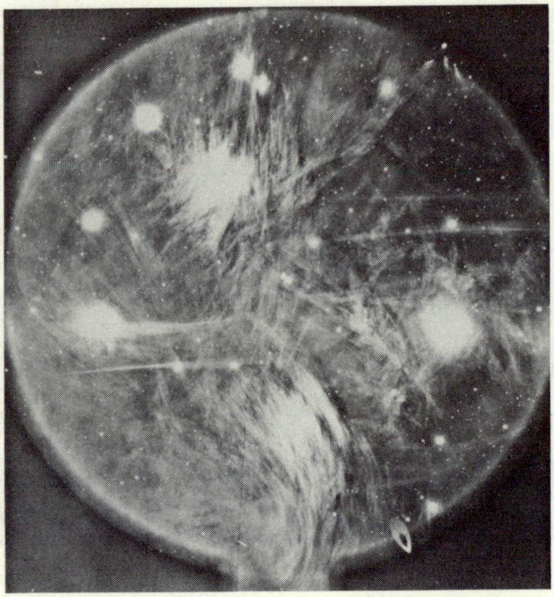

Figure 6. A photograph of the Pleiades showing reflection nebulosity. Merope, the star illuminating the dust cloud at the bottom, has been investigated more than the others because of the hitherto asumed uniformity of the cloud with which it is associated. This new photograph taken with the 4 m telescope at Kitt Peak National Observatory shows a wealth of structure in the cloud indicating a substantial amount of streaming. It should be noted that the Merope cloud probably has a densisty of $n_H \simeq 10^3$ (Greenberg and Roark, 1967) and is therefore a "dark" cloud in this sense.

been determined to be ~ 30 km^{-1} which means that, when the stars were born (some 5×10^7 years ago), the clouds were at least 4500 light years (1500 parsecs) distant from the stars and therefore in an entirely different region of the sky.

Although we can say that the dust and gas have a well-defined general distribution within the galaxy, a brief glance at the night sky reveals a highly inhomogeneous and patchy structure. The interstellar medium actually consists of a chaotic distribution of gas/dust clouds with a variety of densities always in motion and passing from one phase to another in their evolution with the most dramatic phase appearing when new stars are formed. The widest variety of molecules is associated with clouds in intimate association with star formation

d) Quantity of dust and molecules.
Surprisingly, although the qualitative evidence for dust clouds seems obvious to us now, it was at one time denied with learned conviction by those who believed just the opposite; for example, in a textbook written at the turn of the century (1.3) it was possible to find statements like: "The fact is a general one, that in all the forest of the universe there are glades and clearing. How they come to be thus diversified we cannot pretend to say; but we can see that the peculiarity is structural -- that it is an outcome of the fundamental laws governing the distribution of cosmic matter. Hence the *futility* of trying to explain it as of incidental origin, as a consequence, for instance, of the stoppage of light by the interposition of obscure bodies, or aggregation of bodies, invisibly thronging space." It was only with the discovery by Trumpler in 1932 (1.4) that the general reddening (not the Doppler shift) of the light of stars correlated with distance that the evidence for interstellar dust was generally accepted. It is from this kind of quantitative observation of the wavelength dependence of the light blocking power of the dust (greater in the blue than in the red) along with similar observations of polarization and scattering by dust that we can arrive at a physical description of the dust grains. We find that the principal particles are nonspherical with a mean size of the order of 0.2 to 0.3 μm thick. From the combination of observations of the mean amount of starlight extinction per unit distance and from the mean size of the dust grains we can derive a mean spatial number density of about one particle per 100 meter cubed; i.e. about one particle in a cube whose sides are larger than a football field! In dense clouds this density may be higher by a factor of 10^4 to 10^6 and, in protostellar nebulae, much higher. Going with the dust, or vice versa, is a mean hydrogen density of about $n_H = 1\ cm^{-3}$. We see that, by terrestrial standards, even in a dense interstellar cloud the pressure is $p \simeq 3n_H\ 10^{-20}$ mbar which is a very good vacuum. However the vastness of space more than compensates for the low density when one considers the total mass of the interstellar material. For example, although the one hydrogen atom per cubic centimeter gives an average spatial density of $1.66 \times 10^{-24} g\ cm^{-3}$, the mean stellar number density gives only a factor of ten greater; i.e. the mass of gas (mostly hydrogen) relative to stars is about 10%. Similarly mass density of the dust relative to gas turns out to be about 1%, so that relative to the stars it is 1 ÷ 1000. This is a large amount when compared with estimates of the total planetary mass. In our solar system, for example, the mass of all planets together -mostly in Jupiter- amounts to about 1 ÷ 1000 relative to the sun's mass (1.5). Therefore, even if every star in our Milky Way has a planetary system like ours, their mass would not exceed that of the interstellar dust. One can therefore state with confidence that

the overall mass of the small dust particles in space probably exceeds that of all possible earth-like planets by many order of magnitude. We shall come back to this later because a mere comparison of relative masses of two species of objects does not necessarily define their relative importance in a given context.

The molecules are generally concentrated in the denser regions in space although evidence for some molecules like HC_5N, H_2CO and certainly CO is seen in relatively tenuous regions called diffuse clouds with hydrogen densities of $n_H \simeq 10$ cm^{-3}. Although the total number of molecules identified is large -about 50- the absolute number is, with the exception of CO, quite small with respect to the cosmic abundance availability of their constituents. In fact, if we count all the molecules seen and compare their total mass with that of the dust it is less by a factor of at least 100 unless we include CO in which case it may be as large as, but generally considerably less than, 1/2.

e) The interstellar environment.
The physical conditions in interstellar space are dominated by the effects of stars both distant and nearby to the regions in question. We shall not concern ourselves here with the mechanisms by which these conditions are created but will rather limit ourselves to a description of those properties which will be needed in the ensuing discussions.

Although we refer to varying densities of gas and dust as clouds, we should bear in mind that these objects are not always simply described geometrically. They can be highly irregular in shape and are known to have substructures of different densities from the mean density. Nevertheless, for convenience, we categorize regions of different densities by rather specific namens. The term diffuse clouds refers to spatial densities of hydrogen of the order of $n_H = 10$ cm^{-3}. At about ten times this density, $n_H = 100$ cm^{-3}, we find that most of the hydrogen has become molecular (H_2) and furthermore we find relatively abundant quantities of CO such that the number of density of CO relative to hydrogen is $[CO]/[H] \simeq 10^{-4}$ (1.6). At somewhat higher densities $n_H \geq 10^3$ cm^{-3} we find such molecules as H_2CO and many of the others listed in Table 2 but in any one region the number of kinds of molecules detected is normally only a few of the 50. Broadly speaking, clouds with densities of $n_H \geq 10^2$ cm^{-3} are called molecular clouds.

Between the clouds is a far more tenuous medium with a mean hydrogen density of $n_H \simeq 0.1$ cm^{-3}. It turns out that at this time our solar system is passing through such an <u>intercloud</u> medium.

The typical gas kinetic temperatures of diffuse clouds is T = 100 K.

For molecular clouds the temperature range is $10\ K \lesssim T \lesssim 50\ K$. Thus in a diffuse cloud a hydrogen atom has a typical speed of about $v_H \simeq 1.5\ km\ s^{-1}$ and in a molecular cloud $T \simeq 10-50\ K$ so that a CO molecule has a gas kinetic speed of about $v_{CO} \simeq 0.1\ km\ s^{-1}$.

Close to hot young stars the temperature of the gas can be as high as 10,000 K in a region of fully ionized hydrogen, thus an HII region. In the cool clouds, the hydrogen is un-ionized. We refer to these as HI regions.

The interstellar medium is far from thermal equilibrium. Curiously, although both the gas temperature and the dust temperature are basically governed by the radiation from stars, the dust is generally much colder than the gas- the only exceptions being in regions of very high gas density. The mean temperature of the dust grains, which is reached by a balance between absorption of radiation from the ambient field and emission of radiation by the small dust grains, is only about 10 K (1.7). It is only near energetic sources of radiation that grain temperatures get as high as 50 or 100 K. Of course, very close to hot stars temperatures in the thousands- leading to complete evaporation- occur. But this affects only a very small fraction of all the dust grains at any one time. In Table 3 the mean properties of the interstellar medium are summarized.

Table 3. Average Interstellar Medium

	Gas	
$0.1 \leftarrow$	$\langle n_H \rangle = 1\ cm^{-3}$	$\rightarrow 10^5$
	$\langle n_{O+C+N} \rangle \simeq 10^{-3} n_H$	
	$O : C : N \simeq 6.8 : 3.7 : 1$	
	$\langle n_{Mg+Si+Fe} \rangle \simeq 10^{-4} n_H$	
	$Mg : Si : Fe \simeq 1 : 1 : 1$	
$10,000\ K \leftarrow$	$T_{gas} = 100\ K \rightarrow$	$10K$
	Radiation in Ultraviolet	
	$\langle n_{\lambda < \lambda_t} \rangle = 3 \times 10^{-3}\ cm^{-3}$	
	$\lambda_t = 2000\ Å : h\nu_t = 6\ eV$	
	Dust	
	$\langle n_d \rangle \simeq 10^{-12}\ n_H$	
	$T_{dust} = 10\ K$	
	$\bar{a}_d \simeq 0.12\ \mu m$	
	$[a_{core} = 0.05\ \mu m,\ \bar{a}_{mantle} = 0.12\ \mu m]$	
	$a_{bare} \simeq 0.005\ \mu m$	
	$n_{bare} \simeq 10^3\ n_d$	

2. PHOTOCHEMICAL PROCESSING OF DUST.

Where does the dust come from and what is it made of? Although the major components of the interstellar material -the hydrogen- is a result of promordial processes, a large fraction of the elements heavier than hydrogen, which we now find in the gas and dust in space, has been cycled through stars since the birth of the solar system. An estimate of the rate of star formation indicates that in each passage through a spiral arm approximately 1/50 of the medium goes to form new stars (2.1). Since the mean interval between spiral arm traversals is about 10^8 years, we conclude that all the interstellar material is recycled each $50 \times 10^8 = 5 \times 10^9$ years, which is roughly the age of the solar system. This means that if the dust we see today is in steady state it must be created at a rate fast enough to equal its loss rate into stars. Although we have no direct way of knowing whether such a steady state exists, we can derive values for the current production rate of dust and infer that at least at the time of the birth of the solar system, the dust was pretty much the same as it is now.

a) Production and growth of dust grains.
It is apparently impossible for dust particles to condense directly from the gas in interstellar space. In order for condensation to take place there must first be nucleation cores. We have rather convincing evidence for the production of such nucleation cores which we derive from the infrared spectra of evolved cool oxygen rich stars. It is well known that such stars are undergoing mass loss by ejection of gas and that their spectrum shows an infrared excess (2.2) at about 10 μm characteristic of small solid particles of a silicate type material (the 10 μm band is typical of the Si-O stretch in silicates). Combining these two ideas we conclude that solid silicate particles are ejected into space (2.3). Current mass loss estimates for cool evolved stars (M supergiants) lead to a production rate for silicate particles which is sufficient to roughly balance the loss of such particles in the star formation process (2.4). It is certainly clear that such particles exist in space because we see their 10 μm (actually 9.7 μm) absorption in the light from stars which has passed through interstellar clouds (see Fig. 7). In Figure 7 not only the silicate absorption, but also a strong absorption at 3.07 appears (2.5). This band is caused by the O-H stretch in H_2O ice and shows that along with the refractory silicates we have other frozen materials of a much more volatile nature. This suggests that somehow volatiles can condense on the already present silicate particles. However a simple numerical estimate, based on the observed abundance of H_2O in the gas, shows that the solid H_2O of the dust grains is far too much to have come from the gas (2.6).

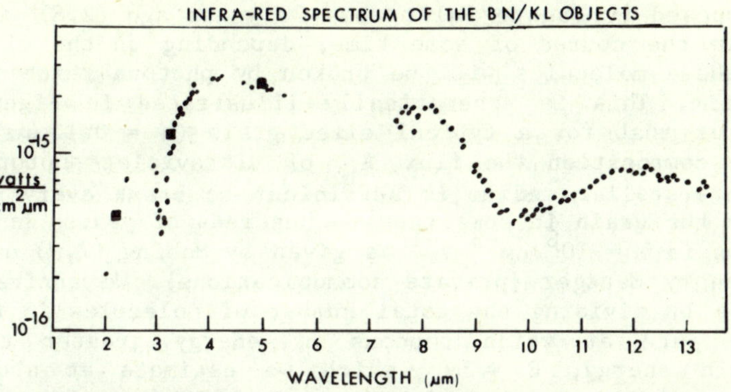

Figure 7. An early infrared absorption spectra of the BN source.

It must have been made on the dust grain by condensation of a much more abundant species -in this case oxygen- and subsequent combination with the ubiquitous hydrogen. This implies that under some conditions, the elements heavier than hydrogen which includes carbon and nitrogen as well as oxygen- must condense on grains. However, although the objects shown in Figure 7 exhibit an H_2O ice band, it is well known that the majority of stars which are obscured by dust do not have a distinguishable ice band even though oxygen is abundant enough to produce it if it were in its saturated form of H_2O. This apparent puzzle must be answered by considering, along with condensation on grains, all the physical processes involved in the evolution of a grain -from the birth of the cores through accretion of mantles and eventually to total annihilation in the birth of new stars.

b) Model for interstellar grains.

A working model for interstellar grains (2.7) consists primarily of two characteristic sizes of particles: (1) core-mantle particles of size $\simeq$ 0.12 µm (radius) with cores of silicates of radius $\simeq$ 0.05 µm and mantles of accreted molecules made up mainly of oxygen, nitrogen, carbon and hydrogen; (2) Bare particles of size $\lesssim$ 0.01 µm (radius) probably made of individual particles of silicates, carbon and perhaps large organic molecules. The bulk of the mass of the dust is in the core-mantle particles and we shall limit our discussion primarily to their chemical evolution.

c) Effect of ultraviolet on a dust grain.

While a core-mantle particle floats about in a cloud of gas it is continually being bombarded by energetic ultraviolet photons. These photons originate either in distant or nearby stars or arise from shocks produced by stellar winds within the clouds. If we were to consider the composition of a grain mantle as consisting, for example, of a frozen mixture of such simple molecules as water (H_2O), methane (CH_4) and ammonia (NH_3), as was proposed by Van de Hulst some 30 years ago (2.8) we realize that, in the course of some time, depending on the ultraviolet flux, these molecules will be broken by photons which penetrate the grain. This is schematically illustrated in Figure 8. It turns out that for a typical sized grain ($a = 0.12\ \mu m$) of such initial composition the flux, $\Phi_{h\nu}$ of ultraviolet photons in the mean interstellar medium is sufficient to break every molecular bond in the grain in some tens to hundreds of years; hundreds if the flux in $\Phi = 10^8\ cm^{-3}\ s^{-1}$ as given by Habing (2.9) or tens if as given by Metzger (private communications). We arrive at this estimate by dividing the total number of molecules in the grain by the rate at which photons of energy greater than some threshold energy, $E = h\nu_t$ - which we estimate at about 6 eV -impinge on the grain. From here on I shall drop the subscript t in $h\nu_t$. See Table 4 for some photodissociation energies.

Table 4. Some approximate molecule and radical bond dissociation energies adopted from Calvert and Pitts, 1966, Photochemistry, J. Wiley, and Sons, N.Y.

Bond Broken R-R'	ΔE (eV)	Bond Broken R-R'	ΔE (eV)
OH - H	5.14	CH_2CH - H	4.54
O - H	4.40	CH_3 - CH_2O	0.54
O - O	5.16	OCH_2 - H	1.02
S - O	5.44	HCO - H	3.79
OS - O	5.66	H_2C - O	7.59
H - H	4.50	CH_3CO - H	3.79
C - O	11.17	H - CO	0.755
OC - O	5.48	C_2H_5 - H	4.27
CH_2 - CO	2.32	CH_3 - CH_3	3.62
CH_3 - OH	3.84	N - O	6.52
CH_3 - H	4.40	NH_2 - H	4.47
C - H	3.53	NH - H	4.14
CH - H	5.48	N - N	9.78

The collision rate is

$$\frac{d(h\nu)}{dt} = \Phi_{h\nu} \pi a^2 \qquad (1)$$

where $\Phi_{h\nu}$ is the flux of photons with energies greater than $h\nu$ and the number of molecules (or bonds) is roughly

$$N_{mol} = \frac{4/3\ \pi a^3 s}{\overline{M} m_H} \qquad (2)$$

where s is the specific gravity, $\overline{M}$ is the mean molecular weight of an H_2O, CH_4, NH_3 mixture in relative cosmic abundance, and m_H is the mass of hydrogen. Thus the time scale for total photon bombardment is obtained by dividing Equation 2 by Equation 1.

$$\tau_{tot} = N_{mol}/d(h\nu)/dt = \frac{4\ as}{3\bar{M}m_H\ \Phi\ h\nu} \qquad (3)$$

$$= 5.7 \times 10^9\ \mathrm{sec} \simeq 200\ \mathrm{yrs}.$$

where we have used s = 1, $\bar{M}$ = 17, $\Phi = 10^8\ cm^{-2}\ s^{-1}$ (see Table 3).

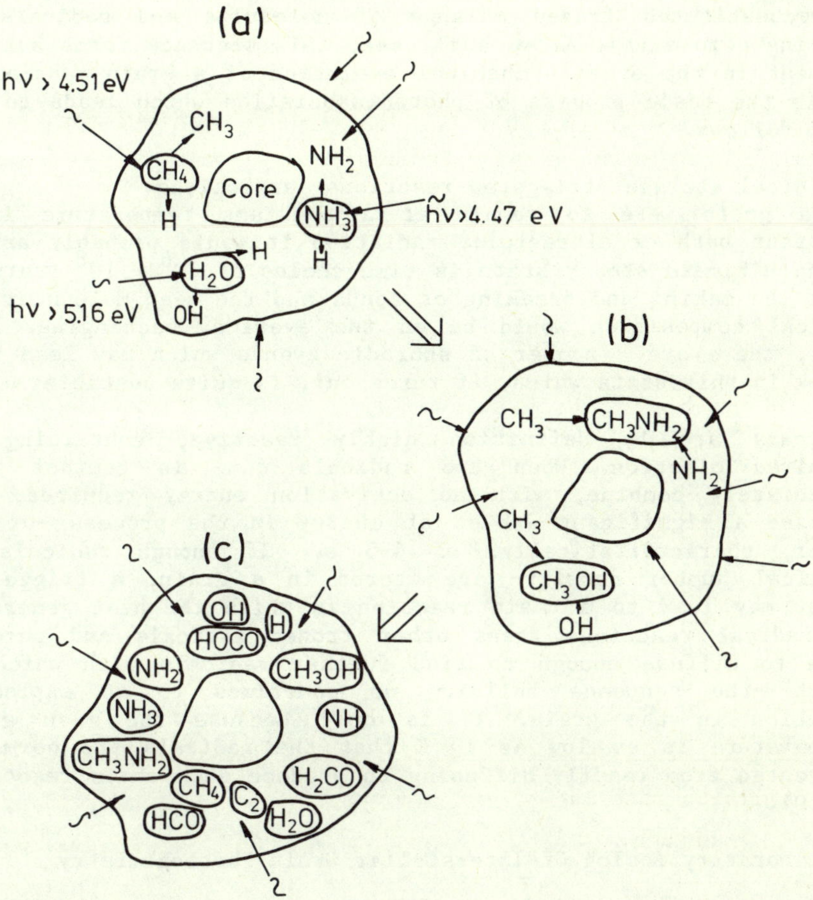

Figure 8. Schematic evolution sequence for a grain mantle at 10 K subjected to ultraviolet photolysis. The processes illustrated are photodissociation, radical-radical combination, production of new molecules and radicals.

Since the time scales which prevail in the interstellar medium are generally in the range of 10^6 - 10^8 years we see that a grain has little chance of remaining chemically static.

What can occur within an individual grain as a result of ultraviolet photons is pictured in Figure 8. In the first instance ultraviolet photons may break up the molecules which are in the grain, leaving the radical pieces frozen in. Sometimes these pieces may recombine to produce the original molecule. Sometimes, as pictured in the second step, adjacent radicals recombine to form new molecules. And sometimes, a radical may remain without reacting. At some stage, pictured in the last step of Figure 8, we envisage a grain which consists of a reconstituted frozen melange of molecules and radicals of varying complexity. As we shall see, this sequence forms but one element in the overall chemical evolution of a grain mantle but it is the basic process of photodissociation which leads to all that follows.

d)Radical storage -triggered reactions- explosions.
If a grain were to remain at a constant temperature in a constant bath of ultraviolet radiation it would probably arrive at an ultimate steady state in time scales of $10^6 - 10^8$ years so that the making and breaking of bonds and the mean molecular and radical composition would be on the average, unchanging. However, there are a number of sporadic events which may lead to a break in this state which, it turns out, is quite unstable.

Radicals are by definition highly reactive, containing an unpaired electron. When two radicals come in contact they immediately combine, with no activation energy required, and release a significant amount of energy in the process -of the order, characteristically, of 4-5 eV. If enough radicals -a critical number density- are stored in a grain, a triggering event may lead to a chain reaction in which the heat generated by radical reactions frees other frozen radicals and permits them to diffuse enough to find further radicals with which to react -the sequence building up sometimes to an explosive reaction in the grain. It is only because the mean grain temperature is as low as 10 K that the radicals are normally prevented from readily diffusing to produce such chain reactions (2.10).

3. Laboratory Analog Of Interstellar Grain Photochemistry

The Astrophysical Laboratory at the University of Leiden is the first to succeed in simulating the essential interstellar space conditions as they affect the evolution of interstellar grains. Although the impetus of this work arose exclusively out of an attempt to answer purely astronomical questions, its extension to a possible connection with problems of prebiological interest was quite natural, even inevitable.

In the 1940's the "dirty ice" model was proposed for the

interstellar dust. In this model atoms of oxygen, carbon, nitrogen and hydrogen were assumed to accrete on the particle surfaces and combine with hydrogen to form a frost of the saturated molecules water (H_2O), methane (CH_4) and ammonia (NH_3) in relative abundances proportional to their cosmic abundances, thus leading to a particle dominated by H_2O ice by with other ingredients; thus the name "dirty ice".

With the introduction of new infrared detection techniques to astronomical problems it became possible in the mid-1960's to search for the expected strong H_2O ice absorption bond at 3 microns due to the O-H stretch (3.1). The initial results of such observations were negative in that they indicated H_2O ice to be far less abundant than predicted, this brought the dirty ice model into question. An alternative suggestion which allows for abundant oxygen in grains without much H_2O is that the general ultraviolet radiation field in space is capable of breaking the H_2O and other molecules in the solid particles so that in recombining, the oxygen need not return to the H_2O form -and therefore not exhibit the 3 μm feature. This proposal was given a strong additional push with the announcement of the discovery in space of the (then) surprisingly complex molecule formaldehyde (H_2CO) by radio astronomers (3.2). The obvious extension of the photodissociation-recombination proposal for reducing the H_2O abundance was that a simple molecule like H_2CO would be one of the most likely molecules produced by recombinations. The next step -to consideration of much more complex molecules- followed easily. Indeed, there was an analogy between the photochemical processing of the small interstellar grains (3.3) and the energetically (lightning, etc.) induced chemical processing in the primitive earth's atmosphere as simulated by Miller and Urey (3.4).

However, though some of the processes which could be induced by ultraviolet radiation in interstellar grains must resemble those which were presumed to have occurred in the atmosphere, there are also some basic differences: one is that at the very low temperatures of grains the broken molecular species are inhibited from moving freely to find a partner to recombine with; and secondly, that the molecules in the solid are at all times in direct contact with each other rather than widely separated as in a gas, so that relaxation times are orders of magnitude smaller.

How the mixing and recombination occurs in the dust involves some processes, in addition to the basic ultraviolet photodissociation, which occur as the grains move about in space. This will be discussed in section 4.

a) Laboratory methods.

A schematic of the main elements of the experimental set-up is shown in Figure 9. The two key components are the low temperature and the ultraviolet. The low temperature is achieved by

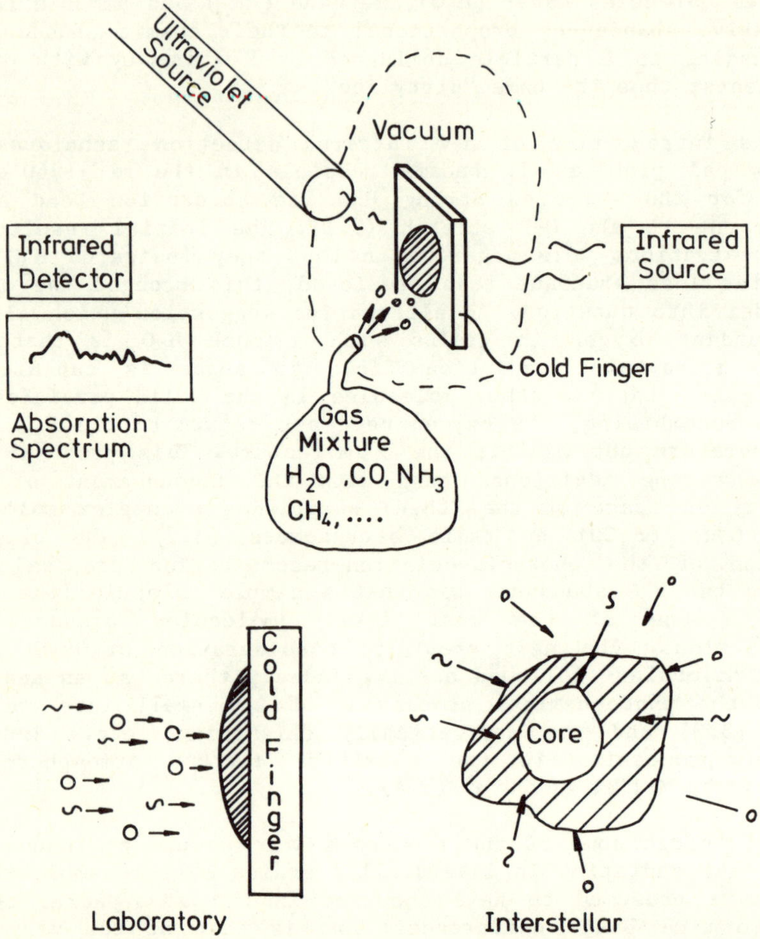

Figure 9. Schematic of the laboratory analog method for studying interstellar grain evolution. Molecules are deposted as a solid on a cold finger in a vacuum chamber and irradiated by ultraviolet photons. The infrared absorption spectrum shows the appearance and disappearance of various molecules and radicals. The cold finger may be an aluminium (~ 3 cm) or a glass, sapphire or LIF window.

means of a closed cycle helium cryostat within which one gets temperatures as low as 10 K on a "cold finger" which can variously be an aluminium block or a transparent window mounted on a metal ring. Various gases may be controllably allowed to enter the vacuum chamber of the cryostat (pressure down to 10^{-8} torr) via a capillary tube. These gases condense as a frost on the cold finger which acts then like the core of the interstellar grains. On one port to the chamber is mounted a source of vacuum ulraviolet radiation which until now has almost exclusively been a microwave stimulated flow hydrogen lamp which has emission peaks at 1216 Å (Lyα) and about 1600 Å. The normal flux of vacuum UV photons by these lamps is ~ 10^{15} cm^{-2} s^{-1}. Through another port (or pair or ports) we may direct the beam of an infrared spectrometer which measures the infrared absorptions in the sample on the cold finger in the range 2.5 μ - 25 μ (4000 cm^{-1} to 400 cm^{-1}). This is the "finger-print" region for identifying molecules by their stretching, bending and rocking modes of oscillation in a solid. Other measurements made are of pressure, chemiluminescence, mass spectra, and visible absorption. Further descriptive details of the equipment as it existed several years ago may be found elsewhere (3.5).

Table 5. Comparison between laboratory and interstellar conditions

	Laboratory	Interstellar
Grain mantle		
- initial composition	CO, H_2O, NH_3, CH_4..	All condensible species
- thickness	0.5 μm t 10 μm	≃ 0.1 μm
- temperature	≥ 10 K	≥ 10 K
Gas: pressure of condensible species	10^{-7} mbar	$3n_{[H]}$ x 10^{-20} mbar
Ultraviolet flux λ < 2000 Å	10^{15} cm^{-2} s^{-1}	10^{8} cm^{-2} s^{-1}
Time scales		
- Diffuse clouds	1 hr.	10^3 yrs.
- Molecular clouds	1 hr.	~ 10^4 - 10^6 yrs.

A comparison between laboratory and interstellar conditions is seen in Table 5. The most important -but necessary- difference is in the time scales for photoprocessing. One hour in the laboratory is equivalent to one thousand years in the diffuse cloud medium and as much as or greater than ten thousand to one million years in the depths of dense molecular clouds.

b) Types of experiments.

The basic mode of operation consists of deposition of mixtures of simple volatile molecules -CH_4, CO, CO_2, O_2, N_2, NH_3, H_2O-

and simultaneous irradiation as they freeze on the cold finger. Sometimes irradiation is continued after deposition is stopped. We simulate in this way the accretion and photoprocessing of grains in molecular clouds. The principal laboratory sequence and operations are:

1. Infrared absorption spectra of unirradiated pure samples and mixtures at 10° K. Infrared studies also made of warmed and recooled samples.
2. Infrared spectra following irradiation to detect presence of radicals and new molecules.
3. Infrared spectra of irradiated material following warmup to detect disappearance of frozen radicals and formation of new molecules.
4. Visible absorption spectra of irradiated and warmed up samples.
5. Simultaneous measures of chemiluminescence (visible) and vapor pressure during warmup of irradiated samples.
6. Production of explosion in warmup period by thermally insullating sample from cold finger.
7. Infrared and mass spectrometric studies of complex non-volatile residues remaining after warmup to room temperature.
8. Visible absorption spectra of non-volatile residue.

Detailed chemical analyses of the nonvolatile residues are planned for the future in collaboration with Professor J. Ferris of Rensselaer Polytechnic Institute and Professor H.J. de Jong at Leiden University.

c) Some Experimental Results.
The basic justification for the experiments which show how interstellar grains evolve towards a chemical composition of prebiological interest must come from <u>comparison</u> with astronomical observations. In the following we shall show how certain key experiments have demonstrated the validity of the assumptions in the laboratory analog method.

1: When the shape of the H_2O ice band was first observed in astronomical spectra (3.6), it was quite different from the absorption of H_2O ice as then measured in the laboratory (3.7). In Figure 10, a comparison is shown between pure amorphous ice deposited at 10° K (3.8) and crystalline ice. It is seen that the astronomical ice band is much closer in shape and in peak position with the amorphous ice band. Only the long wavelength wing is lacking. This wing is, however, produced in appropriate mixtures and a detailed laboratory study of effects of particle sizes and

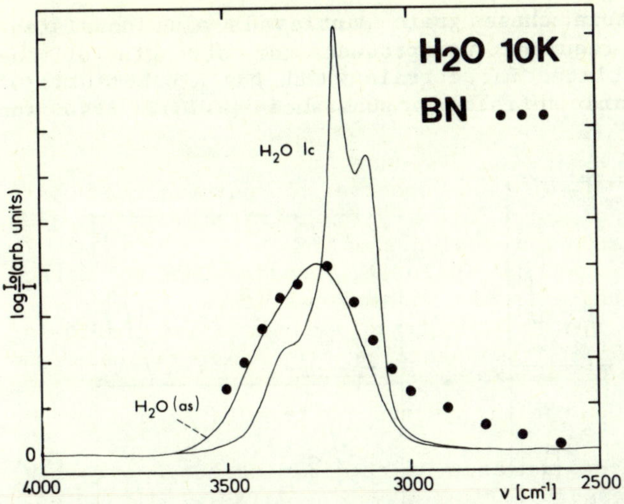

Figure 10. Absorption of amorphous ice H_2O(as) at 10 K and crystalline ice H_2O (Ic). The shape of the ice absorption in BN is shown by the dots.

temperatures shows how the laboratory spectrum can provide a good explanation (3.9) of the shape of the ice band as it appears both in absorption and polarization (Figure 11).

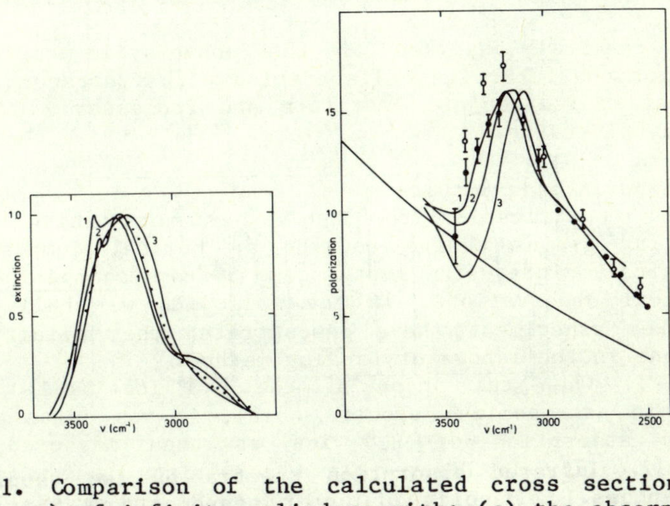

Figure 11. Comparison of the calculated cross sections (3250 cm^{-1} feature) of infinite cylinders with: (a) the absorption and (b) the polarization towards BN.

1. Unannealed mixtrue of H_2O/NH_3 (3/1) (T = 10 K). Mean size 0.15 μm.
2. Partially annealed mixture of H_2O/NH_3 (3/1) (deposited at 10 K warmup to 50 K and recooled to 10 K). Mean size 0.15 μm.
3. Unannealed H_2O(as) (10 K). Mean size 0.4 μm.

Dots (Gillett et al., 1975); open circles (Kobayashi et al., 1980); filled circles (Capps et al., (1978).

The fraction of H_2O in these grain mantles is also consistent with the fraction required to produce the strength of the 3 μm feature as calculated for a grain which has a sub-mantle of the important organic refractory substance (3.10). Also see section 4.

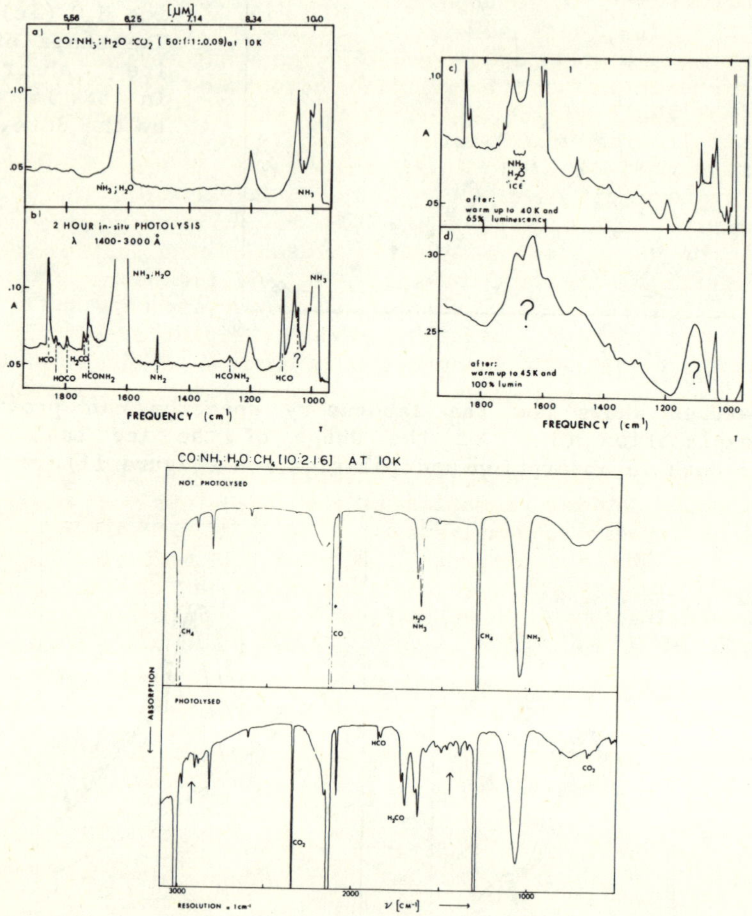

Figure 12. Infrared absorption spectra of two sample analog grain mantles. Left side of upper sequence and the two lower spectra show first the features in unirradiated samples and then the spectra of the irradiated samples showing the appearance of new molecules and radicals produced by photoprocessing. Upper right hand pair of spectra clearly indicate the presence of complex molecules (unidentified) which appear as the more volatile species are evaporated away by warming up.

2,3: In Figure 12 two sequences of irradiation of mixtures are shown by their infrared absorption spectra. In both sequences, some of the new species of molecules and radicals which appear following ultraviolet photolysis are labeled. In the upper sequence, an affect of particular note is that, following warmup, there appears a great change in the character of the spectrum (here shown only in the range < 1800 cm^{-1} but similarly at higher frequencies). The features which appear after evaporation of the volatile components are much broader and are on their way to appearing as they do in the nonvolatile residue following total warmup (see Figure 16). In the lower pair of comparison spectra see the growth of the formaldehyde (H_2CO) feature at around 6 μm and compare it with the H_2O and NH_3 features. It appears that H_2CO is produced abundantly in situ within the grain analog material. The formaldehyde molecule is not terribly abundant in space, but so far no mechanism for its formation in the gas phase has proven adequate to account for what is observed consistent with its depletion by accretion on grains (3.11). Consequently, its copious production within the grains and a mechanism for molecule ejection is required. The latter is shown below in (6:).

Finally, the frozen radicals -exemplified here by the formyl radical HCO- appear immediately in the infrared absorption spectra of irradiated samples.

4: Visible absorption by the HCO radical is seen in Figure 13. These visible absorptions disappear as the sample is warmed showing that HCO diffuses and combines with other species.

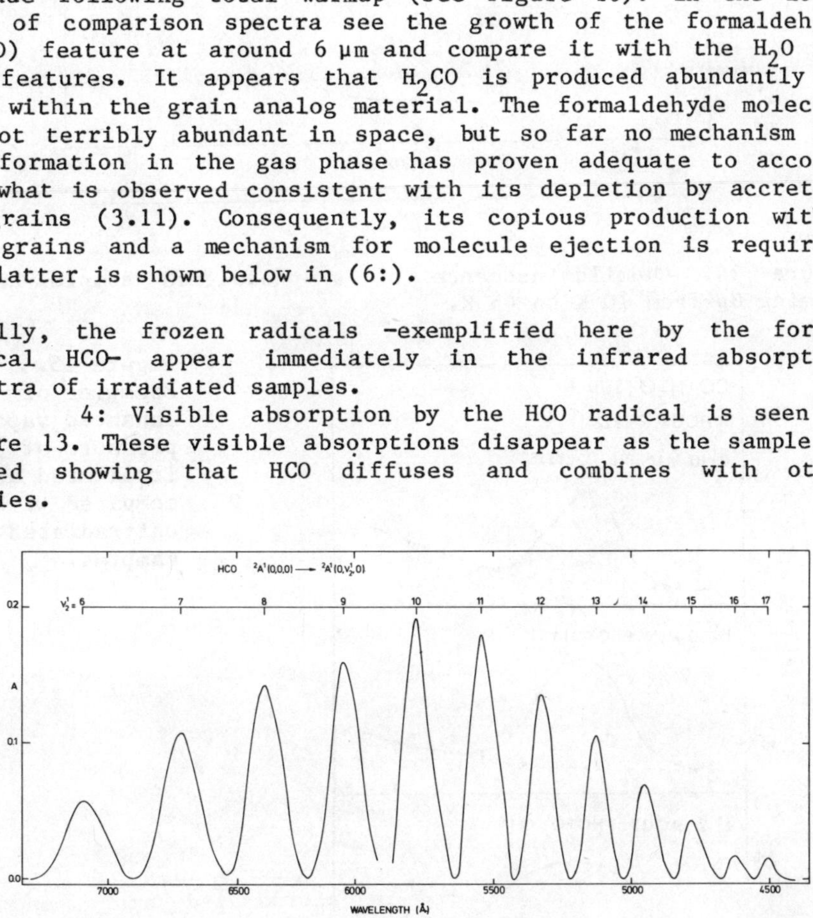

Figure 13. The visible absorption spectrum (1 Å resolution) obtained form a $CO:H_2O$ (40:1) mixture deposited with simultaneous photolysis for 22 h. The features are all due to the HCO radical. Without photolysis, no absorptions are observed.

5: A very nice example of the physical evidence for storage and release of frozen radicals is shown in Figure 14.

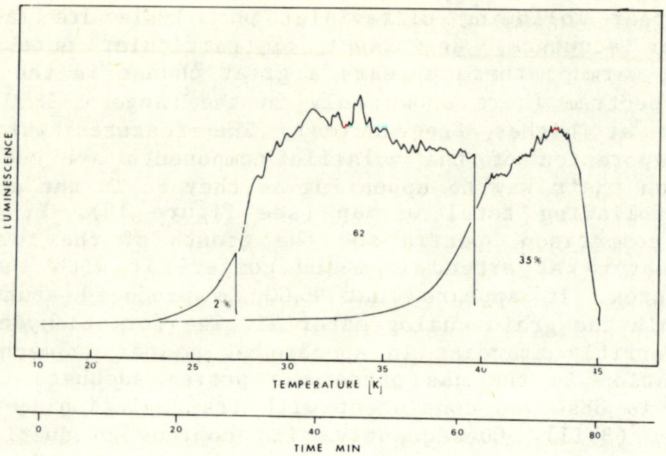

Figure 14. Chemiluminescence of a photolyzed sample during warming up from 10 K to 45 K.

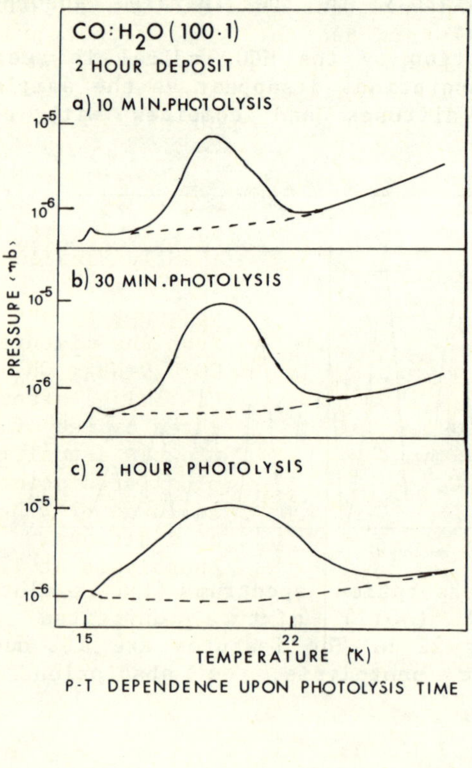

Figure 15. Examples of enhanced vapor pressure of irradiated as compared with unirradiated samples.

Upon warmup, above 10° K the sample begins to emit visible radiation -chemiluminescence arising from molecules produced in excited states by exoenergetic combinations of radicals (3.12). The chemiluminescence stops immediately if the warmup is stopped by turning on the cryostat and recooling. This was done at two temperatures, ~ 27° K and ~ 40° K, and in both cases when the sample was again allowed to warm up (by turning off the cryostat) the luminescence did not reappear until the turnaround temperature was reached (or almost reached at 40° K). It is clear from this example that the freeing of radicals is a diffusion process in which the energy binding sites form a continuous spectrum up to the temperature (E = kT) at which luminescence finally stops entirely. That the energy is released as heat as well as in chemiluminescence is seen by the increase in vapor pressure of an irradiated sample compared with the unirradiated mixture (Figure 15). Only about 1 part in 10^5 of the energy is released in the form of visible light.

6: A special case of energy release by heating is the production of explosive reactions. Such reactions were predicted for interstellar grains (3.13) which would explain not only the source of complex molecules in the interstellar gas but their ability to maintain them against loss by depletion on the grains (3.14). The experiments and application are being intensively pursued (3.15). It has turned out that when the samples are in some way thermally insulated from the cold finger they produce not only gradually varying luminescence and pressure but also spikes, indicating violent chain reactions. The major explosion always occurs at about 27° K when the material is literally blasted off the cold finger.

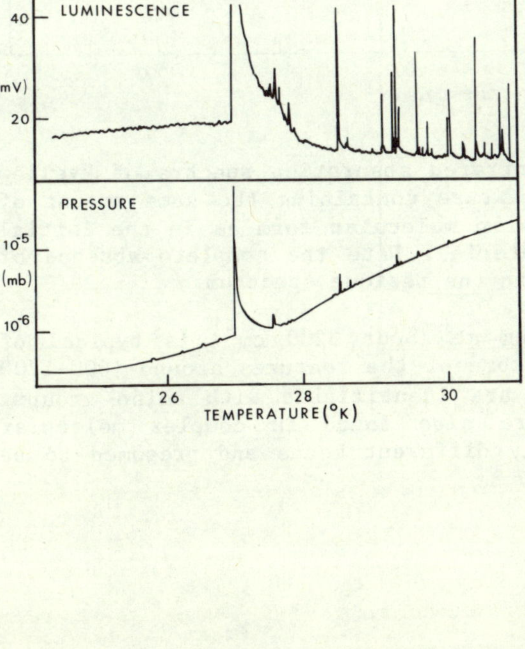

Figure 16. Correlation of chemiluminescence spikes with pressure bursts from the mixture $CO:H_2O:NH_3:CH_4$ (10:1:1:4) after three cycles of 10 minutes deposition plus irradiation followed by 2 hours of additional photolysis at 10 K and subsequent warm-up

An example of the simultaneous light and pressure spike is shown in Figure 16. The critical fractional density of radicals required to produce explosions is estimated to be ~ 10^{-2} and this occurs when the relative flux of photons to accreting molecules is about 1:10. This implies an efficiency for radical production (net number of radicals/number of incident photons) of about 10% which is approximately ten times higher than the conservative pre-experimental estimates (3.14).

7: Subsequent to heating of irradiated samples, there always remains on the cold finger a yellow (near ultraviolet absorbing) nonvolatile residue. This residue remains indefinitely at room temperature and high vacuum (10^{-6} mbar). It is estimated that about 10% (2% to 20%) of the originally deposited volatile materials is converted into this yellow stuff in the interstellar equivalent time of ~ 10^7 years (an astronomically reasonable time). The infrared absorption spectra of a variety of materials produced from complex mixtures are characterized by the general features shown in Figure 17.

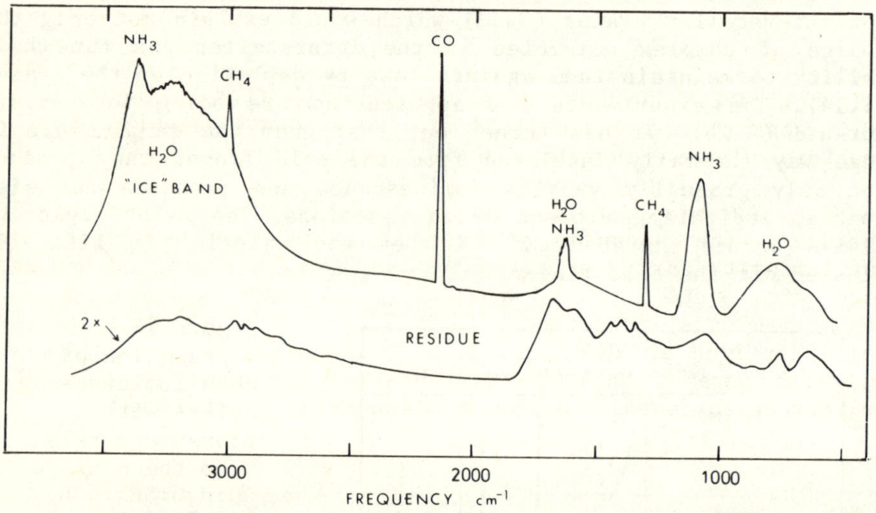

Figure 17. Comparison of infrared absorption spectra of "yellow stuff" residue with 10 K mixture containing the same amount of oxygen, carbon and nitrogen in molecular form as in the initial (pre-irradiated) residue material. Note the complete absence of an H_2O ice band at 3.08 µm in the residue spectrum.

The very broad band peaking at about 3200 cm^{-1} is typical of carboxylic acid groups and some of the features around 1600-1700 cm^{-1} (see also Figure 13) are identifiable with amino groups. These types of spectra are also found in complex molecular mixtures created by entirely different means and presumed to be of prebiological interest. See Figure 18 for a typical example (3.16). The very broad "3 µm" absorption is not to be confused with the H_2O ice band. Some of the smaller features around 3.4 µm have been seen in space (3.17) and these are an indication that what we make in the laboratory is representative of what is made naturally in space and does not require invoking the presence of bacteria or viruses.

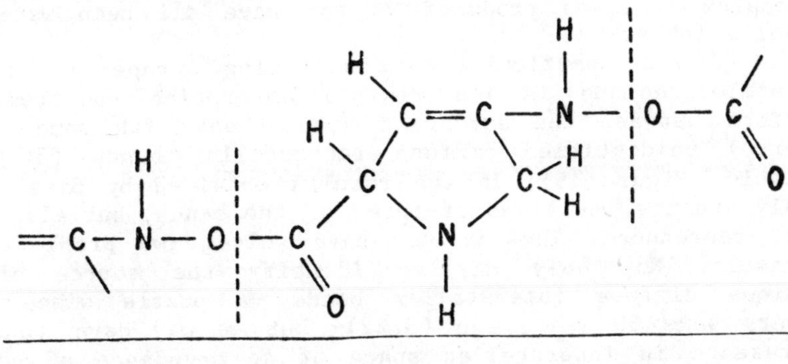

Figure 18. Infrared absorption spectrum of an HCN-oligomerization product after Sephadex G-15 fractionation and HPLC separation. Indentified as 8-carbomoyl adenine (Figure courtesy of A.W. Schwartz and A.B. Voet).

Insofar as the details of the chemical structure of our residues, the amounts we have made so far have not been sufficient to carry out extensive analyses. A beginning has been made with some samples, one of which was examined by high resolution mass spectroscopy (3.18) and found to contain pieces, which appeared at about 400-500° K, identifiable as $C_4H_6N_2$, CO_2 and trace urea ($(NH_2)_2CO$). A tentative structure as part of a polymer which contains amino pyroline rings is given by

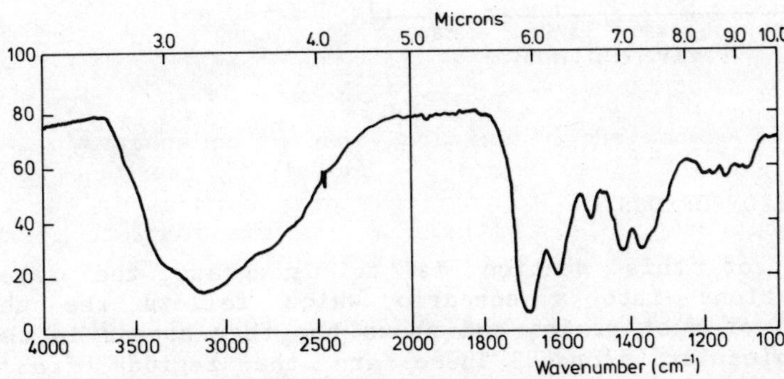

It is interesting to speculate on this basis that we have a natural way of making porphyrins or porphyrin-like

molecules in space as had been conjectured to exist some time ago (3.19). An investigation is about to begin, with support from the National Aeronautics and Space Agency, in which trace quantities (~ 1%) of Fe will be deposited along with the O, C, N bearing molecules. I anticipate the possibility that this will produce some new and exciting results along these lines.

The complex residues produced so far have all been water and methanol soluble.

8: A particularly fascinating property of the nonvolatile residue is its visible absorption spectrum. The comparison between the strongest absorptions with some of the (hitherto) unidentified diffuse interstellar bands (3.20) is remarkable (Figure 19). In the results obtained by Baas (3.21) not only are the positions of three of the bands, but also their shapes, reproduced. Thus we may have solved two problems with one result. Not only may we identify the source of the mysterious diffuse interstellar bands -a puzzle since their discovery some 50 years ago (3.22)- but we may have justified the presence in interstellar space of an abundance of complex organic molecules.

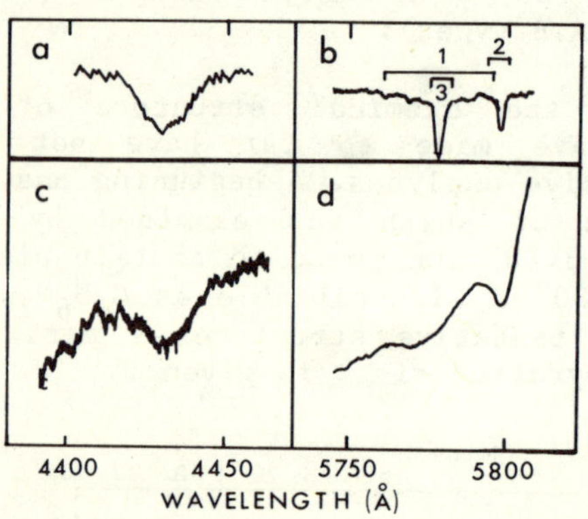

Figure 19. Comparison of diffuse interstellar bands (a and b) with equivalent laboratory spectra from a photolyzed residue of $CO:H_2O:NH_3:CH_4$ (c and d).

4. EVOLUTION OF DUST.

The aim of this section is to integrate the laboratory investigations into a scenario which follows the chemical evolution of dust grains and shows how they appear in the more dense molecular clouds. These are the regions from which intertellar materials may have accreted on the earth in quantities sufficient to be considered to have played a role in the origin of life.

a) From diffuse clouds to molecular clouds to diffuse clouds, etc.

A schematic of several stages in the evolution of a dust grain is shown in Figure 20. We start with the birth of a grain core. An ongoing process is the ejection of small refractory silicate type (probably amorphous silicate) particles which form and are blown away from cool evolved stars (4.1). Other sources of solid refractory ejection are novae and supernovae. These small particles are swept up into the gaseous matter in space to partake in the evolution of the clouds as they pass through the various conditions which we observe as a snapshot at this time.

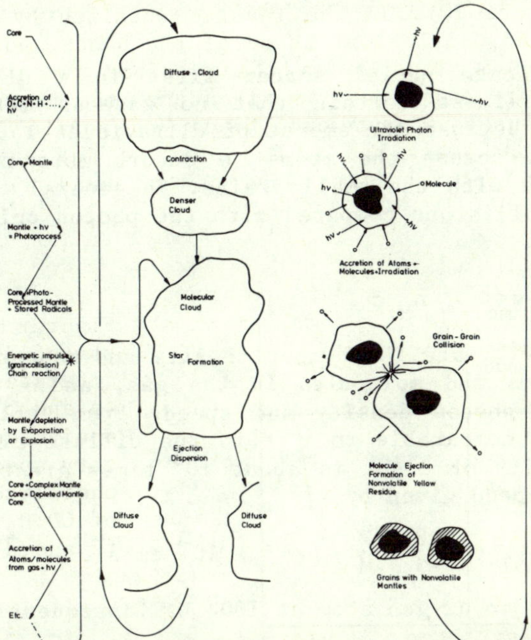

Figure 20. Diagram of grain evolution. The sequence on the left corresponds to one of the subcycles in the dense cloud phase. The sequence on the right is a contraction of how grains evolve through the molecular cloud and star formation phase and back to the diffuse cloud phase.

From the fact that the clouds cannot be static -either because they are observed to be in motion relative to each other, or because we infer or see such dramatic and energetic events as star formation occurring within them- it is obvious that the physical conditions of density and temperature represent different stages in their evolution. The most dense clouds correspond to times just before, during, or just after star formation. Diffuse clouds become denser by some mechanism, perhaps as a result of collisional combinations or some source of external pressure (4.2). Within the dense clouds critical densities may be reached which lead to instabilities and further contraction and finally to star formation (4.3). After the stars form -if they happen to be large hot stars, or if they develop high material ejection speeds by processes other than radiation- the remaining material from which they have been formed is ejected back into the surrounding space (4.4). Much of this material being heated and finding itself in a very tenuous low pressure environment expands to reappear as diffuse clouds. The above sequence is schematically shown in the second column of Figure 21.

Should the silicate nuclei appear first in a diffuse cloud region, it is almost certain that no gaseous material will accrete on them because the amount of ultraviolet radiation will be sufficient to cause the atoms to desorb more rapidly than they can collide with the small grains. In general, the rate of atom/molecule collisions compared with the photon collision rate is given by

$$n_{mol}\, v_{mol} \,/\, n_{h\nu}\, c$$

where n_{mol} and v_{mol} are the number density and mean speed of the condensible atoms and molecules in the gas, and $n_{h\nu}$ and c are the ultraviolet photon density and speed. From Table 3 we see that $n_{at/mol}$ is comparable to $n_{h\nu}$ (in the diffuse cloud medium) while the velocity of light is about 10^6 times greater than the mean molecular speed given by

$$v_{at/mol} = \left(\frac{8}{3}\, k \,\frac{T}{M} \right)^{\frac{1}{2}} \simeq 5 \times 10^4 \text{ cm s}^{-1}$$

corresponding to an oxygen atom at 100° K. Consequently we get

$$(dn_{mol}/dt)/(dn_{h\nu}/dt) = 5 \times 10^{-6}$$

at a characteristic diffuse cloud density of ten hydrogen atoms per cubic centimeter. With an estimated photodesorption efficiency of 2×10^{-5} (two photons in 10^5 which impinge on a grain desorb an atom on the surface), it turns out that accretion begins when $n_H \simeq 50 \text{ cm}^{-3}$ (4.5). When a grain finds itself in such a region, the process of sticking and accretion of atoms and molecules may proceed simultaneously with ultraviolet photolysis. This is where we start at the top of the left sequence in Figure 20. After some period of accretion and photoprocessing, two grains will collide with each other at speeds greater than $\simeq 40 \text{ m s}^{-1}$ which is sufficient to raise the grain temperature to the critical value of 27° K (see Figure

16). Such speeds are highly probable because, as a result of turbulent gas speeds of one to several kilimeters per second in molecular clouds (4.6) grain-grain collision speeds are expected to be > 0.1 km s^{-1} (4.7). Such explosions should prevail over most of the cloud because, in comparison with the diffuse cloud value of $(dn_{mol}/dt)/(dn_{h\nu}/dt) \sim 10^{-6}$, the maximum value allowed for generating critical radical densities is a factor of 10^7 higher. For above critical speeds, all the volatile materials are blown off. But some grains may maintain a fraction of the material which has been converted into nonvolatile residue. Following the explosion -complete or partial- the grain begins again to accrete and be photoprocessed. The overall time for this cycle is governed by the collision time in clouds which is $\approx 5 \times 10^5$ to 5×10^6 years at densities of 10^4 cm^{-3} to 10^3 cm^{-3}. It thus will probably repeat itself many times in the cloud as indicated by the brackets in Figure 20 leading from the first to the second column molecular cloud phase which lasts $\gtrsim 5 \times 10^7$ years. In this way a grain which starts out as a core alone will, after one passage through its molecular cloud phase, emerge with an organic refractory mantle which has the best chance to survive the energetics induced by star formation. Since the overall mean lifetime of a grain is $\sim 5 \times 10^9$ years (4.8), there should arise a steady state distribution of grains with organic refractory mantles distributed in the diffuse clouds. Therefore, although we started originally in our sequence with cores alone, the rate of production of these cores is such that within a fraction of the lifetime (10×10^9 years) of the Milky Way, most of the grains will appear to have developed mantles. This means that when the solar system was born 4.5×10^9 years ago, the grains were as we see them today.

b) Chemical composition of dust in diffuse clouds.

Before discussing the grains in molecular clouds, we will provide the chemical basis for the diffuse cloud grain and gas.

All attempts at observing the H_2O ice band in diffuse clouds have proven negative. This is consistent with these dust grains having had most of their volatiles -including H_2O- sputtered away in the process of their ejection out of the molecular cloud phase. Since the absorption strength of the organic refractory stuff is weak, this in combination with the extreme broadness of its 3 μm "feature" make it rather difficult to detect. However, smaller absorption features appear to have been detected (see 3.17). We conclude that the negative evidence for solid H_2O in diffuse clouds may actually be providing us with positive evidence for the organic refractory materials, particularly if we add in the visible diffuse bands and the infrared bands to strengthen the argument.

Table 6. Elemental composition of diffuse clouds relative to cosmic abundances.

	O	C	N	Mg	Si	Fe	
Gas							
Atoms + Ions	0.75	0.20	0.20	0.03	0.03	0.01	(a)
Molecules	--	--	--	--	--	--	(b)
Dust							
Core + bare	0.09	0.27		~1.0	~1.0	~1.0	(c)
Mantle							
OR	0.11	0.42	0.22				(d)
Volatiles (H_2O, etc.)	~ 0	~ 0	~ 0				
Total Gas	0.75	0.20	0.20				
Total Solid	0.20	0.69	0.22				
Unaccounted	0.05	0.11	0.58				
Available for accretion (Av.)	0.80	0.31	0.78				

a) Depletions in ζ Oph from Morton, 1974, (Astrophys. J. 143, L35), Jenkins and Shaya, 1979 (Astrophys.J. 231, 55) and from de Boer, 1981 (Astrophys.J. 244, 848).
b) Negligible
c) The carbon depletion is based on a 0.025 μm radius graphite particle to produce the 2200 Å bump. Should another type, or smaller size of particle be the actual cause, the carbon depletion in the bare particles will be substantially reduced from the value 0.27.
d) Based on the assumption of a mantle material with relative atomic composition as given by the mass spectrum results for the organic refractory in section 3b.

In Table 6, I have tried to summarize the distribution of O, C, N, Mg, Si and Fe elements as they would appear in the diffuse clouds. It is assumed that an insignificant number of molecules are present in the gas and that the O, C, N, Si, Mg and Fe abundances are as given for the diffuse cloud ζ Ophiucus. The abundances of O, C, Si, Mg and Fe in the core particles and in the bare particles (to which I have only briefly alluded to now) are estimates based on theoretical model calculations (2.7). For the mantle it may be shown that a radius a ≃ 0.12 μm is a good estimate for the organic refractory mantle. If the O, C, and N atoms were in relative cosmic abundance in the OR material the values would all be ~ 0.22. However, we use the mass spectrometer results (for the one sample) in which carbon outnumbers oxygen by about 2:1, to obtain the value of 0.11 for O and 0.42 for C in Table 6. Note that the total amount of oxygen depleted in the grains is probably less than ∿ .20 and that the carbon depleted may be as large as ~ 0.69. This means that when grains grow in molecular clouds they must grow almost exclusively from the oxygen which starts out about 5 times as abundant as carbon in the gas (compared with the cosmic abundance ratio of ~ 2:1). We thus anticipate that a substantial amount of H_2O can grow on molecular cloud dust mantles.

c) Grains in Molecular Clouds.

Although H_2O is seen in the grain absorption in molecular clouds, it has been difficult in the past to estimate its abundance because of lack of data on appropriate indices of refraction. I assume that the dust in a region of accretion may be represented by a silicate core plus an organic refractory mantle and a final extra mantle of accreted molecules in which H_2O is a fractional component.

Table 7. Elemental composition of B.N. type molecular clouds relative to cosmic abundances.

	O	C	N	Si	Mg	Fe	
Gas							
Atoms + Ions	--	--	--	--	--	--	(a)
CO	0.05	0.10	--	--	--	--	
	(~0.01)	(~0.03)					(b)
Other Molecules	<0.01	<0.01	<0.01				(c)
Dust							
Core + bare Mantle	0.09	0.27		~1.0	~1.0	~1.0	(d)
Solid H_2O	0.22						
OR	0.11	0.42	0.22				
Other	0.05	0.11	0.26				
Total Gas	0.06	0.11	0.01				
Total Solid	0.47	0.80	0.48				
Unaccounted	0.47	0.09	0.51				
Available for accretion	0.53	0.20	0.52				

a) Not counting possible significant carbon ions as in Phillips et al. 1980, Astrophys. J. 238, L107.
b) If depletion of gas CO in dense cores of molecular clouds (Rowan-Robinson, 1979, Astrophys. J. 234, 111).
c) Observed (see text).
d) Assumes extra grain mantle of 0.03 μm and the ice absorption for B.N. (see text).

The question is, what is this fraction f? A theoretical

calculation using the measured absorption characteristic at 3 μm for H_2O in mixtures has been used (3.10) to compare with the astronomical observation of the B.N. object. See Table 7 for the element distribution in the dust in a B.N. type molecular cloud. The result is that about 55% of the outer mantle is in the form of amorphous H_2O ice. This result is also consistent with calculations on grain mantle growth by surface reactions alone ignoring the photoprocessing (4.9). Since the B.N. object is in a region of ongoing star formation, what we may be seeing are grains which are not in the molecular cloud steady state accretion-explosion phase, but rather grains which have grown in the last stages of final contraction and which are in a transition state between star formation and final ejection into the diffuse medium.

With or without this speculation, we can provide a substantial body of observational evidence which confirms the general ideas of photoprocessing of mixtures of oxygen, carbon and nitrogen bearing molecules in the various evolutionary phases of the interstellar dust, leading ultimately to a substantial grain component which is a complex organic material. Paraphrasing a nursery rhyme: What are little grains made of? Organics and ice and everything nice, that's what little grains are made of. Typical interstellar grains might be represented as in Figure 21.

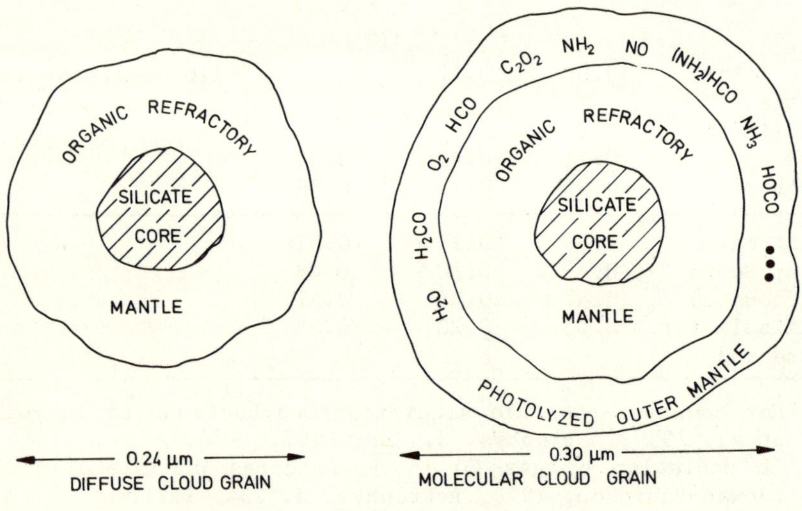

Figure 21. Chemical models of interstellar grains in diffuse and molecular clouds. The molecular cloud mantle size shown here corresponds approximatedly to those in the BN type clouds. In protostellar clouds the outer mantle would be thicker.

5. INTERSTELLAR DUST ON THE EARTH.

a) How Much Organic Material is in a Cloud?

As the earth revolves around the sun, and the solar system as a whole revolves around the center of the Milky Way, we are continually passing through the gas and dust in space. As I have already mentioned, the solar system at the current epoch is in the midst of the most tenuous kind of gas so that the interstellar dust can barely make its presence felt or observed and then only by measurements made outside of the earth's orbital plane. However at quite a few times in the past, the solar system was immersed in dense molecular clouds from which it could have gathered large quantities of dust and gas. Although we know the gas to contain many interesting organic molecules, including H_2CO and HCN, which can play a role in prebiotic chemistry, we also realize that the total fraction of available C, O, and N seen in these molecules is quite small compared to what we infer to be in the dust. Consequently, although we may use the gas phase molecules as an indicator, we will consider that the bulk of molecules are bound in the dust.

In the steady state regime in molecular clouds, the accretion-explosion balance cycles roughly 75% of the cosmically abundant atoms and molecules containing O, C and N as volatiles between the gas and dust. The remaining ~ 25% are in the organic refractory yellow stuff. Of the 75% labile material, approximately 50% is at any moment on the grain so that ~ 65% of the <u>total</u> are in the dust (3.15). From this, we conservatively estimate that 0.1% of <u>all</u> the mass of our entire Milky Way is in the dust. For example, in a molecular cloud of 1 pc radius and a hydrogen density $\eta_H = 10^4\ cm^{-3}$ the complex organic molecules alone -of which about 1/2 are the very complex yellow stuff- account for a mass equal to that of our sun.

However, it is not enough to show that space is an enormous chemical factory producing more complex molecules than all possible planets. For example, if now or at any time since life began, a large quantity of this material were dumped on the planet earth while we passed through a dark cloud, it could hardly provide a competition for the already ongoing life process. If anything it might be poisonous rather than life-giving. As an aside, it is interesting to speculate how the presence of a cloud around the solar system might appear. First of all, if we were in the center of a 1 pc radius, $\eta_H = 10^4\ cm^{-3}$ cloud, the Milky Way and all of the stars would disappear to the naked eye. In fact, all we would see would be the sun (insignificantly affected in brightness), the moon and the visible planets. On the other hand, the sky would light up very brightly -much brighter than the present zodiacal light or Milky Way-after sunset because of the scattered sunlight off the dust particles.

To return to the question of why we consider the prebiotic chemistry of space as a possible alternative for the origin of life to the prebiotic chemistry arising from atmospheric phenomena. Two reasons, as I see it, are that this chemistry could have been important earlier in the life of the earth (see next section); and the state of chemical evolution of the dust (yet to be finally established) could have been such that its molecules were sufficiently key life progenitors that upon deposition they were capable of further development at an already advanced level.

b) Accretion of Interstellar Matter by the Primitive Earth.
There are many ways by which matter from space has been -and still is- deposited on the earth. For example, when the earth was first formed, there were enormous quantities of debris still remaining in the solar system which bombarded the earth's surface (5.1). Furthermore, since the solar system may be presumed to have formed and remained for several million years within a molecular cloud complex it would have continued to accrete large quantities of interstellar matter during this time. However, the current thinking about the state of the earth's surface at such an early epoch (see elsewhere in this volume) makes it highly unlikely that even abundant deposits of prebiotic material could have either survived or found a suitable environment. On the other hand, since there now appears to be evidence for life having already been present on the earth some 3.83 billion years ago (see elsewhere in this volume) we must limit ourselves to the questions of when and how prebiotic matter could have been deposited on the earth's surface during this first 700 million years and actually perhaps in an even narrower time frame between ~ 300 and 700 million years. I shall consider here only the accretion during passage of the solar system through dense clouds. Clouds of comets have been considered elsewhere (5.2).

It turns out that at the distance of the sun from the galactic center, the solar system is rotating about the galactic center at about twice the angular speed of the spiral pattern. Therefore, since the pattern consists of two main spiral arms and since the galactic rotation period of the solar system is about 200 million years (5.3), the sun and earth pass through a relatively high concentraton of dust and gas at the inner edges of spiral arms (see Figure 1) every 110 million years. This is shown schematically in Figure 22. We have thus passed through a spiral arm some 40 times since the formation of the solar system. It can be estimated (5.4) that at the present epoch the collision probability of the earth with clouds of hydrogen density $n_H \sim 2 \times 10^3\ cm^{-3}$ is about 0.25 for each passage through

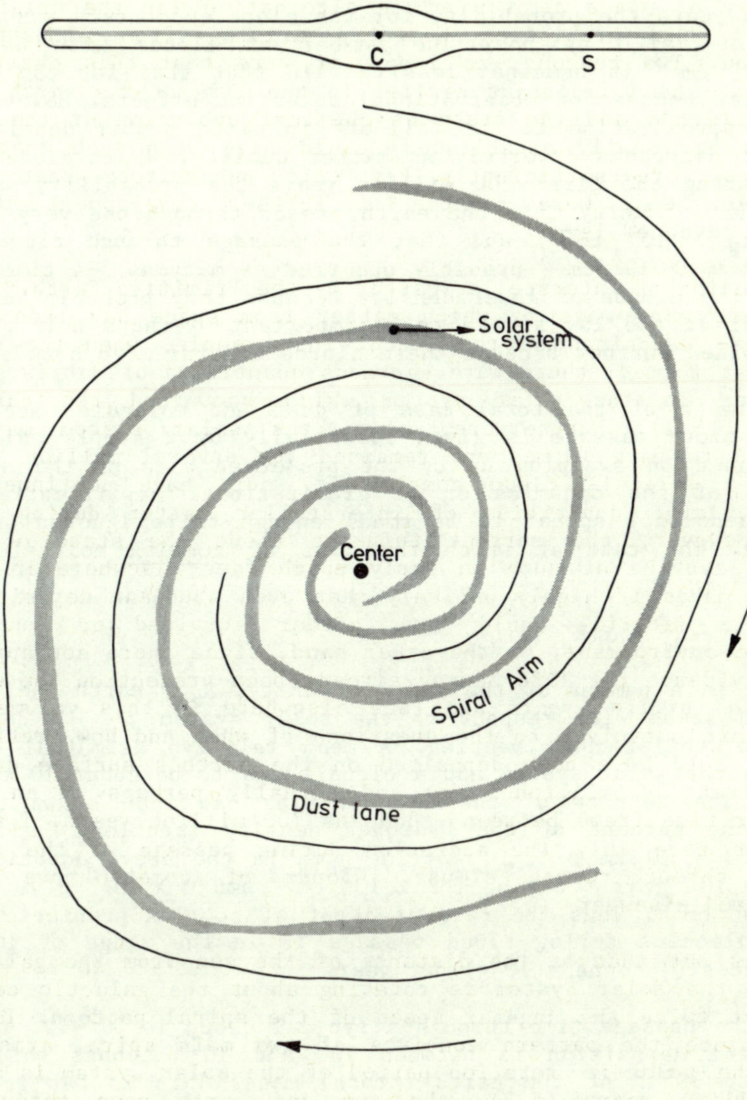

Figure 22. Schematic diagram of a spiral galaxy seen edge-on and face-on showing the concentration of dust. The position and relative velocity of the solar system with respect to the spiral pattern is illustrated at a time when it was passing through a region where the dust clouds are concentrated.

a spiral arm. As a rough approximation we may assume that 4 x 10^9 years ago the amount of interstellar matter was about twice its present value, so that during the first several hundred million years the probability for the cloud encounters was about twice as large as now. The number of clouds with density $n_H \sim 10^4\ cm^{-3}$ is somewhat less certain that that for the lower densities because of observational selection effects. However, a rough approximation to the fall-off in cloud number density is that it decreases inversely as matter density. We conclude then that during the first 700 million years the probability was of the order of unity thta the earth passed through one very dense cloud($n_H > 10^4\ cm^{-3}$) and that the passage through clouds of density $n_H > 10^3\ cm^{-3}$ probably occurred as many as 3-4 times. We ignore the clouds of lower density because they probably deposit material at too low a rate to be important. Perhaps this should be examined further because these clouds are far more abundant.

An estimate of the total mass of dust and molecules accreted during cloud passage is given essentially by a simple calculation based on sweeping up by the projected area of the earth. Effects of gas dynamics or of gravitational amplification on this phenomenon appear to be small enough to be ignored on the average. The rate at which the mass of complex molecules is accreted is

$$\frac{dM}{dt} = \alpha \pi R_e^2\ V \rho_{d-m}$$

where α is a number of the order of unity, R_e = earth radius, V = cloud speed with respect to the solar system, ρ_{d-m} = density of interstellar dust mantles. A mean relative azimuthal speed between the solar system and a cloud seems to be about 10 km s^{-1} (5.5). The density of the dust mantles has been shown to be about one percent of the hydrogen density (section 4). If we insert V = 10 km s^{-1}, $\rho_{d-m} = 10^{-2} n_H m_H$ in the above equation we get $M \gtrsim 10^9$ g yr^{-1} for $n_H \sim 2 \times 10^3\ cm^{-3}$ amd $M \gtrsim 10^{10}$ g yr^{-1} for $n_H \sim 10^4\ cm^{-3}$. Thus the rate of imput of complex prebiotic type dust molecules during cloud passage is in the range of 10^3 to 10^4 tonnes (1 tonne = 10^6g) per year.

Since the passage time through a typical cloud is 10^5-10^6 years, the total deposition is between 10^8 and 10^{10} tonnes which is 10^{-14} to 10^{-12} of the earth's total mass. This is far greater than the current biomass of the earth.

We may probably assume no modification of the chemical composition of the dust as it impinges on the earth. Any existing atmosphere would act like a cushion slowing down these submicron particles without significantly heating them as occurs for much larger meteors and even the interplanetary dust (5.6). However, although the nonvolatile component would be unaffected, the more volatile constituents in the outer mantle of the dust would have partially, if not totally, evaporated because of heating by the sun.

c) Interstellar Dust in Comets.
It is currently believed that comets are a good representative sample of the interstellar matter out of which they form. As such, their chemical composition is basically that of the interstellar dust at the final stage in the prestellar nebula.

Table 8. Suggested chemical mass and volume distribution of the principal condensible atomic consituents in a comet.

Component	Mass Fraction	s	Volume Fraction
Silicates	0.21	3.5	0.086
Carbon	0.06	2.5	0.034
Very complex OR	0.19	~ 1.3	0.21
H_2O	0.19	1	0.27
CO	0.10	1.05	0.13
Other molecules + radicals (CO_2, N_2HCN, H_2CO, HCO,)	0.25	1.3	0.27

This composition is illustrated in Table 8 where it is assumed that all the atoms and molecules in the gas have accreted on the dust. Comet collisions with the earth have undoubtedly contributed substantial quantities of organic material in the past and indeed, many have contributed as much or more than that by the direct accretion of the interstellar dust. However the explosive reaction which occurs on impact of a comet likely would lead to sufficient heating to pyrolize the complex molecules. Nevertheless a relatively high local concentration of organic material should result and perhaps be conducive to rapid continuing chemical evolution.

CONCLUSION

The chemical evolution of interstellar matter via gas phase and solid reactions leads to an enormous reservoir of organic molecules in the space between the stars. The impressive sampling of organic molecules seen by radioastronomical methods in the gas is but a small part of the total of which, by mass and complexity, the major portion floats about in small frozen submicron sized particles of interstellar dust. The accretion of substantial masses of interstellar dust from space by a planet like the earth occurs with high probability in the early stages of crustal development. It is suggested that the high degree of complexity of the dust organics may have been adequate to provide the chemical templates leading to the origin of life at the first opportune time in the earths formation, perhaps previous to very substantial complex molecule formation in a primitive earth atmosphere. The question of whether the ultimate dust composition is so ordered as to give such a head start can only be established by further intensive laboratory studies in solid phase photochemical evolution of interstellar dust particles.

REFERENCES

1.1 Snyder, L.E., Buhl, D., Zuckerman, B. and Palmer, S., Phys. Rev. Letters 22, 679.
1.2 Mann, A.P.C. and Williams, D.A., 1980, Nature, 283, 721.
1.3 Clerke, Agnes, 1903 "Problems in Astrophysics, Black, London.
1.4 Trumpler, R.J., 1930, Lick Obs. Bull., 14, 154.
1.5 Allen, C.W., 1973, Astrophysical Quantities (Athlone, London)
1.6 Wootten, A., Evans, N.J. II, Snell, R. and Vanden Bout, P., 1978, Astrophys. J., 25, L143.
1.7 Greenberg, J.M., 1971, Astron. Astrophys., 12, 240.

2.1 Oort, J.H., 1954, Bull. Ast. Inst. Ned., 12, 177.
2.2 Woolf, N.J. 1973 in Interstellar Dust and Rolated Topics, ed. J.M. Greenberg and H.C. van de Hulst, Reidel, Dordrecht, 485.
2.3 Donn, B., Wickramasinghe, N.I., Hudson, J. and Stecher, 1968, Astrophys J. 153, 451.
Salpeter, E.E. 1974, Astrophys J. 193, 579.
1974, Astrophys J. 193, 585.
2.4 Kwok, S., 1980, J. Roy. Astron. Soc, Can., 74, no. 4, 216.
2.5 Willner, S.P., Gillett, F.C., Herter, T.L., Jones, B., Krassner, H., Merrill, K.M., Pipher, J.L., Puetter, R.C., Rudy, R.J., Russell, R.W. and Soifer, B.T., 1982, Astrophys. J., 253, 174.
2.6 Greenberg, J.M. 1982 in Comets, ed. L. Wilkening (Univ. of Arizona Press).
2.7 Hong, S.S. and Greenberg, J.M., 1980, Astron. Astrophys., 88, 194.
2.8 Van de Hulst, 1949, Rech. Astr. Obs. Utrecht, 11, part 2.
2.9 Habing, H.J., 1968, Bull. Astr. Inst. Ned., 19, 421.
2.10 Pimentel, G.C., 1960, in Formation and Trapping of Free Radicals, eds. A.M. Bass and P. Broida, (Academic Press, N.Y.) chapter 4, 69.

3.1 Danielson, E.R., Woolf, N.J. and Gaustad, E.J., 1965, Astrophys. J., 141, 116.
3.2 See 1.1.
3.3 Greenberg, J.M., Yencha, A.J., Corbett, J.W. and Frisch, H.L., 1972, Mem. Soc. Roy. Sciences Leige, 6e serie, Tome III, 425-463.

Khare, B. and Sagan, C., 1973; in Molecules in the Galactic Environment, eds. M.A. Gordon and L.E. Snyder (Wiley and Sons, N.Y.)
3.4 Miller, S.L., 1953, Science 117, 528.
3.5 Hagen, W., Allamandola, L.J. and Greenberg, J.M., 1979, Astrophys. & Sp. Sci., 65, 215.
3.6 Merrill, K.M. Russell, R.W. and Soifer, B.I. 1976, Astrophys. J. 207, 763.
3.7 Bertie, J.E., Labbé, H.J. and Whalley, E. 1969, J. Chem. Phys. 50, 4501.
3.8 Hagen, W., Tielens, A.G.G.M. and Greenberg J.M., 1981, Chem. Phys., 56, 3, 367.
3.9 Hagen, W., Tielens, A.G.G.M. and Greenberg, J.M., 1982, Astron. Astrophys., accepted for publication.
3.10 Greenberg, J.M., 1982, in Comets, ed. L. Wilkening (Univ. of Arizona Press)
3.11 De Jong, T., Dalgarno, A. and Boland, W., 1980, Astron. Astrophys., 91, 68.
3.12 Van de Bult, C.E.P.M., Allamandola, L.J., Baas, F., van IJzendoorn, L. and Greenberg, J.M. 1980, J. Mol. Struct. 61, 235.
3.13 Greenberg, J.M., 1973, Molecules in the Galactic Environment, eds. M.A. Gordon and L.E. Snyder, J. Wiley & Sons, N.Y., 94.
Greenberg, J.M. and Yencha, A.J., 1973 in Interstellar Dust and Related Topics, Ed. J.M. Greenberg and H.C. van de Hulst, Reidel, Dordrecht, 369.
Greenberg, J.M., 1976, Astrophys. and Sp. Sci., 39, 9.
3.14 Greenberg, J.M., 1979 in Stars and Star Systems, ed. B.E. Westerlund, Reidel, Dordrecht, 177.
3.15 d'Hendecourt, L., Allamandola, L.J. Baas, F. and Greenberg, J.M., 1982, Astron. Astrophys., in press.
3.16 Schwartz, A.W. and Voet, A.B., Private communication
3.17 Allen, D.A. and Wickrsmasinghe, D.T. 1981, Nature 94, 539.
Van de Bult, C.E.P.M., Allamandola, L.J. and Greenberg, J.M. A laboratory analysis in progress.
3.18 Work in collaboration with Professor H.J. de Jong, Leiden.
3.19 Johnson, F.M., 1967, in Interstellar Grains, eds. J.M. Greenberg and T.P. Roark, NASA Sp-140, 229.
3.20 Herbig, G.H., 1975, Astophys. J., 196, 129.
3.21 Baas, F., as reported in Greenberg, J.M., 1981 Ned. Tijds. voor Natuurkunde A47(1), 24.
3.22 Merrill, P.N., 1934, Pubs. Astron. Soc. Pacific. 46, 206.

4.1 Salpeter, E.E., 1977, Annual Review of Astronomy and Astrophysics, Annual Reviews, Palo Alto, 267.
4.2 Field, G.B. and Saslaw, W.C., 1965, Astrophys. J., 142, 568.
Kwan, J., 1979, Astrophys. J., 229, 567.
Oort, J.H., 1954, B.A.N., 12, 177.

Scoville, N.Z. and Hersh, K., 1979,Astrophys. J., 229, 884.
Taff, L. and Savedoff, M., 1972a, Mon. Not. Roy. Astron. Soc., 160, 89.
Taff, L. and Savedoff, M., 1972b, Mon. Not. Roy. Astron. Soc., 167, 357.
4.3 Woodward, P.R., 1978, Ann. Rev. Astron. Astophys., 16, 555.
Bash, F.H., 1979, Astrophys. J., 233, 524.
4.4 Blitz, L. and Shu, F.H., 1980, Astrophys. J., 238, 148.
4.5 Greenberg, J.M., 1982 in Submillimeter Wave Astronomy, ed. J.D. Philips and J.E. Beckman, Cambridge Univ. Press, in Press.
4.6 Larson, R.B., 1981, Mon. Nat. Roy. Astr. Soc., 194, 809.
4.7 Völk, H.J., Jones, F.C;, Morfill, G.E. and Roser, J., 1980, Astron. Astrophys. 85, 316.
4.8 Salpeter, E.E., 1977, Annual Review of Astronomy and Astrophys. Annual Reviews, Palo Alto, 267.
Greenberg, J.M., 1978, In Cosmic Dust, ed. J.A.M. McDonnell, J. Wiley and Sons, 187.
4.9 Hagen, W., Tielens, A.G.G.M. and Greenberg, J.M., 1982, Astron. Astrophys., accepted for publication.
Tielens, A.G.G.M. and Hagen, W., 1982, Astron. Astrophys., submitted.

5.1 Ashworth, D.G., 1978, in Cosmic Dust, ed. J.A.M. McDonnell, (J. Wiley & Sons) 427.
5.2 Ponnamperuma, C., 1981, ed. Comets and the Origin of Life, (Reidel, Dordrecht).
5.3 Bok, B.J. 1981, Scientific American, March.
5.4 Talbot, R.J. Jr. and Newman, M.J., 1977, Astrophys. J. Supplement, 34, 295.
5.5 Spitzer, L. Jr., 1978, Physical Processes in the Interstellar Medium. J. Wiley & Sons, New York.
5.6 Brownlee, D., 1978 in Cosmic Dust, ed. J.A.M. McDonnell, J. Wiley, N.Y., 295.

COMETS, INTERSTELLAR MOLECULES, AND THE ORIGIN OF LIFE

William M. Irvine and Åke Hjalmarson

Onsala Space Observatory
Chalmers University of Technology
S-43900 Onsala, Sweden

1. INTRODUCTION

This review considers two quite distinct subjects: comets, and the relative abundance of molecules in the gas phase in interstellar clouds. The justification for a joint treatment lies in the possibility that comets may consist of a frozen matrix of interstellar volatiles and grains, formed during the early stages of the collapse of the molecular cloud which ultimately produced the Sun and planets. These topics may in turn be related to the origin of life, as the distribution of biologically significant molecules in space bears on the likelihood of chemical evolution throughout the universe, and, more specifically, comets might be a source of planetary volatiles, possibly including complex organic molecules.

In the next section we shall review the observed properties of comets and current models of their nature and composition, stressing also our lack of definitive knowledge in critical areas. Section 3 will then treat the determination of relative molecular abundances in interstellar clouds and the evidence favoring a primarily gas phase chemistry in such regions, concluding with a new tabulation of molecular abundances in the cold cloud known as TMC-1, which has been called a "protosolar nebula". The final section then briefly relates these subjects to a possible role in chemical evolution.

The reader is reminded at the outset that the interstellar solid particles ("dust" or "grains") may well play a critical

C. Ponnamperuma (ed.), Cosmochemistry and the Origin of Life, 113–142.

role in interstellar chemistry and/or contain the bulk of interstellar organic matter. See this Volume's Chapter by M. Greenberg.

2. COMETS

Comets are among those astronomical phenomena for which the oldest written records exist, with descriptions dating back to about 1000 BC in Chinese annals. The periodic appearances of Halley's Comet have been described since classical times.

2.1 Visual Appearance

In visual appearance comets typically exhibit a fuzzy, roughly symmetric head or coma, and a tail whose length increases as the comet approaches the Sun. Two morphologically distinct types of tails occur, referred to as dust and as gas or ion tails, on the basis of their deduced composition; a single comet may show a mixture of both phenomena. Comets differ markedly in their general appearance and temporal behavior, with flares (sudden brightenings), splitting into two or more components, and even disappearance sometimes occurring.

2.2 Orbits

The orbits of comets show them to be members of the solar system, and they may be classified according to their period of revolution about the Sun (truly interstellar comets quite likely exist, but there is no well documented evidence that one has ever been observed). Of the approximately 630 comets with determined orbits, about 100 are referred to as "short period" (periods < 200 years), and the remainder as "intermediate" or "long period" (periods 200 to $> 10^6$ years); more than 80 of the latter are new comets, entering the inner solar system for the first time. It is hypothesized that the Sun is surrounded by a vast cloud of 10^{11} to 10^{12} comets (the Oort cloud) in orbits which extend to distances comparable to that of the nearest stars. Gravitational perturbations by such stars are thought to cause changes in the orbits which may cause a comet to plunge into the inner solar system (a new comet); some such comets are gravitationally "captured" by the major planets (particularly Jupiter) into more elliptical orbits and so become periodic comets. Each time a comet enters the inner solar system, it leaves behind a trail of gas and dust, which produces meteor showers when encountered by the Earth. As a result, periodic comets exist for only a finite time (at least as comets; whether some become inert asteroids is still an open question); it has been estimated that these comets lose some 10^{-3} to 10^{-4} of their mass on each such revolution. It is well documented that cometary orbits are affected by non-gravitational as well as gravitational forces, the nature of which plays an important role in current models of comets (see below).

2.3 Spectra

The evidence concerning the composition of comets comes primarily from their spectra, traditionally observed in the visible region but more recently extended into the ultraviolet, the infrared, and the radio portions of the spectrum. A tabulation is presented below. The ions are observed

TABLE 1

Atomic and Molecular Species Observed in Comets

Organic:	C, C_2, C_3, CH, CN, CO, CS, (HCN), (CH_3CN)
Inorganic:	H, NH, NH_2, O, OH, S, Si, (H_2O)
Metals:	Na, Ca, Cr, Co, Mn, Fe, Ni, Cu, V, K
Ions:	C^+, CO^+, $CO_2{}^+$, CH^+, CN^+, $N_2{}^+$, OH^+, H_2O^+, Ca^+
Dust:	Silicates

in the ion tails, while the metals occur in the coma of those comets which approach the Sun quite closely. The reality of the detections for the molecular species given in parentheses (all observed by radio astronomy) will be discussed subsequently. Note that, apart from the species in parentheses and perhaps CO and CS, all the observed molecules are chemically unstable radicals or ions. It is hypothesized that they are dissociation products of parent molecules which are the true constituents of the comet far from the Sun.

2.4 Model

The observed properties of comets are generally interpreted in terms of the "icy conglomerate" ("dirty snowball") model, originally proposed by F. Whipple. It is believed that the basic cometary entity is a nucleus consisting of a mixture of frozen volatiles (ices) and refractory grains (dust); whether some comets are differentiated radially, perhaps containing an asteroidal interior, is important in discussions of their evolution, but is at present undecided (existing data is most compatible with an undifferentiated nucleus). As the nucleus approaches the inner solar system, solar heating produces sublimation of the volatiles, and subsequent photochemistry, interaction with the solar wind, and possibly gas phase chemistry in the coma lead to the observed composition in the coma and tail.

Outbursts of gas produced by the heating of an inhomogeneous, rotating nucleus are invoked as the cause of a jet-effect, to produce the non-gravitational forces; and the outflowing gas carries with it the particulate matter observed in the dust tails. The size of the nucleus could play an important role in determining its thermal evolution, but direct measurements do not exist; current estimates for most comets are in the range of 1 to 10 kilometers (compared to 10^4 to 10^6 km for the coma and up to 10^8 km for the tail). Such comets would have masses $\lesssim 10^{18}$ grams and gas phase densities near the nucleus of 10^{13} to 10^{14} molecules per cc (the densities in the molecular clouds discussed in Section 3 fall in the range 10^3-10^6 cm^{-3}, which occur in cometary comae at 10^4 km or so from the nucleus).

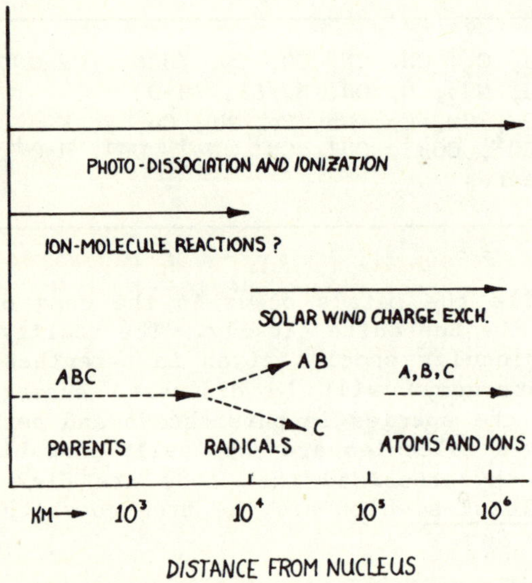

Figure 1: Physical and chemical processes in cometary comae versus distance from the nucleus. Parent molecule ABC first converted into radicals and ions, finally into atoms and atomic ions.

2.5 Composition

Although the relative elemental abundances of H, C, N, and O are roughly known for some recent comets (they are essentially the same as in terrestrial life), there is almost no direct evidence for the actual chemical composition of the nucleus. Consider the possible parent molecules in Table 1: CO might be present in the nucleus, but could also be produced from heavier species; CS, on the basis of its brightness distribution in the coma, is believed to be a dissociation product; while H_2O, HCN, and CH_3CN each have been claimed in only one comet, the detection of H_2O in particular has been severely challenged, and searches in other comets for HCN and CH_3CN have produced negative results.

On the other hand, there is considerable <u>indirect</u> evidence that H_2O is abundant in comets: the existence of H_2O^+ in tails, the relative abundance of H and OH, the observed gas production rate as a function of heliocentric distance (which agrees with the volatility of H_2O ice), and models of coma chemistry which yield the observed radicals in Table 1. Likewise, some nitriles must be present to produce the observed CN in the coma.

Considerable information on the abundances and mineralogy of nonvolatile constituents is available <u>if</u> the so-called <u>Brownlee</u> particles are in fact cometary in origin. These are obtained from both stratospheric sampling and oceanic cores, and are believed to be interplanetary dust captured by the Earth, the dust in turn presumably being cometary debris (<u>cf</u>. the earlier comments on meteor showers). The elemental abundances in the Brownlee particles are very similar to those in chondritic meteorites, although some differences in mineralogy apparently exist. A possible connection between cometary dust and <u>interstellar</u> particles is implied by the presence of a spectral feature at 10 micrometers attributed in both cases to amorphous silicate material.

Recent evidence for several comets shows a noticeable similarity in the relative abundances of such molecules as C_2, CN, C_3, and OH, independent of the period of revolution (and hence "age" spent subject to solar heating and mass loss). This is consistent with a considerable homogeneity within and among cometary nuclei and supports the prevalent idea that cometary material is the least metamorphosed in the solar system and hence may retain the clearest record of conditions during the early stages of the solar nebula. Nonetheless, considerable heterogeneity in the relative abundances of H_2O, CO_2, NH_3, and of course more complex molecules remains a possibility. The relative abundance of dust (non-volatile matter) to volatiles does vary over perhaps an order of magnitude among comets, with a "canonical" value of the mass ratio sometimes taken as 0.7.

Finally, let us speculate on the degree of chemical complexity which might be present in comets. Direct evidence, as noted above, is meager. Laboratory experiments in which simulated cometary ices are irradiated with high energy particles ("cosmic rays") or photons do, however, indicate the synthesis of heavier species. The possibility of aqueous synthesis of complex organics during a relatively long time period also cannot be ruled out; an initial radioactive heat source may have been present, and cometary ice is an excellent insulator (1,59).

The presence of amino acids, purines, pyrimidines, "porphyrin-like" compounds, and yet heavier polymers in carbonaceous meteorites also raises the possibility of similar syntheses in comets.

2.6 Origin

The chemical composition of comets presumably contains important information concerning their place and mode of origin, as well as their subsequent evolution. In the "classical" theory comets were formed by condensation from the cooling solar nebula in the vacinity of the giant planets (in older theories near Jupiter and Saturn, while more recent suggestions might favor Uranus and Neptune). Gravitational perturbations by these planets then threw the comets into their present distribution in the Oort cloud. Such models often discuss a composition predicted by chemical equilibrium for a cooling solar composition nebula (with C in CH_4, N in NH_3, and considerable H_2O), although catalytic processes on particulate surfaces might well produce less reduced compounds (2).

Alternatively, some more recent models favor cometary formation as occurring by condensation of interstellar gas onto interstellar solid grains, perhaps in the early stages of collapse of the molecular cloud which ultimately formed the solar nebula, or perhaps in gravitationally related but distinct cloudlets. Since interstellar conditions are very far from thermodynamic equilibrium, the composition of cometary nuclei might be rather different in this case. There is some slight evidence from theoretical models of coma chemistry that an "interstellar" composition for the nucleus leads to better agreement with the observed distribution of radicals and ions than does an "equilibrium" model.

It is unlikely that the origin of comets can be specified until, first, the chemical composition of comets themselves is better known, probably with the aid of spacecraft sampling; and, second, that the relative abundances in interstellar molecular clouds are likewise more quantitatively determined, which is the subject of the following section of this review. Since there is evidence for considerable compositional differences among molecular clouds in our Galaxy, this discussion raises the intriguing possibility that the study of comets might ultimately enable us to tell the type of interstellar cloud which did in fact form the solar nebula, and hence to identify possible sites of future solar-type star formation.

3. CHEMICAL ABUNDANCES AND PROCESSES IN INTERSTELLAR MOLECULAR CLOUDS

3.1 Observational Methods

Most information concerning the chemical makeup of dense interstellar clouds has until now been derived by the techniques of radio astronomy. This may be understood by recalling the energy level diagram for a typical molecule. The internal energy of a molecule may be divided into electronic, vibrational, and rotational contributions, which are normally of quite different magnitudes. Thus, the difference in energy between electronic states typically corresponds to photons of visible or ultraviolet light, which in temperature units T = (energy)/(Boltzmann's constant) represents several thousand degrees Kelvin or more; each electronic state is divided into vibrational states, whose energy difference tends to fall in the infrared portion of the spectrum (50 K $<$ T $<$ 2000 K); while each vibrational state is further subdivided into rotational levels with excitation energies (except for the lightest molecules) in the radio frequency region, corresponding to temperatures T $<$ 100 K. As a result, at the temperatures characteristic of dense interstellar clouds (see below), the emitted radiation is predominately at radio and far-infrared wavelengths, except near energy sources such as hot stars, shock waves produced by cloud collisions, _etc_. Radio frequency observations have also the technical advantage over shorter wavelength methods that heterodyne techniques allow extremely high spectral resolution, thus permitting accurate line identification, measurement of physical characteristics such as rotational temperature, and high velocity resolution kinematic studies (through the Doppler effect).

Infrared observations are seriously handicapped by absorption due to the terrestrial atmosphere. In the future, however, the use of high altitude and spacecraft instruments, together with improvements in spectral resolution such as those arising from laser heterodyne techniques, will vastly increase the opportunities for this type of research. It is already clear from preliminary studies and from ground based near infrared observations that, particularly in regions of active star formation like the giant molecular clouds, this spectral region holds tremendous promise. Chemical studies will inevitably profit by these developments, since a number of expected constituents of the interstellar gas have infrared but not radio frequency transitions (_e.g._, H_2, HC_2H, CO_2 and other symmetric species; see below).

Molecular radio astronomy, meanwhile, is fortunate in that several new telescopes for millimeter wavelength research have either just come into operation (20 m diameter dish at the

Onsala Space Observatory in Sweden; 14 m dish at the University of Massachusetts, USA; interferometers operated by the University of California at Berkeley, USA; and by the California Institute of Technology , USA) or are in late stages of completion (joint French-German 30 m telescope in Spain; 45 m telescope in Japan). Moreover, receiver development is continuing to achieve major strides in sensitivity at the highest frequencies.

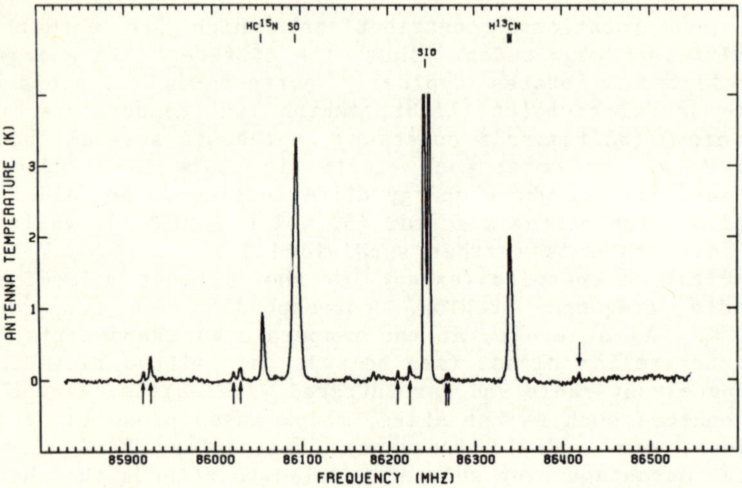

Figure 2: Radio frequency spectrum of the Orion molecular cloud obtained at the Onsala Space Observatory. Upward pointing arrows indicate transitions of methyl formate (HCO_2CH_3), downward arrow indicates an unidentified line, and stronger lines are labeled.

3.2 Abundance Determinations

The determination of relative molecular abundances in interstellar clouds raises complex problems of measurement and interpretation. Rather than presenting the analytic formulation given in the References, we shall give a qualitative description of procedures.

For convenience in calibration, radio astronomers present their measured signal intensities in temperature units (cf. Figure 2). The intensity of a spectral line T_A (for antenna temperature) for a transition from energy level E_{upper} to level E_{lower} for molecular species Y depends on 1.) telescope efficiency, which due to surface imperfections inevitably is a function of frequency; 2.) the angular size of the celestial source relative to the angular region instantaneously sampled by the telescope (the beam size); 3.) the number of molecules of Y in the telescope beam; 4.) the population distribution among the energy levels of Y (characterized by the rotational temperature); and 5.) complex radiative transfer (saturation) effects governing the ease with which an emitted photon will escape from the source and hence be observable. These effects are, moreover, not independent.

Consider, for example, point (4), the population distribution. If the abundance of Y were sufficiently low, the observed intensity would depend only on the population in level E_{upper}, since only molecules in that state emit the photons of interest. Since conditions in interstellar clouds may be far from thermal equilibrium, however, knowledge of the population of one state does not in itself provide the total number of molecules Y, as would be the normal situation in the laboratory (i.e., for which a Boltzmann population distribution at ambient temperature applies). Even if the population distribution is approximately characterized by a Boltzmann distribution, the corresponding temperature T_{rot} (for rotational) is normally not equal to the kinetic temperature characterizing the velocity distribution of the gas molecules or the temperature describing the wavelength dependence of intensity in the ambient radiation field; instead, T_{rot} will be determined by a competition among the microscopic processes of excitation and de-excitation for Y, and hence will be a function of molecular parameters such as the coefficient of spontaneous emission, the number density of molecules colliding with Y in the cloud (mainly H_2 and He), their kinetic temperature, and the ambient radiation field. To accurately estimate T_{rot}, several transitions should be observed, which may introduce problems (1) and (2) above, since the transitions should arise from rather different energy levels, will therefore often occur at quite different wavelengths, and the telescope beam size is a function of wavelength (there are, however, exceptions with many transitions at similar frequencies, such as NH_3 and SO_2). In addition, as soon as the abundance of Y is high enough for absorption of the line radiation by molecules in state E_{lower} to be significant, the population becomes coupled to the radiative transfer problem (5), and hence to cloud geometry and velocity structure, which are typically unknown.

On the other hand, it is the dependence of T_A on these physical conditions which enables molecular emission to be used as a probe of temperature, density and other physical parameters in interstellar clouds, particularly since this dependence differs among molecules as a function of their intrinsic properties (collision cross section, electric dipole moment, etc.). For example, a molecule like CO with a low spontaneous emission coefficient for its $J=1\rightarrow0$ transition tends to adopt a population distribution characterized by its collisional excitation and de-excitation, and hence described by the kinetic temperature of collision partners, therefore becoming a good thermometer; while, at the opposite extreme, a molecule which spontaneously emits a photon following each collision will provide a measure of the number of colliding molecules and hence of the density (under some conditions approximately the case for the low lying rotational transitions of HCN and CS).

In the simplest physical situation, when every emitted line photon escapes the cloud (in which case the cloud is said to be optically thin at the requisite frequency), the cloud is homogeneous and larger than the telescope beam, the rotational temperature $T_{rot} >> 2.7$ K (the cosmic black body radiation temperature), and $T_{rot} >> h\upsilon/k$ (where υ is the frequency of the line and k is Boltzmann's constant), then the signal T_A depends only on known molecular constants (assuming these are known; some molecules identified by radio astronomy have not yet been studied in the laboratory) and on the population of Y in level E_{upper} integrated through the cloud (designated N_{upper}). This latter quantity, in units of molecules per cm^2, is referred to as a column density. A rotational temperature must then be estimated either theoretically or by observations of other transitions in order to determine N(Y), the total column density of species Y (averaged across the telescope beam). This optically thin case is not so rare as might be at first assumed, since for the more abundant molecules rarer isotopic species can be observed (e.g., $^{13}C^{16}O$, $^{12}C^{17}O$, $^{12}C^{18}O$, and even $^{13}C^{18}O$ have been detected), although abundance determinations then require that the isotopic ratios be known and that isotopic fractionation in the source be small or known (a problem for H and ^2H, and to a lesser extent for ^{12}C and ^{13}C).

Relative abundance determinations then present the further problems that abundance ratios cannot be invariant along the line of sight, because cloud chemistry depends on environmental factors such as density, temperature, and radiation field; and even if the relative abundances were constant, differences in molecular parameters would (in the prevailing non-equilibrium conditions) ensure that different molecular transitions sampled different environments. Thus, ratios of column densities N(Y)/N(Z) are not necessarily characteristic of relative volume densities (molecules per cm^3) under typical source conditions, particularly if molecules Y and Z have very different excitation requirements.

Another cautionary note, to be kept in mind when examining the tables presented below, is that many molecules possibly present in interstellar clouds are not observable by radio astronomical means. This includes, first of all, those present in the solid phase in the interstellar grains. A model for grain composition which is rich in complex organic molecules has been presented at this Institute (M. Greenberg), but there is as yet no general agreement among astronomers on this subject. Next there are molecules which, for reasons of symmetry, have no permanent electric dipole moment and hence no strong rotational transitions. Some of these may be relatively abundant in interstellar clouds (e.g., N_2, CO_2, CH_4, HCCH, and O_2; the latter could be observed from above the atmosphere via its strong magnetic dipole transitions). Then there are molecules whose rotational transitions are either at higher frequencies (the light hydrides), or which may not be excited at the temperature of the colder interstellar clouds (e.g., the well known 22 GHz transition of H_2O, observed in regions of star formation but not in cold, dark clouds). Both vibration-rotation bands of symmetric molecules like CO_2 or HC_2H, and the fundamental rotational spectrum of the light hydrides, are becoming observable through advances in infrared technology.

3.3 Interstellar Chemistry

The relative importance of the chemical processes determining the abundances of those gas phase molecules directly observed astronomically is still a subject of considerable uncertainty. M. Greenberg (this Volume), Hayatsu and Anders (2), and Allen and Robinson (3), from quite different points of view, have argued that solid particles play a basic role in this regard. Most other workers in the field have concentrated on purely gas phase processes, at least partly because of the much easier prediction of quantitative abundances. Although the solid grains must interact with the gas phase molecules, we shall here complement Greenberg's Chapter by presenting the evidence supporting a predominantly gas phase chemistry. All workers, however, presently agree on the importance of grain surface catalysis for the production of H_2.

Since only binary reactions play a role at the low densities characteristic of interstellar clouds, and since such reactions between neutral species often have activation barriers greater than or comparable to the available kinetic energies, the major breakthrough in interstellar chemistry has been the realization of the fundamental importance of ion-neutral reactions. The energy sources for these reactions, which typically have no activation barrier, are cosmic rays and ultraviolet photons. For example, ionization of molecular hydrogen can be followed by

$$H_2 + H_2^+ \rightarrow H_3^+ + H , \quad (1)$$

with the H_3^+ ion then being important to the production of heavier observed molecules.

Evidence usually cited for the importance of gas phase ion-molecule chemistry in the interstellar medium includes:
1.) The observed existence of ions. Thus, eq. (1) predicts the production of species such as HCO^+ and HN_2^+ via

$$H_3^+ + CO \rightarrow HCO^+ + H_2 \tag{2}$$

and

$$H_3^+ + N_2 \rightarrow HN_2^+ + H_2. \tag{3}$$

Both these ions were observed in space before their microwave spectra had been measured in the laboratory.
2.) The existence of the isomeric pair HNC and HCN. HCN corresponds to the lower energy state, and, under thermal equilibrium or given the energy available for isomerization following evaporation or explosion from a grain, the percentage of HNC should be negligible. In contrast, in many interstellar clouds the relative abundances of HNC and HCN are comparable, in agreement with predictions from gas phase chemistry (Figure 3).
3.) The strong deuterium fractionation observed for many interstellar molecules. Although the relative abundance of D $\equiv$ ^2H to H in the solar system is $\sim 10^{-5}$, the ratios in interstellar clouds for, e.g., [DCN]/[HCN], $[DCO^+]/[HCO^+]$, $[HDCO]/[H_2CO]$, and [DNC]/[HNC] are orders of magnitude larger. (Henceforth we shall denote the abundance of species Y by square brackets [Y].) This has a natural explanation in ion-molecule chemistry because of the exothermicity of the reaction

$$H_3^+ + HD \rightarrow H_2D^+ + H_2 , \tag{4}$$

which results in

$$[H_2D^+]/[H_3^+] \gg [D]/[H] , \tag{5}$$

which propagates the fractionation to heavier molecules produced by reaction with H_2D^+ and H_3^+.
4.) A similar fractionation, dependent on the ambient ultra-violet radiation density via the free electron abundance, is predicted for ^{13}CO and ^{12}CO in dark clouds:

$$^{13}C^+ + {}^{12}CO \rightarrow {}^{12}C^+ + {}^{13}CO . \tag{6}$$

Differential fractionation in agreement with theory is apparently observed (4).
5.) The electron abundance derived in dark clouds from, e.g., observations of HCO^+ is in rough agreement with that predicted from the measured cosmic ray flux at the Earth, showing a general consistency of the theory.

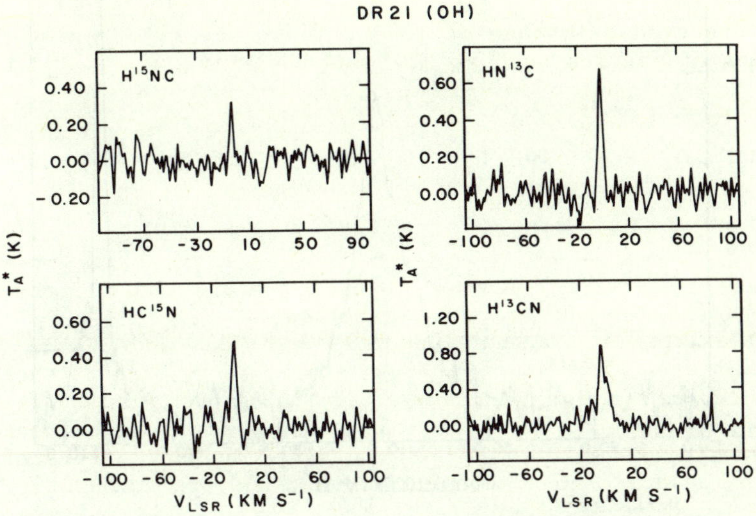

Figure 3: Radio frequency spectra of rare isotopic variants of HCN and HNC in the interstellar molecular cloud DR21(OH), showing comparable abundances for these two isomers (Onsala Space Observatory data; 7). Abscissa converted from frequency to velocity units via the Doppler effect. Transition is J=1→0 in all cases.

It has in addition been argued that the observed abundances are in rather good agreement with predictions from a purely gas phase model of the chemistry ((53,63) cf., however, Section 3.4); although comparison with predictions from grain chemistry is extremely difficult, since the latter processes are poorly understood.

Finally, it is becoming clearer that the chemical processes producing interstellar molecules are quite selective. Thus, no ring molecules have been found, and there seems to be (at least in some sources) a predominance of linear, unsaturated carbon chains over more reduced species (see particularly the new data on TMC-1 discussed below). Quantitative data are needed for more molecular clouds, but such selectivity should ultimately place strong constraints on molecular formation mechanisms.

3.4 The Composition of Interstellar Clouds

It is important to emphasize, however, that there is not one specific chemical composition for interstellar clouds. Such

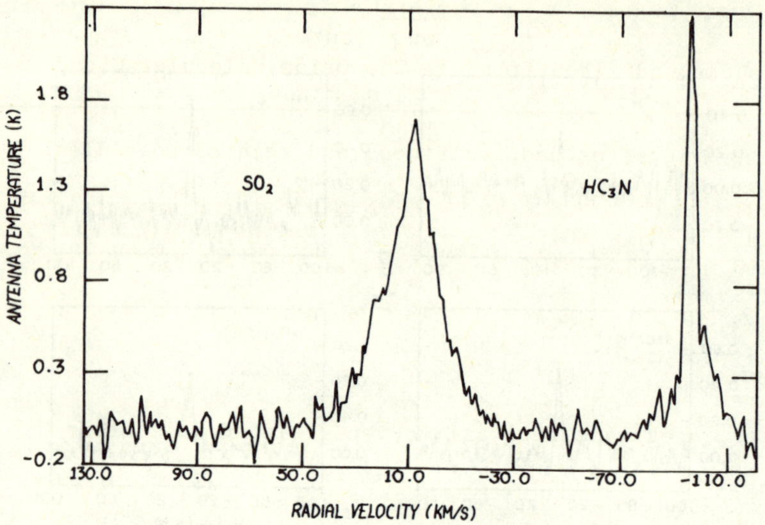

Figure 4: Radio frequency (73 GHz) spectrum of the Orion molecular cloud obtained at the Onsala Space Observatory. Abscissa converted from frequency to velocity units via the Doppler effect. Note the much greater width of the SO_2 relative to the HC_3N line, showing that the latter is present in a more quiescent region along the line of sight. Transitions are $6_{06}-5_{15}(SO_2)$ and $8-7(HC_3N)$.

regions differ considerably in their physical properties, and, it is becoming increasingly clear, also in their chemical makeup (e.g., 6,7). We shall discuss here only the denser, cooler portions of the interstellar medium (which still exhibit significant variation), and not the so-called diffuse clouds, "the intercloud medium", or regions of ionized hydrogen. In the diffuse clouds the simple diatomic molecules CH^+ and C_2, not yet detected in dense regions, are observed in absorption against bright stars.

The "giant" molecular clouds are known to be formation sites of massive stars, and probably for lower mass stars (like the Sun) as well. They are relatively warm (20-100 K), massive (10^3-10^5 M_0, where M_0 is the solar mass), and dense (up to 10^6-10^7 molecules per cm^3). The Galactic Center cloud Sgr B_2 is an example and is the richest source of complex molecules currently known.

Since the Galactic Center is not a typical environment (there are apparent isotopic anomalies, perhaps a result of extensive nuclear processing), and since the large distance from the Sun results in extremely poor spatial resolution and hence

TABLE 2

Compounds Identified In The Orion Molecular Cloud

Simple hydrides, oxides, sulfides, and related molecules:		
H_2	CO	H_3CNH_2
OH	CS	$N{\equiv}N^+H$
CH	OCS	
H_2O	SiO	
H_2S	SO	
NH_3	SO_2	
	O_3 ?	
	CO^+ ?	
Nitriles, acetylene derivatives, and related molecules:		
$C{\equiv}N$	$C{\equiv}CH$	
$HC{\equiv}N$	$H_3C{-}C{\equiv}C{-}H$	
$H_2N{-}C{\equiv}N$?	$HN{=}C{=}O$	
$H_3C{-}C{\equiv}N$	$HN{=}C$	
$H_3C{-}CH_2{\equiv}CN$		
$HC{\equiv}C{-}C{\equiv}N$		
$HC{\equiv}C{-}C{\equiv}C{-}C{\equiv}N$		
Aldehydes, alcohols, ethers, and related molecules:		
$H_2C{=}O$	$H_3C{-}O{-}CH{=}O$	
$H{-}C^+{=}O$	H_3COH	
$H{-}C^+{=}S$	$(H_3C)_2O$	
$H_3C{-}CH{=}O$?		
$H_2C{=}S$		

particularly large uncertainties in abundance ratios for a given set of physical conditions, we shall not consider Sgr B_2 further. The existence there, however, of the molecules (8) NO, NS, HNO, HCO, H_3CSH, HNCS, HCOOH, H_2CNH, H_2CCO, $HCONH_2$, CH_2CHCN, and CH_3CH_2OH, which have not yet been detected in the sources discussed below, suggests that future developments in instrumental sensitivity may uncover such species in nearby clouds.

The Orion molecular cloud is the nearest region of active formation of massive stars, and has become an archetypal giant cloud. There is clear evidence for chemical and physical heterogeneity in the line of sight to its denser regions (e.g., 9). In particular, a region of high velocity mass flow (the "plateau source") can be distinguished from cooler, more quiescent and extended "ridge" clouds, with additional regions ("hot core") also probable (cf. Figure 4). Moreover, Orion is not necessarily typical of other giant molecular clouds, for example in its [HNC]/[HCN] ratio (7). Rather than attempt to give abundance ratios, we have therefore only tabulated the molecular species detected in the Orion Cloud (8,10,11,12,13). Those molecules followed by question marks in Table 2 have been claimed but not confirmed (cf. 14).

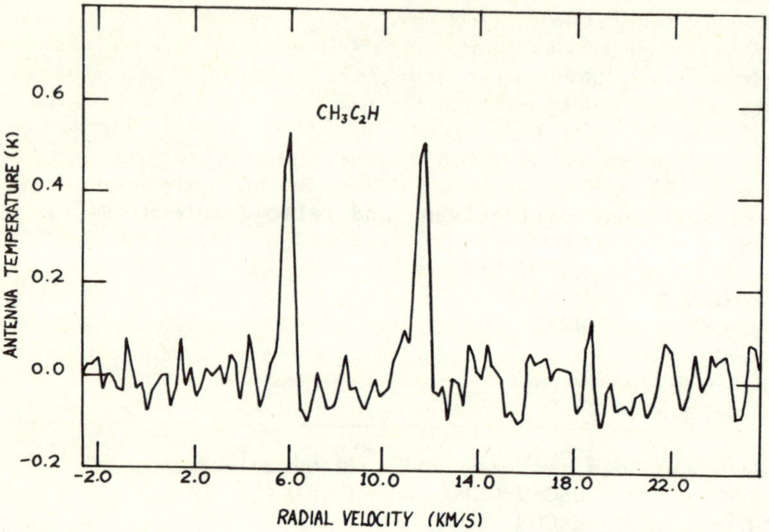

Figure 5: Radio frequency (85 GHz) spectrum of the cold, dark interstellar cloud TMC-1 obtained at the Onsala Space Observatory. The observed transitions are (left to right) the $5_0 \rightarrow 4_0$ and the $5_1 \rightarrow 4_1$ lines of CH_3C_2H (methyl acetylene). Note the narrow width of the lines (in velocity units) in this cold cloud compared with those in Orion (Figure 4).

The second class of objects discussed here are the "dark" clouds, which normally contain no apparent energy sources, are relatively cold (T < 10-15 K) and generally less dense (n $<10^4$-10^5 cm^{-3}) than the giant molecular clouds, and may be formation sites for low mass stars like the Sun. In spite of their apparently similar physical characteristics, significant chemical differences seem to occur among the dark clouds. Two of the most studied examples of this class are Taurus Molecular Cloud One (TMC-1) and L134N (also referred to as L183). We shall concentrate our attention on the former, while tabulating also some properties of L134N.

TMC-1 is estimated to be at a distance of about 140 pc (1 pc ≡ parsec is 3.26 light years) from the solar system. It appears in projection as a narrow ridge, roughly 0.06 pc x 0.6 pc, when viewed in the emission from several high dipole moment molecules (15). Observations of a number of molecular transitions suggest that it has a kinetic temperature of about 10 K, and a molecular hydrogen density which reaches a maximum of 10^4 to 10^5 per cm^3. TMC-1, and to a lesser extent other portions of the Taurus dark cloud complex, possess a strikingly high abundance of triply-bonded carbon compounds relative to other known interstellar clouds. The reasons for this are unknown, although it has been suggested that we may be observing these clouds at a particular stage in their chemical evolution. It is of course even more uncertain whether the Sun and the comets formed from such a cloud, but the possible presence of carbynes in meteorites (16,17; the identification has been challenged, 18) may provide a plausible link to such a chemistry, and at least some authors have referred to TMC-1 as a "proto-solar nebula". Table 3 lists those molecules which have been detected in TMC-1 as of summer, 1981 (8,10,11,19).

TABLE 3

Compounds Identified In The Taurus Molecular Clouds

Simple hydrides, oxides, sulfides, and related molecules:		
OH	SO	$N\equiv N^{+}H$
CH	CO	
NH_3	CS	
Nitriles, acetylene derivatives, and related molecules:		
$C\equiv N$		$C\equiv CH$
$HC\equiv N$		$HN=C$
$C\equiv C-C\equiv N$		$C\equiv C-C\equiv CH$
$HC\equiv C-C\equiv N$		$H_3C-C\equiv CH$
$HC\equiv C-C\equiv C-C\equiv N$		
$HC\equiv C-C\equiv C-C\equiv C-C\equiv N$		
$HC\equiv C-C\equiv C-C\equiv C-C\equiv C-C\equiv N$		
Aldehydes and related molecules:		
$H_2C=O$		
$H-C^{+}=O$		
$H-C^{+}=S$		

Since the physical environment is relatively well defined, TMC-1 provides perhaps the best current interstellar "laboratory" to test against models of chemical processes (how typical its composition is remains, it must be recalled, still uncertain). We have therefore tabulated current estimates for molecular column densities N in units of molecules per cm^2 averaged over the telescope beam. For the reasons discussed in the previous section, the telescope beam size, assumed or deduced values of physical parameters, and the Einstein spontaneous emission coefficient $A_{u\ell}$ are also included.

TABLE 4

Molecular Column Densities in TMC-1*

Molecule	Beam (Arc-min)	Transition	Frequency (GHz)	$A_{u\ell}$ (10^{-7} s^{-1})	N (10^{13} cm^{-2})	Notes**
C_3N	2	$N=3 \rightarrow 2$	30	8.4	0.7	b
HC_3N	2	several			6	b,c
	1x3	several			13	d
	2	$5 \rightarrow 4$	45	83	1	e
	1.5	$1 \rightarrow 0$	9	0.4	4	f
C_4H	2	$N=3 \rightarrow 2$	29	0.9	15	g
CH_3C_2H	2	$2_K \rightarrow 1_K$, K=0,1	34	1.1	6	g
C_2H	3	$N=1 \rightarrow 0$	87	20	8	h
HCN	2	$1 \rightarrow 0$, $H^{13}CN$	86	240	12	v
CN	1	$N=1 \rightarrow 0$	113	120	<0.2	i
					1.3	
HNC	2	$1 \rightarrow 0$, $HN^{13}C$	87	240	8	j
N_2H^+		$1 \rightarrow 0$	93			k
NH_3	2.5	J,K=1,1	24	1.7	8	l
SO	2	1_0-0_1 (SO, ^{34}SO)	30	2.4	5	a
CS	2	$1 \rightarrow 0$, $2 \rightarrow 1$	49,98	18,170	2	m
HCO^+	2	$1 \rightarrow 0$, $H^{13}CO^+$	87	240	5	w
CO	1	$1 \rightarrow 0$, ^{13}CO, $C^{18}O$	110	0.7	50000	p
Pyrolle	3	$4_{14}-3_{13}$; $4_{04}-3_{03}$	41		<0.4	s
Imidazole	2	$3_{03}-2_{02}$	33	21	<0.3	t
Glycine	2.5	$3_{12}-2_{11}$	23	11	<0.2	u
CH	15	Λ-doublet	3.3	0.002	20	n
OH	16	Λ-doublet	1.7	0.0008	200	o
H_2					10^9	q
H_2CO	11	$1_{11} \rightarrow 1_{10}$	5	0.036	6	r
HNCO	1.4	$1_{01}-0_{00}$	22	0.6	0.2	***
Glycine I	1.3	several			†	

* Observations at or near the cyanopolyyne peak ($\alpha=04^h38^m38.6^s$, $\delta=25°36'18''$; 1950), unless noted. Column densities N are average over beam. Upper limits to N from ½ peak-to-peak noise.

** Notes to Tables given in Section 6.

*** Brown, R.L.: 1981, Astrophys. J. (Letters) 248, L119.

† See the discussion of alternative possible conformers for gas phase glycine in Hollis, J.M., Snyder, L.E., Suenram, R.D., and Lovas, F.J.: 1980, Astrophys. J. 241, 1001.

Determinations of abundance ratios through observations with the same instrument at similar frequencies should give more reliable values than individual measurements (although data over a wide frequency range may be important for sorting out questions of excitation and saturation for a given molecular species). Table 5 gives such ratios, where these have been determined by a given author(s) (except for ^{13}CO), presumably in a consistent manner; some isotopic ratios are included, since isotopic ratios in comets provide in principle a valuable clue to their origin. Table 6, based primarily on unpublished Onsala observations, lists some upper limits on abundance ratios for undetected molecules relative to detected species of similar molecular weight (and hence, hopefully, similar collision cross sections); the comparisons are made for transitions with similar Einstein spontaneous emission coefficients, to minimize differences in excitation.

TABLE 5

Abundance Ratios in TMC-1 (unless noted)

Molecules	Ratio	Reference	Position or Source
HC_3N/C_3N	8	27	See Table 4
$HC_3N/HC_5N/HC_7N/HC_9N$	100/38/9/2	28	
	100/79/36/-	29	
	100/159/40/-	15	
	100/36/20/8	12	
	-/100/14/-	44	
NH_3/HC_3N	20	15	TMC-1 at (-2,+3)arcmin
	1000	49	L134N
NH_3/HC_5N	5	45	
	600	46	L134N
	1000	50	B335
HC_3N/DC_3N	12→50	22	
C_4H/CH_3C_2H	2→4	19	
$CO/^{13}CO$	75±8(1σ)	47	B5,B335,L1262
e/H_2	$<3(10)^{-8}$	48	
$HC_5N/^{13}CO$	$2(10)^{-3}$	29	
	$3(10)^{-4}$	29,46	L134N

Finally, Table 7 lists relative abundances of species probably present but currently unobservable in dark clouds, as calculated from models of gas phase chemistry. The large uncertainties in some cases, dependent partly on unmeasured reaction rates, are obvious.

It has not been possible for us to compare in detail the procedures used by individual authors to obtain the numbers

TABLE 6

Upper Limits to Abundance Ratios in TMC-1*

Molecule	Rotational Transition	E_ℓ (K)	A,B,C (K)	A(10^{-7} s^{-1})	T_A^* (K)	N/N_{std}	Notes
HC_3N	3-2	1.31	0.218	6.0	3.4	1	a
CH_3CH_2CN	$3_{12}-2_{11}$	2.43	1.33 0.227 0.204	5.9	<0.14	<0.2	b
CH_2CHCN	$3_{03}-2_{02}$	1.37	2.40 0.239 0.217	6.7	<0.10	<0.1	c
HC_2CHO	$3_{03}-2_{02}$	1.34	3.27 0.232 0.216	6.2	<0.11	<0.1	d
HCO_2CH_3	$3_{13}-2_{12}$	2.32	0.961 0.332 0.255	4.7	<0.06	<0.4	e
CH_3C_3N	8_0-7_0	5.56	7.60 0.0993	27	<0.07	<0.03	f
CS	2-1	2.36	1.178	170	1.31	1	g
SiO	2-1	2.09	1.047	290	<0.6	<0.2	h
CS	1-0	0	1.178	18	1.70	1	i
SO_2	$1_{11}-0_{00}$	0	2.922 0.496 0.423	35	<0.3	<1	j
NH_3	(1,1)	23		1.7	0.5	1	k
CH_2NH	$1_{11}-2_{02}$	9.21	9.416 1.663 1.409	0.9	<0.1	<0.4	l

*E_ℓ is energy of lower state in transition relative to ground state, (A,B,C) are rotational constants in temperature units, A is Einstein spontaneous emission coefficient. Antenna temperature T_A^* is corrected for antenna, radome, and atmospheric losses (upper limits are peak-to-peak values). Observations made at the Onsala Space Observatory unless noted, with the equipment described in (19) and (14). Values assume all lines are optically thin with equal velocity width, beam filling and rotational temperature (10K). Upper limits on column densities N are normalized relative to values N_{std} computed for the "standard" molecules HC_3N, CS, and NH_3.

quoted in the Tables 4 and 5. It is clear, however, that molecules like OH, CH, CO, and H_2CO are observable throughout a much wider spatial volume than TMC-1 itself, so that the column densities for these species can only with caution be compared to those for molecules requiring high H_2 densities for excitation (the cyanopolyynes, for example), if relative volume densities are desired. We note that differences in column density of an order of magnitude are apparent among different authors who have

TABLE 7

Molecular Species Unobservable in Dark Clouds (Theoretical Calculations)

Molecule	Reference	Abundance**	Notes*
H_2O	51	$1(10)^{-5}$	a,e,f
	52	$8(10)^{-6}$	b,e,f
	53	$6(10)^{-6}$	d,f,h
CO_2	51	$5(10)^{-9}$	a
	52	$3(10)^{-7}$	b
	53	$5(10)^{-8}$	d
CH_4	51	$1(10)^{-10}$	a
	52	$8(10)^{-11}$	b
	54	$2(10)^{-7}$	c
N_2	51	$2(10)^{-5}$	a
	52	$2(10)^{-5}$	b
	53	$6(10)^{-6}$	d
O_2	51	$7(10)^{-5}$	a,h
	52	$3(10)^{-5}$	b,h
	53	$5(10)^{-5}$	d,h
HC_2H	52	$1(10)^{-11}$	b
	21	$1(10)^{-6}$	g

* Section 6.2.
** Relative to H_2.

observed the same molecule. This is probably a rough measure of the uncertainty in most of the tabulated abundances in Table 4. Relative abundances (Table 5) are in somewhat better agreement, but discrepancies of factors of 2-4 are present.

Several interesting points are apparent from the Tables. First, there appear to be real differences in the chemical composition of the clouds TMC-1 and L134N, in spite of their apparently similar temperature and density; in particular, the abundance of the acetylene and polyacetylene derivatives seems to be much higher in TMC-1 (the possibility that at least part of these apparent differences arise from excitation effects, and not abundance, must be borne in mind, however (12)). It seems likely that chemical heterogeneity between such dark clouds and warmer molecular clouds like Orion is even larger.

Secondly, the predominance of linear, unsaturated molecules in TMC-1 is striking. There are problems of observational selection, since the partition functions for the linear chains are small relative to those for more saturated analogs, so that individual rotational transitions are considerably stronger for the cyanopolyynes. Nonetheless, the data show almost an order of magnitude difference between HC_3N and the upper limits on the more saturated analogs CH_2CHCN and CH_3CH_2CN. Although the uncertainties are larger, there also seems to exist an underabundance of aromatic molecules and branched chains (glycine) relative to the cyanopolyynes of comparable molecular weight.

This kind of selectivity agrees with models of gas phase chemistry in which condensation reactions involving the C^+ ion are a primary means of building heavier organic molecules (5,61); other specific pathways have also been suggested (21). Moreover, the relative abundance of the deuterated species DC_3N and DC_5N supports the idea of gas phase formation for at least HC_3N and HC_5N (22,23). On the other hand, the rather slow decline in abundance with molecular weight for the cyanopolyynes has been used as an argument in favor of their formation by fragmentation of grains or very large molecules (20), or through catalytic surface reactions (2); perhaps followed by isotopic fractionation in the gas phase (62). The question is not yet settled, and quantitative laboratory experiments of the type being planned by Greenberg (this Volume) are awaited with interest.

Of specific relevance to observed cometary radicals are the relative abundances of molecules containing (e.g.) CN, C_2, and C_3. If the situation for TMC-1 is at all analogous to the composition of cometary nuclei, then it could well be that HCN contributes only ~1/2 the observed CN, while there are possibly several parent molecules which may contribute to C_2 and C_3.

Finally, we may comment on the possible existence of more complex molecules than those presently confirmed in interstellar clouds. We must stress that our tabulations are by no means complete, even apart from species such as those in Table 7. Compounds can be detected only if they have been sought, and they may only be sought if their microwave spectra are known, if the existing technology is sufficiently sensitive to provide meaningful limits, and if they exist in the gas phase.

Laboratory spectra have often not been measured for molecules much more complex than those heretofore detected, partly because low vapor pressure may make it difficult to obtain gas phase samples without dissociating the molecules. The rotational spectrum of glycine was only determined in 1978, and no other amino acids have been similarly measured; likewise, no radio frequency data are available for, e.g., adenine. Moreover, compounds much less abundant than those in our Tables could simply not be detected with currently available equipment. And the most complex molecules may simply be locked up in or on grains (preceding Chapter), and hence not directly observable.

Thus, although the basic building blocks of biologically interesting monomers are abundant in interstellar clouds (HCN, H_2CO, etc.), and somewhat more complex molecules clearly exist (such as CH_3CH_2OH, and HCO_2CH_3), the presence of such monomers as amino acids, purines, or pyrimidines is still hypothetical. Nonetheless, the presence of the latter types of molecules in carbonaceous meteorites, the synthesis of heavy molecules in laboratory simulation experiments, and the likely presence of organic mantles on interstellar grains in dense clouds all support the plausibility of the existence of complex organic molecules throughout our Galaxy.

4. RELATION TO THE ORIGIN OF LIFE

Even if biologically interesting molecules are present in space, could they have played a role in the origin of terrestrial life?

There is no doubt that significant quantities of cometary material have been and are today being accreted by the Earth as it encounters interplanetary dust. That portion of this matter with sizes less than a few tens of micrometers is not significantly heated in passing through the atmosphere, and thus is relatively pristine when it reaches the surface. It has even been suggested that the bulk of the Earth's volatiles might originally have been derived from cometary impact, although this is disputed and the impact of a cometary nucleus would largely pyrolize complex organics (2,24,25). The possibility of direct accretion of interstellar particles when the solar system passes through a dense interstellar molecular cloud is also not negligible.

Whether any such extraterrestrial molecules were necessary for chemical evolution is another matter. It may be argued that synthesis is so efficient and pervasive that no non-terrestrial contribution could have played a significant role. This cannot, however, be established until the primitive environment can be more precisely specified. The oxidation state of the primitive atmosphere versus time, the temperature history of the Earth's crust, and the origin of the hydrosphere and ocean-land interfaces are all matters of considerable uncertainty, but are of critical importance to chemical evolution (in addition to the other Chapters in this Volume, see (26)). It is still conceivable that certain critical molecules, serving as catalysts, templates, or reaction partners, were not produced in the terrestrial environment, but were instead supplied from outside, after synthesis under the very different conditions existing in interstellar space or in comets.

ACKNOWLEDGEMENTS

In addition to the references listed in the following section, we have drawn upon unpublished colloquia presented at the Onsala Space Observatory by V. Vanysek (Charles University, Prague, Czechoslovakia), W. Langer (Princeton University, USA), and M. Guelin (Observatoire de Paris, France). We are grateful to Dr. Langer and our colleagues at Onsala for permission to use unpublished data in our tables. We are particulary thankful to T. Grzybowski for preparing this manuscript. Onsala Space Observatory, Chalmers University of Technology, is operated with financial assistance from the Swedish Natural Science Research Council. William Irvine has been partially supported by NASA grant NGL 22-010-023.

5. REFERENCES

Background references are listed under appropriate subject headings, while references for specific points and to the Tables are given in the following category labeled General. Research described in the listed conference proceedings or major reviews has normally not been separately referenced in the text.

5.1 Comets

The proceedings of two recent conferences contain reviews and very recent research results:

Ponnamperuma, C. (ed.): 1981, Comets and the Origin of Life, D. Reidel.
Wilkening, L.L. (ed.): 1981, Comets, Univ. Arizona Press.

Older useful reviews include:

Delsemme, A.H. (ed.): 1977, Comets, Asteroids, and Meteorites, Univ. Toledo Bookstore.
Donn, B., Mumma, M., Jackson, W., A'Hearn, M., and Harrington, R. (eds.): 1976, The Study of Comets, NASA SP-393.

Whipple, F.L. and Huebner, W.F.: 1976, "Physical Processes in Comets", in Ann. Rev. Astron. Astrophys. 14, pp. 143-172.
Whipple, F.L.: 1978, "Comets," in Cosmic Dust (ed. J.A.M. McDonnell), J. Wiley, pp. 1-74.

The possible interrelationships of comets, interstellar molecules, and the origin of life have also been discussed by:

Brown, R.D.: 1981, "Interstellar Molecules and the Origin of Life," in Origin of Life (ed. Y. Wolman), D. Reidel, pp. 1-10.
Irvine, W.M., Leschine, S.B., and Schloerb, F.P.: 1981, "Comets and the Origin of Life," in Origin of Life (ed. Y. Wolman), D. Reidel, pp. 27-32.
Chang, S.: 1979, "Comets: Cosmic Connections with Carbonaceous Meteorites, Interstellar Molecules, and the Origin of Life," in Space Missions to Comets (ed. M. Neugebauer, D.K. Yeomans, J.C. Brandt, R.W. Hobbs), NASA Conf. Publ. 2089.

5.2 Interstellar Chemistry of Dense Clouds

Discussions of gas phase interstellar chemistry are given by:

Watson, W.D.: 1980, "Molecule Formation in Cool, Dense Interstellar Clouds," in Interstellar Molecules (ed. B.H. Andrew), D. Reidel, pp. 341-353 (and references therein).
Dalgarno, A. and Black, J.H.: 1976, "Molecule Formation in the Interstellar Gas," Rep. Progress Physics 39, pp. 573-612.

A list of observed interstellar molecules by source and frequency, including some information on abundances, is given by Mann, A.P.C. and Williams, D.A.: 1980, Nature 283, pp. 721-724.

The review papers presented at a recent Royal Society Discussion on "Molecules in Interstellar Space" will be published in Proc. Royal Soc. London (A).

The composition of interstellar grains is discussed from a somewhat different perspective to that presented by M. Greenberg in this Volume by:

Whittet, D.C.B., 1981: "The Composition of Interstellar Grains," Quart. J. Roy. Astron. Soc. 22, pp. 3-21.

The physics of interstellar molecules and related radio observations are discussed by:

Winnewisser, G., Churchwell, E., and Walmsley, C.M.: 1979, "Astrophysics of Interstellar Molecules," in Modern Aspects of Microwave Spectroscopy (ed. G.W. Chantry), Academic Press, pp. 313-505.
Somerville, W.B.: 1977, "Microwave Transitions of Interstellar Atoms and Molecules," Adv. Atomic Molec. Physics 13, pp. 383-437.

Problems in the determination of abundances, as applied to isotopic ratios, are discussed by:
Penzias, A.: 1980, Science 208, pp. 663-669.

5.3 General

(1) Klinger, J.: 1980, Science 209, 271.
(2) Hayatsu, R. and Anders, E.: 1981 in Topics in Current Chemistry 99, in press.
(3) Allen, M. and Robinson, G.W.: 1977, Astrophys. J. 212, 396.
(4) Langer, W.D., Goldsmith, P.F., Carlson, E.R., and Wilson, R.W.: 1980, Astrophys. J. (Letters) 235, L39.
(5) Fertel, J.H. and Turner, B.E.: 1975, Astrophys. Letters 16, 61.
(6) Turner, B.E. and Thaddeus, P.: 1977, Astrophys. J. 211, 755.
(7) Goldsmith, P.F., Langer, W.D., Ellder, J., Irvine, W.M., and Kollberg, E.: 1981, Astrophys. J., in press.
(8) Mann, A.P.C. and Williams, D.A.: 1980, Nature 283, 721.
(9) Rydbeck, O.E.H., Hjalmarson, Å., Rydbeck, G., Ellder, J., Olofsson, H., and Sume, A.: 1981, Astrophys. J. (Letters) 243, L41.
(10) Irvine, W.M., Hjalmarson, Å., and Rydbeck, O.E.H.: 1981, in Origin of Life (ed. Y. Wolman), D. Reidel, pp. 11-17.
(11) Thaddeus, P., Guelin, M., and Linke, R.A.: 1981, Astrophys. J. (Letters) 246, L41.
(12) Bujarrabol, V., Guelin, M., Morris, M., and Thaddeus, P. 1981, Astron. Astrophys., in press.
(13) Erickson, N.R., Snell, R.L., Loren, R.B., Mundy, L., and Plambeck, R.L.: 1981, Astrophys. J. (Letters) 245, L83.
(14) Ellder, J., Friberg, P., Hjalmarson, Å., Hoglund, B., Irvine, W.M., Johansson, L.E.B., Olofsson, H., Rydbeck, G., and Rydbeck, O.E.H.: 1980, Astrophys. J. (Letters) 242, L93.
(15) Tölle, F., Ungerechts, H., Walmsley, C.M., Winnewisser, G., and Churchwell, E.: 1981, Astron. Astrophys. 95, 143.
(16) Whittaker, A.G., Watts, E.J., Lewis, R.S., and Anders, E.: 1980, Science 209, 1512.
(17) Hayatsu, R., Scott, T.G., Studier, M.H., Lewis, R.S., and Anders, E.: 1980, Science 209, 1515.
(18) Smith, P.P.K.: 1981, Nature 291, 15.
(19) Irvine, W.M., Höglund, B., Friberg, P., Askne, J., and Ellder, J.: 1981, Astrophys. J. (Letters), in press.
(20) Sagan, C. and Khare, B.: 1979, Nature 277, 102.
(21) Mitchell, G.F., Huntress, W.T., and Prasad, S.S.: 1979, Astrophys. J. 233, 102.
(22) Langer, W.D., Schloerb, F.P., Snell, R.L., and Young, J.S.: 1980, Astrophys. J. (Letters) 239, L125.
(23) Schloerb, F.P., Snell, R.L., Langer, W.D., and Young, J.S.: 1981, Astrophys. J. (Letters).

(24) Chang, S.: 1979, in Space Missions to Comets (eds. M. Neugebauer, D.K. Yeomans, J.C. Brandt, and R.W. Hobbs), NASA Conf. Publ. 2089, pp. 59-112.
(25) Oro, J., Holzer, G., Lazcano-Araujo, A.: 1980, in COSPAR: Life Sciences and Space Research (ed. R. Holmquist) 8, pp. 67-82.
(26) Brooks, J.: 1981, Proc. Roy. Soc. (London).
(27) Friberg, P., Hjalmarson, Å., Irvine, W.M., and Guelin, M.: 1980, Astrophys. J. (Letters) 241, L99.
(28) Avery, L.W.: 1980, in Interstellar Molecules (ed. B.H. Andrew), D. Reidel, pp. 47-58.
(29) Snell, R.L., Schloerb, F.P., Young, J.S., Hjalmarson, A., and Friberg, P.: 1981, Astrophys. J. (Letters) 244, 45.
(30) Wootten, A., Bozyan, E.P., Garrett, D.B., Loren, R.B., and Snell, R.L.: 1980, Astrophys. J. 239, 844.
(31) Frerking, M.A. and Langer, W.D.: 1981, private communication.
(32) Churchwell, E.: 1980, Astrophys. J. 240, 811.
(33) Frerking, M.A., Langer, W.D., and Wilson, R.W.: 1979, Astrophys. J. (Letters) 232, L65.
(34) Turner, B.E. and Zuckerman, B. 1978, Astrophys. J. (Letters) 225, L75.
(35) Rydbeck, O.E.H., Sume, A., Hjalmarson, A., Ellder, J., Rönnäng, B.O., and Kollberg, E.: 1977, Astrophys. J. (Letters) 215, L35.
(36) Linke, R.A. and Goldsmith, P.F.: 1980, Astrophys. J. 235, 437.
(37) Guelin, M., Langer, W.D., and Wilson, R.W.: 1981, Astrophys. J., submitted.
(38) Rydbeck, O.E.H., Irvine, W.M., Hjalmarson, Å., Rydbeck, G., Ellder, J., and Kollberg, E.: 1980, Astrophys. J. (Letters) 235, L171.
(39) Rydbeck, O.E.H., Kollberg, E., Hjalmarson, Å., Sume, A., Ellder, J., and Irvine, W.M.: 1976, Astrophys. J. Suppl. 31, 333.
(40) Turner, B.E.: 1973, Astrophys. J. 186, 357.
(41) Allen, M. and Knapp, G.R.: 1978, Astrophys. J. 225, 843.
(42) Frerking, M.A., Langer, W.D., and Wilson, R.W.: 1981, Astrophys. J.
(43) Sume, A., Downes, D., and Wilson, T.L.: 1975, Astron. Astrophys. 39, 435.
(44) Rodriguez, L.F. and Chaisson, E.J.: 1980, Mon. Not. Roy. Astron. Soc. 192, 651.
(45) Little, L.T., MacDonald, G.H., Riley, P.W., and Matheson, D.N.: 1979, Mon. Not. Roy. Astron. Soc. 189, 539.
(46) Benson, P.J. and Myers, P.C.: 1980, Astrophys. J. (Letters) 242, L87.
(47) Wilson, R.W., Langer, W.D., and Goldsmith, P.F.: 1981, Astrophys. J. (Letters) 243, L47.
(48) Langer, W.D., Wilson, R.W., Henry, P.S., and Guelin, M.:

1978, Astrophys. J. (Letters) 225, L139.
(49) Ungerechts, H., Walmsley, C.M., and Winnewisser, G.: 1980, Astron. Astrophys. 88, 259.
(50) Hjalmarson, Å. and Friberg, P.: 1980, in Interstellar Molecules (ed. B.H. Andrew), D. Reidel, pp. 69-70.
(51) Mitchell, G.F., Ginsburg, J.L., and Kuntz, P.J.: 1978, Astrophys. J. Suppl. 38, 39.
(52) Prasad, S.S. and Huntress, W.T.: 1980, Astrophys. J. 239, 151.
(53) Brown, R.D. and Rice, R.U.: 1981, Proc. Roy. Soc. (London),
(54) Huntress, W.T. and Mitchell, G.F.: 1979, Astrophys. J. 231, 456.
(56) Kutner, M.L., Machnik, D.E., Tucker, K.D., and Dickman, R.L.: 1980, Astrophys. J. 242, 541.
(57) Irvine, W.M., Ellder, J., Hjalmarson, Å., Kollberg, E., Rydbeck, O.E.H., Sørensen, G.O., Bak, B., and Svanholt, H.: 1981, Astron. Astrophys. 97, 192.
(58) Brown, R.D., Godfrey, P.D., Storey, J.W.U., Bassez, M.P., Robinson, B.J., Batchelor, R.A., McCulloch, M.G., Rydbeck, O.E.H., and Hjalmarson, Å.: 1978, Mon. Not. Roy. Astron. Soc. 186, 5P.
(59) Irvine, W.M., Leschine, S.B., and Schloerb, F.P.: 1980, Nature 283, 748.
(60) Castor, J.I.: 1970, Mon. Not. Roy. Astron. Soc. 149, 111.
(61) Woods, R.C.: 1981, in Molecular Ions: Geometric and Electronic Structure (ed. J. Berkowitz)
(62) MacLeod, J.M., Avery, L.W., and Broten, N.W.: 1981, Astrophys. J. (Letters).
(63) Graedel, T.E., Langer, W.D., and Frerking, M.A.: 1981, "The Kinetic Chemistry of Dense Interstellar Clouds," preprint.

6. NOTES TO TABLES

6.1 Table 4

The initial number given in parentheses for each note is a literature reference to Section 5.3. This is followed by assumed or deduced physical conditions, where LVG refers to the large velocity gradient or Sobolev approximation (e.g., 60); LTE signifies local thermodynamic equilibrium; and we use the symbols T_x ≡ excitation temperature, n ≡ number density of H_2, T_{kin} ≡ kinetic temperature, τ ≡ optical depth.

a. (38) $T_x(^{32}SO)=T_x(^{34}SO)$, terrestrial $[^{32}S/^{34}S]$
b. (27) Stat. Equil. LVG: $n=10^3-10^4 cm^{-3}$, T_{kin}=10 K; μ=2.2 D
c. (12)
d. (28) Stat. Equil. LVG: $n=9(10)^3-6(10)^4 cm^{-3}$, T_{kin}=10-20 K
e. (29) Stat. Equil. LVG: $n=2(10)^4 cm^{-3}$, T_{kin}=10 K

f. (15) Stat. Equil. LVG: $n \sim 3(10)^4 cm^{-3}$, T_{kin}=10 K
g. (19) LTE, $\tau \ll 1$, T=10 K
h. (30) LTE, $\tau \ll 1$, T=15 K
i. (32) Hyperfine components in LTE; second value at (-4,6) arcmin
j. (33) Derived by present authors from data in (33), assuming $\tau(HN^{13}C) \ll 1$, terrestrial [$^{12}C/^{13}C$], LTE at 10 K
k. (34) No value of N given because of possible self-absorption
l. (35) Hyperfine components in LTE, T_{kin}=10 K
m. (36) Stat. Equil. LVG: $n=2(10)^5 cm^{-3}$, T_{kin}=10 K, no self-absorbtion
n. (39) $T_x=T_{kin}$=5.5 K or -15 K, $T_{rot} \ll 25$ K
o. (40) LTE
p. (41,42) Terrestrial isotope ratios, LTE
q. (42) Same H_2/extinction ratio as diffuse clouds
r. (43) T_x=2.3 K, T_{rot}=7 K
s. (56) LTE, $\tau \ll 1$, T_{kin}=10 K
t. (57) LTE, $\tau \ll 1$, T_{kin}=8 K
u. (58) LTE, $\tau \ll 1$, T_{kin}=7 K
v. (31) Stat. Equil. LVG: $n \sim 10^4 cm^{-3}$, T_{kin}=10 K
w. (37) Stat. Equil. LVG: $n \sim 10^4 cm^{-3}$, T_{kin}=10 K

6.2 Table 6

Number in parentheses is transition frequency in MHz.

a. (27294) Hyperfine component F=4→3, deconvolved from F=3→2
b. (27562) F=4→3
c. (28441) F=4→3
d. (27972)
e. (34157) Total for E plus A species
f. (33051) Assumes electric dipole moment equals that for HC_3N
g. (97981) T_A^* from reference (36)
h. (86847)
i. (48991) T_A^* from reference (36)
j. (69576)
k. (23695) Column density from reference (35)
l. (33705) F=2→3

6.3 Table 7

Assumptions used in the calculations were

a. $n=5(10)^4$ cm^{-3}, T=50 K, elemental abundances as observed towards the star ζ Oph
b. $n=2(10)^4$ cm^{-3}, T=50 K, elemental abundances as towards ζ Oph
c. Assumes large rate for the radiative association reaction $CH_3^+ + H_2 \rightarrow CH_5^+$+photon
d. $n=10^4$ cm^{-3}, T=30 K, cloud age 10^7 years
e. For comparison, these calculations predict $n(CO)/n(H_2) \approx 10^{-4}$,

very close to the observed value for TMC-1 (Table 4).
f. H_2O has been observed in the Orion molecular cloud (Table 2)
g. Assumes high CH_4 abundance (see note c.).
h. Recent calculations (63) for T=10 K show no significant change vs. temperature.

IMPACT OF SOLAR SYSTEM EXPLORATION ON THEORIES OF CHEMICAL EVOLUTION AND THE ORIGIN OF LIFE

DONALD L. DeVINCENZI

National Aeronautics and Space
Administration
Washington, D.C. 20546

ABSTRACT

The impact of solar system exploration on theories regarding chemical evolution and the origin of life is examined in detail. Major findings from missions to Mercury, Venus, the Moon, Mars, Jupiter, Saturn, and Titan are reviewed and implications for prebiotic chemistry are discussed. Among the major conclusions are: prebiotic chemistry is widespread throughout the solar system and universe; chemical evolution and the origin of life are intimately associated with the origin and evolution of the solar system; the rate, direction, and extent of prebiotic chemistry is highly dependent upon planetary characteristics; and continued exploration will increase understanding of how life originated on Earth and allow better estimates of the likelihood of similar processes occurring elsewhere.

INTRODUCTION

Research in disciplines as diverse as molecular biology and radioastronomy have contributed greatly to the understanding of processes which led to the origin of life on Earth. It is now generally accepted that the origin of life was preceded by a period of chemical evolution where the biogenic

C. Ponnamperuma (ed.), Cosmochemistry and the Origin of Life, 143–174.

elements and molecules on the primitive Earth, or in the vicinity of the acreting and evolving planet, were converted through a series of chemical and physical reactions into small organic molecules essential for life, molecules such as amino acids, nucleic acid bases, and simple sugars. In the presence of suitable energy sources and catalytic surfaces, these small molecules reacted further to form the key catalytic and informational polymers, that is, proteins and nucleic acids. This was followed by a series of reactions of increasing complexity which resulted in the formation of such biological structures as membranes and subcellular organelles and such biological systems as metabolic pathways and the genetic system. These formations culminated in the origin of a cellular entity capable of growth, reproduction, and transfer of genetic information.

Scientific evidence for this process has been accumulated from a variety of sources including 1) simulations of primordial Earth and planetary chemistry, 2) examination of chemical and biological species in ancient terrestrial rocks, 3) analysis of meteorites and observations of interstellar molecules, and 4) unmanned exploration of planetary environments. The unmanned explorations have provided the opportunity to test the more general aspects of hypotheses that have been heavily influenced by the only models we have available for direct study--Earth as the planet and Earth life as the form.

Exploration of the planets, satellites, and moons of the solar system by unmanned spacecraft over the last 20 years has provided a new perspective to understanding processes relevant to the origin of life. It is becoming increasingly apparent that the origin and evolution of life on Earth is intimately associated with the origin and evolution of the planet itself as well as the solar system as a whole. Both the rate and direction of chemical and biological evolution have been shaped by the physical and chemical environments in which these processes occurred. The recently acquired ability to explore extraterrestrial environments, uncomplicated by life processes that have drastically altered the geological record on Earth, has provided a wealth of new data crucial to understanding the chemical mechanisms leading to life.

In the context of the Institute on Cosmochemistry and the Origin of Life, this paper will deal with results of solar system exploration missions which are pertinent to the process of chemical evolution. This planet-by-planet review will include 1) a brief description of the general properties of each planet, 2) the type of mission conducted, 3) a more detailed summary of the major scientific findings, and (4) a discussion of their implications for our understanding of chemical evolution pathways. It is hoped that this approach will create an effective bridge between the preceding papers, which have emphasized extrasolar processes and systems, and the following papers which explore in detail the pertinent information contained within the terrestrial record.

SOLAR SYSTEM EXPLORATION

Mercury

Mercury, the planet nearest the sun, was surveyed in 1974 by the Mariner 10 spacecraft as it flew by the planet. On-board cameras revealed a heavily cratered surface, much like that of the moon (Figure 1). Since Mercury has no detectable atmosphere, its surface features have remained unweathered from the time of its formation. Because of Mercury's proximity to the sun, surface temperatures exceed 300° C and the surface is continuously bathed with intense ultraviolet light.

Although these properties and the absence of reactive atmospheric species make Mercury hopeless from a chemical evolution standpoint, this hostile environment does provide some useful information regarding the limits or extremes of planetary conditions in which the chemistry of life can occur. For example, although Mercury is smaller than the Earth, its density is approximately the same since it is richer in heavier elements because its proximity to the sun resulted in the volatilization of lighter elements. Furthermore, absence of an internal heat source prevented plate tectonic activity and the various differentiation processes which were so important in providing conditions

Figure 1. The surface of Mercury, photographed by the Mariner 10 spacecraft, shows a rugged surface much like that of the Moon.

conducive for chemical evolution on the primitive Earth. These examples illustrate how overall planetary properties (e.g., size, mass, density, and distance from the sun) can be critical in determining whether a planet has the proper initial inventory of biogenic elements and molecules or the internal sources of energy and volatiles that are needed to initiate the process of chemical evolution.

Venus

Venus (Figure 2), the cloud-shrouded planet most like Earth in size and mass, was examined in

Figure 2. Venus cloud formations are shown in this photo taken by the Pioneer orbiter.

great detail during the Pioneer Venus orbiter and multiprobe missions following encounter late in 1978. There were 12 scientific experiments on the orbiter, 2 on the carrier bus, 7 on the large probe, and 3 identical experiments on each of the three smaller probes. This mission provided direct measurements of cloud structure, atmospheric composition and structure, and surface properties. Of particular interest are the results from the large probe's neutral mass spectrometer and gas chromatograph instruments which were designed to measure the composition of the lower atmosphere of Venus.

Table 1 summarizes several key findings from the Pioneer Venus mission and highlights the vast differences in temperature, pressure, and atmospheric constituents between Venus and Earth. The significance of Venus in understanding the process of chemical evolution must then lie in questions dealing with why and how Venus diverged so greatly from Earth after presumably similar beginnings.

It is generally agreed that high surface temperatures on Venus, higher than expected solely on the basis of distance from the sun, are due to the so-called greenhouse effect. According to this model, visible light penetrating the clouds is absorbed by the surface material which then radiates infrared light. Atmospheric carbon dioxide then traps the infrared light very effectively and produces a tremendous heating of the planet's surface and lower atmosphere. This greenhouse effect is especially dramatic on Venus for two reasons: 1) carbon dioxide is the main atmospheric constituent, and 2) the atmosphere is 90 times more massive than the Earth's.

Venus, then, represents another case where gross planetary characteristics (e.g., distance from the sun) can produce conditions that prohibit development of the kind of chemistry that is a necessary prerequisite for life. A planet this close to the sun cannot maintain liquid water on its surface, and without liquid water, there is no mechanism by which carbon dioxide can be removed from the atmosphere and converted into carbonate rocks as is the case on Earth (a process that is, incidentally, greatly accelerated by life).

Table 1

Significant findings from Pioneer Venus spacecraft

Temperatures	
	Cloud top: -33° C
	Surface: 475° C
Atmospheric composition	
	Major constituents (% composition):
	CO_2 (96)
	N_2 (3.5)
	Minor constituents (ppm range)
	H_2O
	O_2
	CO
	SO_2
	Others
Surface pressure	
	90 atm

Another series of measurements conducted by the Pioneer Venus spacecraft which is important to understanding the planet's history deals with isotope abundances of the rare gases argon and neon, and how their abundance values on Venus differ from their values on Earth. Because argon, in particular, is believed to be a measure of the primordial solar-nebula material from which the planets condensed, the large difference in argon values between Venus and Earth challenges the assumption that the terrestrial planets all had a common origin, and thereby complicates attempts to understand how and why their environments diverged so radically. An alternative explanation, however, is that this is a localized phenomenon, where evolution of planetary atmospheres is highly dependent upon initial fractionation of elements, including rare gases, because of proximity to the sun.

Moon

From 1969 to 1972, manned exploration of the Moon (Figure 3) during the Apollo program provided the first opportunity for exhaustive study of the history of another planetary body by direct examination of samples returned from its surface. Over 2000 samples weighing nearly 400 kg were collected from nine different sites during the six Apollo lunar landings. Scientists continue to study these samples in an attempt to recreate the origin and history of the Earth's nearest neighbor and decipher the clues the Moon contains regarding the early history of the solar system.

Careful examination of lunar samples shows that the Moon is deficient in such volatile elements as hydrogen, carbon, and nitrogen, the very elements essential for life as we know it. Because of the Moon's smaller size (about 1/2 that of the Earth) and reduced gravity (about 1/6 that on Earth), both of which also preclude the possibility of sustaining an atmosphere necessary to initiate chemical evolution reactions, these volatile elements probably have been deficient since the Moon was formed.

Lunar samples also have contained no water, either free or combined, and only traces (i.e., ppm range) of carbon or carbon-containing compounds. Aside from questions about terrestrial contamination during sample analysis, the presence of trace amounts of organic matter probably results from deposition during meteorite bombardment and/or interaction of elemental components of the solar wind with surface material. Finally, examination of lunar crustal material gives no evidence for tectonic activity, that is, plate movements which are major forces in the geological activity of the Earth.

Bleak as the Moon may be with regard to its own potential to foster chemical reactions important for life, other considerations highlight its

Figure 3. The heavily cratered surface of the Moon indicates its crust must have formed early in the history of the solar system. The dark regions (*maria*) are low areas covered with basaltic lava. The light regions (*highlands*) are higher areas that are older than the *maria*.

significance for the origin and evolution of life on Earth. For example, the tidal action exerted by the Moon on the Earth could have provided environments conducive for organic reactions of increasing complexity, such as tidal basins where dilute solutions of abiotically synthesized organic matter could be concentrated and activated for reaction by periodic cycles of wetting and drying of catalytic clay surfaces. In addition, the stabilizing effect the Moon has on the orientation of the Earth's rotational axis may be responsible for long-term climate stability, a factor crucial for development, evolution, and persistence of life on a global scale.

Mars

Of all the planets in the solar system, Mars (Figure 4) exhibits a range of environments and planetary characteristics that make it the best candidate for occurrence of processes similar to those that led to life on Earth. Although Mars is considerably smaller than Earth, is farther away from the sun, and has only about 1/3 the gravitational field, it has several important similarities to Earth: 1) a geologically diverse surface, 2) a dynamic atmosphere, 3) waxing and waning polar caps, 4) day-night cycles nearly identical to those of Earth, and 5) similar seasonal changes. More intensive investigations from Earth-based observatories and flyby and orbiter spacecraft, however, highlighted very significant differences: 1) much lower temperatures, including daily freeze-thaw cycles even at the warmest locations; 2) an atmosphere with carbon dioxide as the major constituent and only traces of oxygen; 3) an atmospheric pressure approximately 1/200 that of Earth; and 4) intense bombardment of the surface by ultraviolet light from the sun. Despite this harsh environment, evidence from microbiological studies on the adaptation of organisms to extreme terrestrial environments and from chemical studies of the potential of simulated Martian environments to support chemical evolution reactions made the search for life on Mars a valid scientific objective.

The Viking missions to Mars in 1976 were designed to acquire detailed information about the atmosphere and surface of the planet with special

Figure 4. This view of Mars is a composite of pictures taken by Viking 2. To the right of the morning terminator is a giant volcano highlighted by plume like water-ice clouds. To the south is the frost-filled Argyre basin.

emphasis on the question of life. Extensive scientific investigations were conducted on each of the two missions using the 12 instrument packages listed in Table 2. There was one spacecraft in orbit and a soft lander on the surface in each mission. Only the biology instrument was designed to search for life processes directly, but almost all of the other instrument packages, especially those for molecular and inorganic analysis, water and thermal mapping, and meteorology, contributed directly to assessments of the capacity of the Martian

TABLE 2

Scientific payload for the Viking missions to Mars

Investigation	Instruments
Orbiter	
Visual imaging	Television cameras
Water vapor mapping	Infrared spectrometer
Thermal mapping	Infrared radiometer
Lander	
Imaging	Facsimile camera
Biology	Analyses for metabolism, growth, or photosynthesis
Molecular analysis	Gas chromatograph-mass spectrometer
Inorganic analysis	X-ray fluorescence spectrometer
Meteorology	Pressure, temperature, and wind velocity gauges
Seismology	Seismometer
Magnetic properties	Magnet on sampler
Physical properties	Engineering sensors
Radio Science	
Atmospheric, planetary, interplanetary, and relativity data	Orbiter and lander radio and radar systems

environment to sustain life or the chemistry preceding life. Table 3 summarizes some results important in this regard from several Viking instruments.

The remainder of this section will deal with findings from three of the other experiments: inorganic analysis, molecular analysis, and biology.

TABLE 3

Significant findings from 1976 Viking Mars missions

Temperature ranges:	
Planetwide	-143° C to 17° C
Landing sites	-83° C to -33° C
Atmospheric pressure:	8 millibars
Permanent polar cap:	H_2O ice
Atmospheric composition - gas and (% composition):	
CO_2	(95)
N_2	(2.7)
Ar	(1.6)
O_2	(0.1)
CO	(0.07)
H_2O	(0.03)
O_3	(0.03 ppm)

Inorganic analysis. Elemental abundances in the Martian soil were determined by X-ray fluorescence spectrometers which analyzed several different soil samples in the vicinity of each lander. Table 4 shows the results of analyses carried out at the two landing sites which were separated by several thousand kilometers. In general, the data indicate that elemental compositions at each site are virtually identical, and that the abundances are quite similar to those found in several common terrestrial soil and rock types. After careful examination of the Mars data, and exhaustive comparisons of spectra with various terrestrial and artificially created model soils, it was concluded that iron-rich clays of the nontronite class represent the best fit to the data. This point will be important when the results of the biology experiments are discussed. It should also be noted that this experiment, as designed, could provide only elemental abundance data and not information on the oxidation state of minerals in the soil. Consequently, there is a crucial gap in

TABLE 4

Elemental composition of Mars soil

Element	% of total composition (average of both sites)
Silicon	21
Iron	13
Magnesium	5
Calcium	4
Sulfur	3
Aluminum	3
Chlorine	0.5
Titanium	0.5
All others (thought to be mostly oxygen)	50

the data because most interpretations of the unusual chemical activity of the Martian soil discovered by Viking instruments invoke the presence of oxidizing soil constituents for which there was no means of positive identification.

Molecular analysis. The purpose of the molecular analysis experiment was to identify and quantitate organic compounds expected to be present in the Martian soil. The approach involved pyrolysis of soil samples to vaporize the organic constituents, separation on a gas chromatographic column, and identification by mass spectrometry. The results of this experiment have had far-reaching implications for assessing the state of chemical evolution on Mars.

In summary, the results show that approximately 0.1 to 1 percent water was released from soil samples at each site following heating to 350° C or 500° C. Presumably, this water was released from hydrated minerals, although exact identification of the mineral source is difficult to infer from the available data. In addition, small amounts of carbon dioxide were released from the heated samples although positive identification of the carbonate source is likewise lacking.

The most dramatic discovery from this experiment, however, was the complete lack of detectable organics in all samples and at both sites, down to the parts-per-billion level. There is no question that the molecular analysis instrument performed properly since it did detect trace organic contaminants which had also been detected during the cruise phase of the mission. Furthermore, extensive ground-based tests with a flightlike instrument demonstrated its ability to detect organics in a wide variety of samples, including meteorites and sterile antarctic soils.

These findings suggest the presence of active destructive processes operating in the surface materials which have obliterated any record of deposition of organics by previous eras of meteorite bombardment. Apparently these processes are also active enough to prevent the accumulation of abiotically synthesized organics which can be shown to occur in simulated Martian environments (carbon dioxide, carbon monoxide, water, and ultraviolet radiation) and which may have been much more prominent during earlier periods of more favorable environmental conditions.

It should be noted, however, that caution must be exercised in extrapolating these results to all locations on the planet. For example, only samples from the top 6 cm of soil were analyzed. Because of the atmospheric and geological conditions on Mars, this layer is most subject to the destructive processes mentioned earlier. It is entirely possible that substantial quantities of organic matter are present at depths beyond the reach of Viking's sampler arm or in other areas on the planet where conditions might be more conducive to preservation of organic matter (i.e., at or near the polar caps). Moreover, the Viking instrument would have been unable to detect organic material if it were in a highly polymerized state, although one would expect smaller, detectable fragments to be associated with such complex compounds. Finally, organic material embedded within rocks, as is commonly found in rock samples from extreme terrestrial environments like the antarctic dry valleys, would have escaped detection because Viking had no provision for analyzing rock samples.

Biology. Perhaps the most intriguing data from the mission were produced by the biology instrument, a complex device containing three separate experiments to search for active metabolic processes in soil samples. These experiments were designed to detect microbial life forms based on the assumption that any Martian biota would employ a chemistry based on carbon and water.

The pyrolytic release (PR) experiment was designed to measure the presence of photosynthetic activity in soil by measuring the ability of samples to incorporate radioactively labeled carbon dioxide or carbon monoxide gas into organic matter. The gas exchange (GEX) experiment employed a gas chromatograph to measure biologically induced changes in metabolically important gases, (e.g., hydrogen, nitrogen, oxygen, and carbon dioxide) following incubation of soil samples with a rich nutrient medium. The labeled release (LR) experiment measured the ability of components in the

Martian soil to catalyze the breakdown of radioactively labeled organics to carbon dioxide gas during the course of incubation.

Table 5 summarizes the major findings of the three life detection experiments. Qualitatively at least, all three registered initial responses that could be interpreted as being consistent with biological processes. On closer scrutiny, and in light of results from other lander experiments, biological interpretation of these results gave way to speculation about inorganic explanations for the reactivity of the samples. In fact, following extensive experiments on Mars over several months, and numerous simulations and analyses in terrestrial laboratories over several years, it is now generally concluded that the Viking results are best interpreted as being due to the activity of various inorganic species in the Martian soil.

TABLE 5

Summary of results of Viking biology experiments

Experiment	Findings
Pyrolytic release	Small incorporation of CO/CO_2 into organics
Gas exchange	Initial rapid release of O_2; slow release of CO_2, N_2, ${}^2Ar/CO$
Labeled release	Initial rapid release of labeled gas; continuing slow release

This inorganic explanation, however, is quite complex and requires the postulation of several different mechanisms to be consistent with all of the findings. For example, the rapid evolution of oxygen in the GEX experiment can be shown to be simulated by the addition of water to labile chemicals induced in soils by photochemical or other treatments. Similarly, soils containing iron-rich clays or peroxides of various kinds will catalyze the production of carbon dioxide at a rate, and in amounts, similar to those observed in the LR experiment. Finally, small amounts of carbon dioxide gas can be fixed into organic matter, as observed in the PR experiment, by several minerals whose presence is compatible with the findings of the inorganic analysis experiment. Although this explanation is complicated, it is the most consistent with all available data, mission as well as ground-based.

Although the Viking biology results do not indicate active biology in the soil samples analyzed, extrapolation to the planet as a whole again must be done with caution. As was the case with the molecular analysis experiment, no data are available from samples at depth, at or near the polar caps, or inside rocks. However, it is known that in the closest terrestrial analogue to the Martian environment, the antarctic dry valleys, microbial life is abundant several millimeters below the surface of rocks where the microbes themselves have created a hospitable environment. Furthermore, the Viking experiments were not designed to explore the question of life that might have existed earlier in the planet's history although other Viking data indicate there might have been much more water available, a denser atmosphere, and more favorable temperature regimes.

Mars, then, remains an enigma with respect to chemical evolution and even life. Although Viking increased our knowledge about the planet and its atmospheric and surface properties by orders of magnitude, a whole new series of questions has emerged, among which are the following:

Are there organics, as might be expected, in other locations on the planet?

How far has chemical evolution progressed?

Were atmospheric conditions in the planet's history more conducive to prebiotic chemistry?

What is the current state of water on Mars, where is it, and was it more plentiful in the past?

What is the exact mineralogical composition of the soil?

Are the soil minerals playing any role in chemical evolution on Mars as we suspect clays played on the primitive Earth?

Are the activities observed in the Viking biology experiments related in any way to prebiological processes?

Is there evidence for extant life, fossil life, or organic molecules in Martian rocks?

The list of questions could go on. It is clear that answers will come only from continued exploration, either with more sophisticated spacecraft sent to Mars or with soil samples returned from Mars for analysis in terrestrial laboratories.

Jupiter

Jupiter (Figure 5), the largest planet in the solar system, has a mass approximately 300 times that of Earth but only about 1/4 the density. It represents a clear and intriguing demarcation in the transition from the dense and rocky inner planets to the large gaseous giant planets of the outer solar system.

Three features of the planet have made it of intense interest to scientists concerned with chemical evolution and the origin of life. First, the atmosphere is reducing; that is, it contains methane, ammonia, and water, with a large excess of hydrogen and helium. Because elemental abundances on Jupiter are similar to those of the sun, it is assumed to reflect the elemental composition of the presolar nebular gas cloud from which the planets condensed. If, in fact, all the planets had a common origin, Jupiter's present atmosphere may reflect the composition of the Earth's primordial atmosphere about 4 billion years ago. Because reduced atmospheres containing methane, ammonia, water, and hydrogen are known to produce biologically important molecules when appropriate energy sources are applied, Jupiter is believed to be a site where

Figure 5. Voyager 1 took this photo of Jupiter at a range of 32.7 million km (20 million miles). The banded atmospheric patterns parallel the planet's direction of rotation. The Great Red Spot, 25,000 km long, is a feature that has persisted for at least several centuries.

extensive synthesis of organic matter important for chemical evolution is taking place at the present time.

The second feature of interest is that although Jupiter is five times farther away from the sun than the Earth is from the sun, it generates more energy than it receives. This suggests the presence of an internal energy source which may be important for synthesis of organics.

Third, the distinctive colored bands as well as the Giant Red Spot have spectral features similar to several classes of organic compounds synthesized in terrestrial simulations. Even though there are equally plausible inorganic explanations for the origin of the colors, the consistency with organic spectra is a property that makes Jupiter of interest from the chemical evolution perspective.

This gaseous giant, together with its many moons and satellites, has been extensively surveyed by the earlier Pioneer 10 and 11 missions and more recently by the Voyager spacecraft encounters in 1979. Voyager studies of the Jovian atmosphere emphasized atmospheric dynamics, composition, structure, and magnetospheric effects; the satellite studies stressed both the comparative and detailed geology of the Galilean satellites. The investigations of particular significance to atmospheric composition included infrared radiation, radio science, and ultraviolet spectroscopic measurements.

Table 6 summarizes some of the key findings from the Voyager 1 mission. In addition to confirming the presence of reduced gases, spectroscopic and radiometric investigations demonstrated the occurrence of both acetylene and ethane in the Jovian atmosphere. Temperature measurements at various atmospheric depths were low, but Voyager provided the unexpected finding of cloud-top lightning bolts. This, coupled with the equally surprising discovery of active volcanism on the satellite Io, indicates the energetic and dynamic nature of the Jovian system. The discovery of ethane and acetylene was subsequently verified by Voyager 2, but their ratio varied at different latitudes and was twice as high as the ratio from the Voyager 1 data.

Although the Voyager missions confirmed theories that Jupiter contains both the atmospheric constituents and energy sources necessary for prebiotic synthesis of classes of molecules important for life, many questions still remain unanswered. For example:

> Are hydrocarbons larger than two carbon units, or different classes of organics such as amino acids present; or is more complex chemistry not possible in an atmosphere dominated by hydrogen?

Despite the measurement of ethane and acetylene, is their synthesis balanced by a continuous destruction process resulting from convection currents which cause degradation deep in the atmosphere where temperatures are much higher?

Is this same process responsible for destruction of more complex organic molecules?

The absence of microenvironments, at moderate temperatures and containing liquid water in proper abundance, may indeed have prevented chemical evolution from proceeding beyond this primitive stage on Jupiter.

The Jovian discoveries also affect theories about chemical evolution in other interesting ways. For example, they raise the question of whether a

TABLE 6

Significant findings from Voyager 1 mission to Jupiter

<u>Atmospheric constituents:</u>
Hydrogen
Helium
Methane
Acetylene
Ethane
Ammonia
Phosphine
Water
Others
<u>Temperature:</u>
-173° C at 0.1 atm
-113° C at 1 atm

reduced atmosphere might have bathed the Earth for a length of time sufficient to allow for more complex and varied organic synthesis. Theorists have suggested that on the primitive Earth such a reduced atmosphere may have been a rather transient phenomenon, being supplanted rather quickly by an atmosphere which was formed by outgassing from the Earth's interior, an atmosphere dominated by more oxidized species such as carbon monoxide, carbon dioxide, nitrogen, and water. Terrestrial simulations of organic syntheses in these atmospheres using a variety of energy sources have, to date, been largely unsuccessful in producing organics of equal complexity and similar abundances as observed under more reducing atmospheric conditions.

On a broader scale, the Jovian system can be viewed as a miniature solar system. In fact, density calculations for some of the Galilean satellites are consistent with similar determinations for the solar system as a whole, with the density decreasing from the inner to the outer satellites. This is believed to indicate that, under the nebular theory of planetary formation, all planetary systems should show the same variation from dense inner planets to gaseous outer planets. The vastly different capacities of the various planets in the solar system to sustain prebiotic chemistry is another reminder that planetary characteristics such as size, density, gravity, distance from the sun, and incident or internal energy may each be a critical factor that determines the rate, extent, direction, and duration of chemical evolution reactions.

Saturn

As with Jupiter, atmospheric studies of Saturn (Figure 6) are important because they can provide significant information about the processes that governed the origin and evolution of the solar system and about primordial organic chemistry.

Because the Saturnian system is approximately a billion miles farther from the sun than the Earth is from the sun, Saturn receives only a small fraction of the solar energy incident on the Earth. Saturn's mass is 95 times that of the Earth and it has an atmosphere consisting mainly of hydrogen and helium. The planet is obscured by clouds moving at velocities

Figure 6. Voyager 1 recorded Saturn and its rings from a distance of 18 million km. A series of dark and light cloud bands are apparent in the planet's northern hemisphere through a high-altitude atmospheric haze.

that vary with latitude and is surrounded by an intricate complex of rings, satellites, and moons, each with its own set of unique and distinguishing features.

From the perspective of information on the nature and extent of primordial organic chemistry occurring on Saturn, the same objectives and instruments that were important during the Voyager encounters with Jupiter were also of prime importance when these same spacecraft reached Saturn in 1980 and 1981. In particular, the infrared radiation and ultraviolet spectroscopy investigations provided information about atmospheric dynamics, composition, structure, and magnetospheric effects on Saturn.

Results from infrared observations of Saturn's atmosphere by Voyager 1 are summarized in Table 7. The similarity to Jupiter's atmosphere is quite apparent and, in fact, the concentrations of methane, acetylene, and ethane are approximately equal in both atmospheres. In addition to these components, other spectral data provide suggestive evidence for the presence of more complex molecules such as methylacetylene and propane. Further analysis is necessary to confirm these identifications.

TABLE 7

Atmospheric composition of Saturn from Voyager 1

Gas	Mole Fraction
Positively identified	
Hydrogen	0.94
Helium	0.06
Ammonia	2×10^{-4}
Phosphine	1×10^{-6}
Methane	8×10^{-4}
Ethane	5×10^{-6}
Acetylene	2×10^{-8}
Tentatively identified	
Methylacetylene	
Propane	

In assessing the importance of these findings to theories about prebiotic chemistry, the conclusions drawn earlier for Jupiter apply to Saturn as well. Both planets have overall elemental abundances that reflect solar abundances more than the inner planets. Both have apparently retained much original matter from the presolar nebula in contrast to the inner planets whose thin atmospheres are probably of secondary origin. And both planets appear to sustain, on a continuing basis, synthesis of small organic molecules. This is consistent with both chemical evolution theory and Earth-based simulations of planetary atmospheric chemistry.

One other set of Saturnian observations by the Voyager spacecraft is worthy of note. During close approach to several heavily cratered icy moons, precise measurements of their size were made, which allowed more accurate determination of densities. The calculated densities are very similar to densities predicted for cometary nuclei. These findings have led to speculation that the moons may have formed from cometary material outside the Saturnian system. This is an interesting possibility because, as previous papers in this series have shown, cometary spectra show the presence of several reactive species which may be fragments of parent molecules known to be key intermediates in chemical evolution pathways. It also highlights the need to understand the role, if any, of comets, meteorites, and interstellar dust grains in the delivery of preformed organic matter to an acreting planet.

Titan

Of all the outer planet satellites and moons, Saturn's moon Titan (Figure 7) is the only one that contains both an atmosphere and solid surface. These features, together with its size (slightly larger than Mercury), reddish color, and a methane atmosphere, make Titan of intense interest for chemical evolution studies.

Although the Voyager spacecraft was not able to visualize the surface through the dense, hazy atmosphere that envelops Titan, it did obtain new information on the composition, structure, and dynamics of the atmosphere that significantly altered earlier conclusions based on observations from Earth. For example, infrared measurements show that nitrogen is the dominant atmospheric constituent. Further proof is suggested by detection of compounds containing nitrogen, carbon, and hydrogen, and the discovery of nitrogen emission spectra by the ultraviolet observations. Additional components of Titan's atmosphere are shown in Table 8.

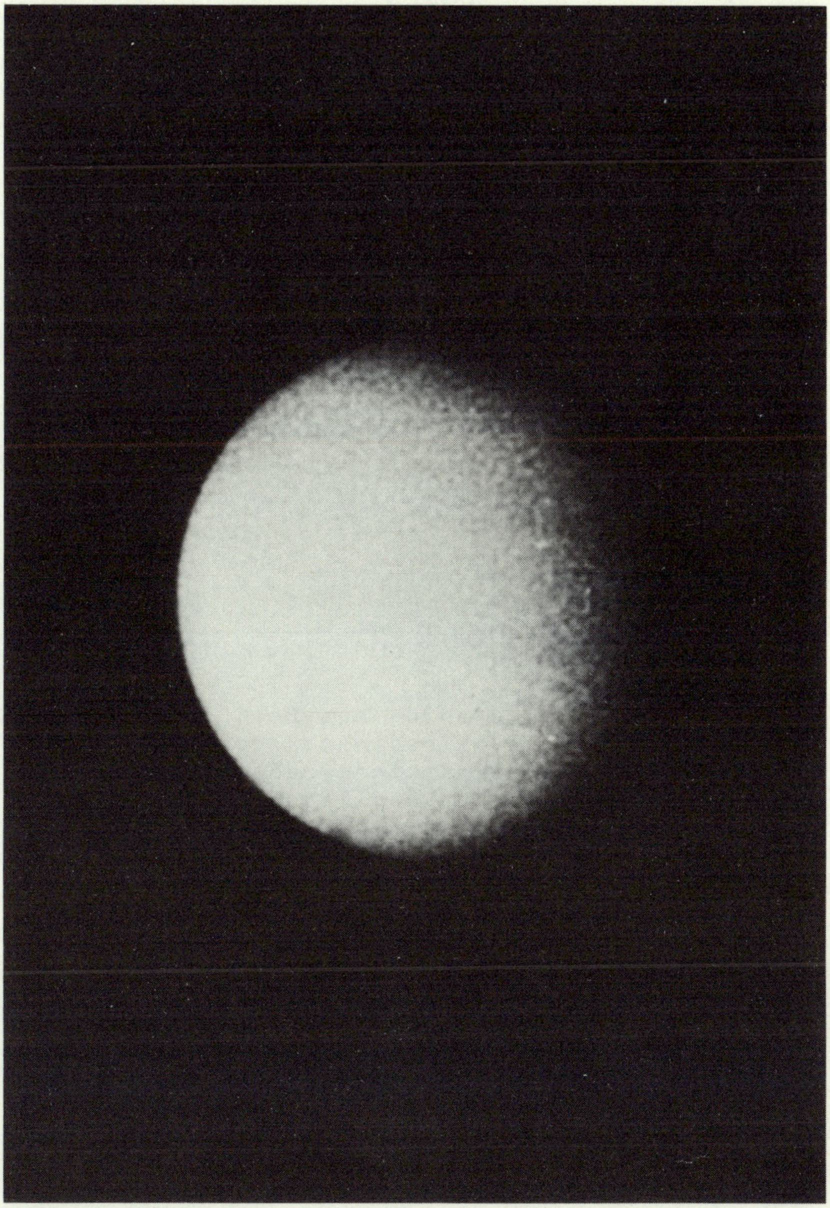

Figure 7. Voyager 1's cameras show Titan's surface to be totally obscured by a thick layer of atmospheric haze.

TABLE 8

Atmospheric composition of Titan from Voyager 1

Gas	Approx. mole fraction
Positively identified	
Methane	1×10^{-2}
Ethane	2×10^{-5}
Acetylene	3×10^{-6}
Ethylene	1×10^{-6}
Hydrogen cyanide	2×10^{-7}
Tentatively identified	
Methylacetylene	
Propane	

Although hydrogen cyanide has been detected in the interstellar medium by radioastronomical observations and has been synthesized in prebiotic simulations of outer planet and Titan atmospheric chemistry, this is the first direct proof of its existence in a planetary atmosphere. This finding is of major significance because in prebiotic simulations hydrogen cyanide is a key reactive intermediate in the synthesis of more complex biologically relevant compounds such as amino acids and purine and pyrimidine bases. Hence, the synthesis of precursors to biochemicals essential for life appears to have occurred on Titan as well as on Earth.

Temperature and pressure measurements also provide new data on Titan's present state and have implications for its evolutionary history. Although it is not clear that Voyager sensors were able to penetrate to the surface of Titan, temperatures of 90° K were recorded at what is believed to be near-surface levels of the atmosphere. The corresponding pressures in this area were approximately 1.5 times higher than at the Earth's surface. When these data are viewed in light of an atmosphere dominated by nitrogen and containing small amounts of methane, they suggest that Titan's present atmosphere may be reflective of the primitive atmospheres of the inner planets, but is preserved in this state because of the low temperatures. Therefore, although the current atmosphere of Titan is very different from the atmospheres of the inner planets, it may contain clues about the chemical and physical processes that shaped their development and subsequent evolution.

Several laboratories have conducted simulations of Titan's atmospheric chemistry by subjecting gaseous mixtures of nitrogen and methane to various energy sources, including ultraviolet radiation. The results are generally in excellent qualitative agreement with Voyager findings, depending on the energy source, and demonstrate the synthesis of hydrogen cyanide as well as numerous other saturated hydrocarbons. Voyager also discovered that the abundances of ethane and acetylene on Titan are significantly higher than on Saturn (by factors of 4 and 150, respectively) because of physical processes which result in a nearly irreversible conversion of methane to these species. Therefore, Earth-based and mission data support the concept that substantial quantities of organic matter may be synthesized in the Titan atmosphere, transported downward, and then condensed, precipitated, and accumulated on the surface.

CONCLUSIONS

The solar system is replete with physical and chemical information that broadens our perspective about processes affecting chemical evolution and the origin of life. For the first time, exploration of the solar system by means of unmanned spacecraft has provided direct measurements in extraterrestrial environments to support theories that previously were based upon limited and rather geocentric terrestrial simulation experiments and ground-based planetary observations.

When planetary data are assimilated and analyzed, an integrated scheme begins to emerge which indicates that chemical evolution and the origin of life on Earth are intimately associated with the origin and evolution of the planet itself, and even the solar system and universe. Planetary chemistry, coupled with analyses of meteorites, observations of comets and interstellar molecules, and simulations of interstellar chemistry, show that the synthesis of precursors to molecules essential for life is a universal phenomenon. It is also clear, however, that the nature and extent of this prebiotic chemistry is highly dependent upon the same factors and processes that affect solar system and planetary evolution and development. It has been during this period of planetary exploration that we have gone far beyond asking simply what is synthesized in a particular atmosphere when it is subjected to various energy sources and we now are questioning the nature of the primordial atmosphere itself, its origin and duration, the effect of planetary size and distance from the sun on atmospheric evolution and chemistry, and the extent to which this chemistry occurs as a function of the planet's gravity which affects both the initial inventory of the biogenic elements as well as their rate of escape.

Throughout this planet-by-planet survey, some of the many questions that still remain unanswered have been listed. Some answers can be obtained by more advanced terrestrial simulations and ground-based observations. Others, however, can only be obtained by continued exploration of planetary environments where measurements can be made in situ or on samples returned to terrestrial laboratories. Advanced planetary missions to Jupiter, with both an orbiter and atmospheric probe, and to Saturn, with an orbiter and probes into the atmospheres of both Saturn and Titan, will greatly increase our understanding of the detailed chemistry occurring in those environments. The state of chemical evolution on Mars and conditions during its past would be clarified by more sophisticated lander missions or by a returned sample mission. Comet rendezvous or sample return missions, both technologically feasible at present, would provide detailed information regarding solar and extrasolar chemistry in these relics of the primordial solar nebula and allow a better assessment of their role in delivering an initial inventory of biogenic elements and molecules to acreting planets by impact. Finally, an initial reconnaissance of asteroids will complete the inventory of major classes of solar system bodies, allow an analysis of the nature of these objects in comparison with meteorites and comets, and shed light on their role in the physical and chemical evolution of the solar system.

In view of the fact that these kinds of missions are complex and costly, it should be noted that scientific information of fundamental importance for chemical evolution and the origin of life can also be obtained from the smaller Pioneer and Mariner class missions. Advanced studies currently underway indicate that with careful planning, clear definition of science requirements, and innovative approaches to instrumentation, these missions can provide the opportunity to investigate a smaller set of scientific questions in flyby, probe, and even limited lander mission modes.

Finally, the space telescope, to be placed into orbit by the space shuttle in 1985, will provide the capability to observe deep-space phenomena which cannot be observed through the Earth's atmosphere or which are degraded by terrestrial interference. With suitable adjustments, it is possible that this instrument could be used to detect the presence of planets around other stars and, for the first time, directly demonstrate the validity of the nebular theory that predicts planets should be a frequent, if not absolute, accompaniment to the formation of stars. With further refinements, it is conceivable that spectroscopic techniques can be used to explore their overall properties and even identify atmospheric constituents on these planets.

As summarized in this paper, knowledge gained from planetary exploration has greatly expanded our understanding of the detailed prebiotic chemistry occurring in the Earth's solar system environment. It has also contributed to a more expanded view of life as a planetary phenomenon. Continued exploration will increase our knowledge of how life originated on Earth and allow better estimates of the likelihood of such processes occurring elsewhere beyond the solar system.

BIBLIOGRAPHY

1. Beatty, J.K., O'Leary, B., and Chaikin, A.: 1981, The New Solar System, Sky Publishing Corp., Cambridge, Massachusetts.

2. Coldsmith, D. and Owen, T.: 1980, The Search for Life in the Universe, Benjamin/Cummings Publishing Co., Inc., Menlo Park, California.

3. J. Geophys. Res., special Pioneer at Venus issue, Vol. 82, No. 28, 1977.

4. J. Geophys. Res., special Viking at Mars issue, Vol. 82, No. 28, 1977.

5. Klein, H.P.: 1979, 'The Viking Mission and the Search for Life on Mars', Rev. Egophys. Space Phys. 17, 1655-1662.

6. Morrison, D. and Samz, J.: 1980, Voyage to Jupiter, National Aeronautics and Space Administration, Washington, D.C. (NASA SP-439).

7. Science, special Pioneer at Venus issue, Vol. 203, No. 4382, 1979.

8. Science, special Voyager 1 at Jupiter issue, Vol. 204, No. 4386, 1979.

9. Science, special Voyager 2 at Jupiter issue, Vol. 206, No. 4421, 1979.

10. Science, special Voyager 1 at Saturn issue, Vol. 212, No. 4491, 1981.

THE CHEMICAL COMPOSITION AND CLIMATOLOGY OF THE EARTH'S EARLY ATMOSPHERE

A. Henderson-Sellers

University of Liverpool, Liverpool, UK
Goddard Institute for Space Studies, New York, N.Y.

ABSTRACT

The unique system which comprises the Earth's climate is examined and seven fundamental characteristics derived. The surprisingly robust nature of the climate which is revealed seems to be a direct result of the existence of a global hydrosphere. Analysis of the chemical and physical processes operating within the Earth's atmosphere--surface environment around the critical period of the origin and earliest evolution of life suggests that the likely global conditions result in hospitable physical environment whilst also permitting local chemical inhomogeneities. The length of time required for the earliest stages of evolution may be critical.

1. INTRODUCTION--THE ROLE OF THE ATMOSPHERIC SCIENTIST IN COSMOCHEMISTRY AND THE ORIGIN AND EVOLUTION OF LIFE

"The Earth--a strange and beautiful anomaly in our solar system" (56)

Advancement in science often depends upon accurate, or at least adequate, resolution of data and theories in both the time and space dimensions. Such definition is critically important for questions regarding the cosmochemical setting for the origin and evolution of life on the Earth. Atmospheric science can contribute to these fundamental discussions by the interpretation (and possibly extrapolation) of climate model results which use the physical and chemical parameters appropriate for the early Earth. The importance of the atmosphere is undeniable; its very existence is probably a fundamental requirement for the evolution of

C. Ponnamperuma (ed.), Cosmochemistry and the Origin of Life, 175–212.

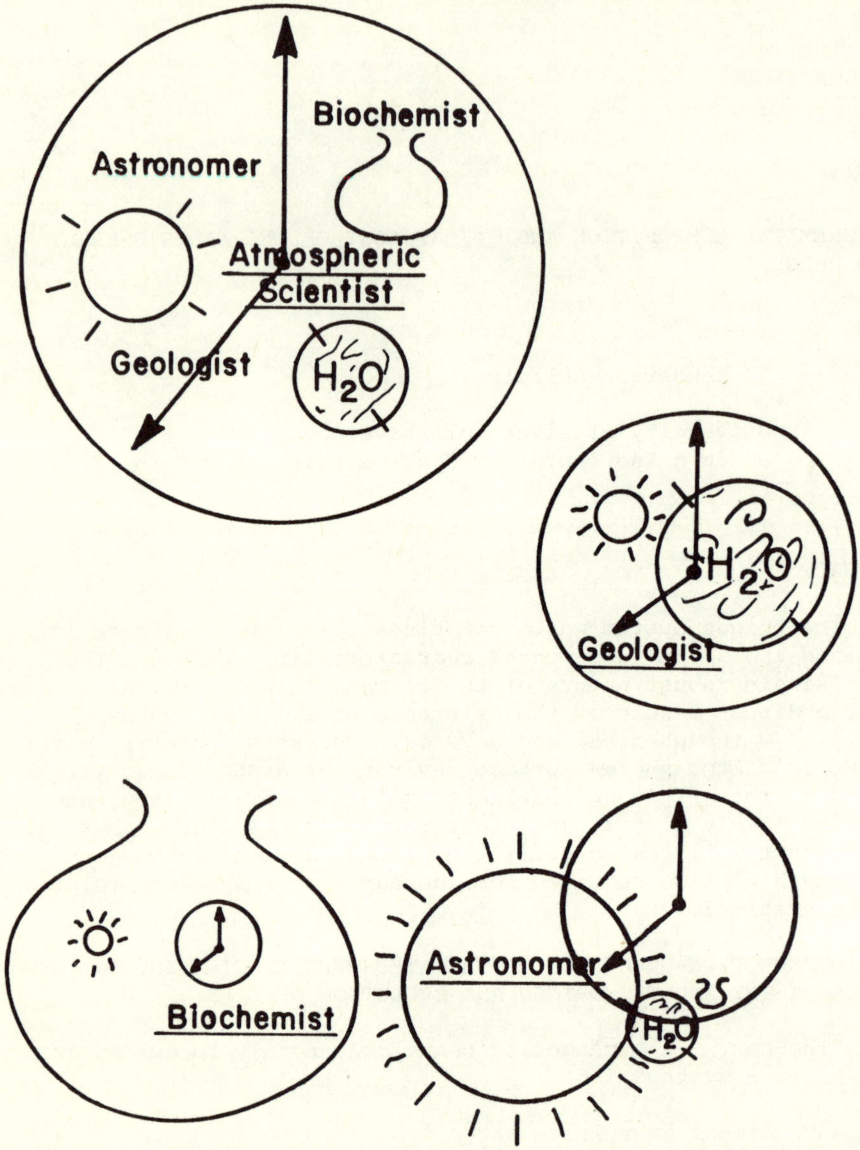

Figure 1. The fundamental importance of time and space resolution can be overlooked within specific research fields. The atmospheric scientist must draw data from the three major research disciplines: geology, biochemistry and astronomy. The evaluation of the relative importance of TIME and data, here represented by PLACE (Earth), ENERGY (Sun) and EXPERIMENTS (flask), seems to differ.

life. The atmosphere is predominant in its control of the chemical and physical conditions at the surface of the Earth. For instance, surface type is a result of atmospheric process--oceans (over 70 percent of the surface), deserts and ice fields are a result both of specific local atmospheric processes (i.e., precipitation, insolation, wind, etc.) and also a direct function of global scale energy circulation which controls local and regional temperatures. The ability of plants to photosynthesise and life to respire depends upon the chemical composition of the atmosphere. Indeed, if the evolution of man depended upon fire and later the engines of the industrial revolution, then it too has been abetted by the atmosphere since oxygen levels below 12% make combustion impossible. Similarly, if, as seems likely, aerobic respiration preceeded oxygen releasing photosynthesis, there must have been at least a small mixing ratio of free O_2 in the early atmosphere.

Figure 1 illustrates schematically the primary areas of interest of the three principal scientific disciplines contributing to the discussions on cosmochemistry and the origin of life. The cartoon clock encompassing all the groups is to underline the fundamentally important nature of time scales and time resolution. Despite the obvious need for communication between scientific disciplines, one is often hampered by the need to draw data from other areas of expertise while feedback in the form of results is neither common nor adequate. For instance, it is necessary for the atmospheric scientist to use data from geology, biochemistry and astronomy. Each of these disciplines has a slightly (or grossly) different perspective upon the relative importance of the data sources, schematically shown in Figure 1 by the sun, a laboratory flask and the earth itself. The three remaining cartoons clearly illustrate the author's perception of colleagues' prejedices(!). The need for the intensive multi-disciplinary discussions such as have taken place at this Study Institute is clear.

The importance of time control is seen in the need to date events accurately but, perhaps more importantly, in the <u>length of time</u> required to accomplish each step in the origin of evolution of life. It is in this sense that the perceptions and areas of interest of different disciplines become confused. For instance, it is now generally agreed (42,66) that the atmospheric composition of the terrestrial planets has been primarily a mixture of CO_2 and H_2O vapour throughout geologic history. Analyses of Precambrian rocks are continuing to push backwards the time period before which evidence supporting this basic premise does not exist (61, 76). However, the majority of experiments on the synthesis of organic materials are conducted in flasks containing highly reduced mixtures of gases. Chang et al. (18) remark upon this anomoly in their summary, "the connection between the model environmental conditions implicit in many of the syntheses and the geologic and meteorologic realities of the prebiotic Earth may be tenuous...." These two situations may not, in fact, be mutually exclusive, but the space scales and/or time factor may be critical. It is not unreasonable to presume that reducing conditions could exist in a restricted local environment. Indeed, such environments exist today on the ocean floor (23) and are known to have existed at times during the geological record (87, 76). Alternatively, the earliest conditions on a homogeneously accreted Earth may approach those in the experiments. It is important to note that organics synthesized at this stage would have to survive the considerable thermal trauma of core formation and atmospheric removal followed by degassing. Therefore, the important questions for the biochemist become: i) whether complex molecules readily produced can survive these astrogeological upheavals and perhaps more likely, with regard to localized synthesis, ii) what is the necessary time period required to complete adequate synthesis so that the building processes will not be totally destroyed when the local scale conditions change, thus removing the sustaining environment. If this time period is relatively short and survival likely in the different global conditions then the apparently exclusive experiments and theories could turn out to be mutually reinforcing.

Time periods over which events occur are also going to be vital for the atmospheric scientist since changes in or recovery of the global climate state will have to be achievable within time periods not incompatible with the data and the known characteristics of the system. For instance, the occurrence of snow and ice at a particular latitude in a winter season may be a weather anomoly but persistence of packed snow and ice over a number of years implies a climatic perturbation and may be itself a reinforcing agent (the "ice-albedo feedback mechanism"--see Section 3). As well as focussing upon specific climate anomolies, the atmospheric scientist's contribution to cosmochemical questions must take on a global dimension. It is difficult to consider the environmental conditions at a particular location, whether on the surface, in the ocean or within the atmosphere itself without immediate reference to the global atmospheric state. This is because the net global situation of emitted thermal energy just balancing absorbed solar energy masks complex energy transfer processes both horizontally and vertically. The local temperature and the vertical lapse rate as well as wind and precipitation patterns are a direct function of atmospheric mass and chemical composition. This is simply illustrated by consideration of the diurnal surface temperature ranges on Mars, the Earth and Venus. Thus the global atmospheric configuration must be adequate to support both mean conditions and local extremes as interpreted from data. A further facet of the global atmosphere is that of continuity. If an external mechanism such as the surface heating likely to be associated with late core formation (49) or sweeping by the solar wind (76) removed or considerably modified the majority of the atmosphere, then surface conditions will suffer an enormous upheaval. If such an event took place after the origin of life, the likely preservation of climatic environment at at least adequate levels to permit survival of the life forms or their building blocks seems to be required. The issues raised here are considered in the next three sections as follows: i) a detailed consideration of the best available data and likely conditions which leads to a characterization of the atmospheric and environmental state; ii) consideration of the problems and possibilities of planetary climate sensitivity underlines the uniqueness of the Earth's system; iii) finally, these results are applied to the areas of continuing research into cosmochemical questions of the origin of life, thus providing a first attempt at constructive feedback within this diverse subject.

2. THE EVOLUTIONARY ENVIRONMENT--THE ATMOSPHERIC COMPOSITION AND ENVIRONMENTAL CHARACTERISTICS ON THE EARLY EARTH

The physical and chemical characteristics of the Earth's atmosphere around the time of the origin and earliest evolution of life are considered. This evolutionary environment is defined by seven fundamental parameters (37). Data are drawn from the geological record as far as possible but often inferences have been made from available astrophysical and climatological models. Some difficulties of apparently incompatible data sets seem to be a result of inadequate definition of the timescales at which and over which mechanisms operate. In this discussion of the Earth's atmosphere, the analyses will be for the time period over which the evolution of life and the evolution of the environment coincided. The time that atmospheric and surface properties define the regime within which biological evolution must have progressed, whilst at the same time the enormous strides of evolutionary processes may have impinged upon global, chemical and climatological conditions.

The time domain (see Figure 2 from (37)) lies between the earliest period of planetary formation, termed the astrophysical environment, and the period around the mid Precambrian when, after the appearance of large quantities of free oxygen in the Earth's atmosphere, the regime may be termed the modern environment. Positioning on the time axis of even these end points is extremely difficult. Probably the start of the modern environment lies between 2.5 and 1.6 x 10^9 yrs BP (55), but Schidlowski's (74) recent suggestion that carbon isotope ratios in the Isua rock formation indicate photosynthetic activity prior to their formation may imply a very much earlier start for the modern environment, as indicated in Figure 2. The earliest history of the

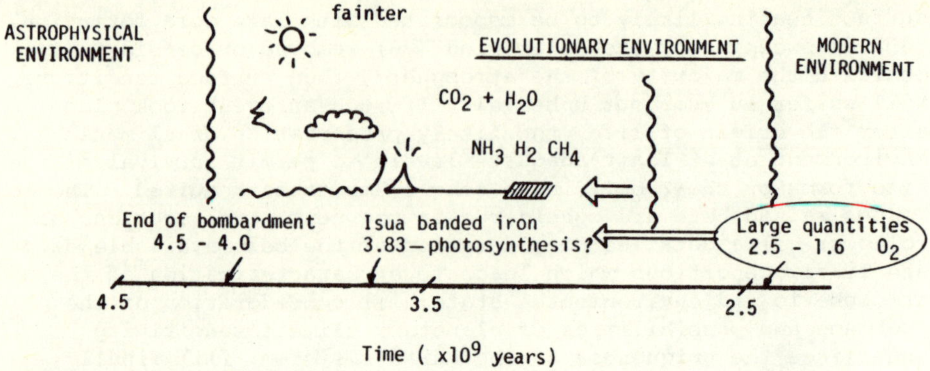

Figure 2. Schematic diagram of the evolutionary environment (from (37)). The time period extends from the cessation of planetary scale astro-geophysical upheavals until the geological era when the ocean-continent regime is established and oxygen is a significant constituent of the atmosphere.

Earth's environment will have been dominated by processes associated with planetary formation. The cessation of these major disruptions is again difficult to locate upon the time axis. Certain models of planetary formation can build the Earth and its atmosphere within 10^5 to 10^6 years (93), whilst other evidence seems to indicate a much later end to astrophysical disruptions. The time period is defined to begin between 4.5×10^9 and approximately 4.0×10^9 and to terminate between approximately 3.9 & 2.5×10^9 yrs BP. It is in this time period, termed the evolutionary environment and having a length of between 400 and 600×10^6 years (see Figure 2) that consideration will be made of the interactions between the evolutionary processes of life and the composition of the atmosphere. The characteristics of this period are summarized in Table I. The data and reasoning behind their derivation is described below.

2.1 A Carbon Dioxide and Water Vapour Atmosphere (Plus Neutral and Trace Gases)

The classic work of Rubey (70) is now underlined by a comparison between the total volatile inventories of the Earth and Mars and Venus (65). Support for the view long held by geologists (1) that the primitive atmosphere was an almost neutral mixture of the volatiles CO_2, H_2O, N_2 and CO is now evident in the literature (42, 66). Noble gas data from the Viking 1 and 2 missions to Mars and estimates for Venus from Venera 9 and 10 reveal a very similar pattern of relative abundances as found on the Earth. These ratios are also found in the 'planetary components' of meteoritic gases re-inforcing the view that

TABLE I

Characteristics of the Evolutionary Environment
(Time span 400 - 600 x 10^6 years--see Figure 2)

1. CO_2 + H_2O atmosphere (plus neutral and trace gases)
2. Atmospheric mass between 500-1000 mb
3. Extensive surface water + global hydrosphere
4. Lower incident solar flux
5. Hospitable mean surface temperatures
6. Convectively active troposphere (lapse rate ≃5-7K km^{-1})
7. Trace gases common and variable.

fractionation occurred prior to planetary formation. The following propositions of Owen (65) are the "new orthodoxy" (66).

(i) 'The atmospheres of the inner planets are derived from volatile-rich veneers that coated these planets in the last stages of accretion.'

(ii) 'The primitive atmospheres of the inner planets were not highly reduced mixtures of CH_4, NH_3 and H_2, captured from the primordial solar nebula but only weakly reducing containing CO_2, CO, N_2 and a small amount of H_2, produced by degassing during and after accretion.' It is particularly interesting to note that all likely mechanisms of atmospheric evolution result in an atmosphere dominated by CO_2 and H_2O, further, that this composition does not conflict with Owen's final proposition that:

(iii) 'The mean temperature of the early Earth was about 273K, even though the solar luminosity was low and significant amounts of NH_3 were not present in the atmosphere.' (see 2.5 below.)

The lack of free iron in the upper mantle (85) requires degassing of predominantly neutral or oxidized rather than reduced gases for all models of planetary formation. Although NH_3 cannot persist in any globally significant quantities in the early atmosphere (51), it may be important on a local scale (41).

It is by no means certain precisely how or at what rate the

atmosphere evolved (see Section 2.2). However, the time span of the evolutionary environment is long enough to make consideration of chemical and geophysical processes important. The removal of gases (e.g., CO_2) from the atmosphere not only affects the total atmospheric mass, but also decreases the greenhouse effect. Chemical reactions (especially photochemical) may considerably perturb levels of trace gases in the primitive atmosphere and small amounts of H_2, NH_3 and CH_4 may have been transiently present as a result of meteoritic bombardment and local scale activity (41).

Despite this general consensus about the chemical nature of the primitive atmosphere, biologists concerned with the origin of life still often work with mixtures of reduced gases. These experiments are interesting and valuable but should be supplemented by others using a CO_2 + H_2O mixture + trace gases.

2.2 Atmospheric Mass Between 500 and 1000 mb

Planetary surface temperatures are a function of both the atmospheric mass and its chemical composition; but the former, in turn, depends on the (highly uncertain) degassing rate. Henderson-Sellers et al. (42) have therefore looked at two extreme cases: (1) when the degassing rate is rapid (i.e., effectively instantaneous compared with the lifetime of the Earth); (2) when the degassing rate is slow (i.e., taking 10^9 yrs or more to produce an atmosphere nearly equal in amount to that present now).

Rapid atmospheric production by impact vaporization during the final stages of planetary accumulation (2) is important when impact velocities of a few kilometres a second are expected in the accretional process. Benlow and Meadows (6) have shown that for a typical impact velocity of 5 km s^{-1} there is a critical planetary mass of ~5 x 10^{23} kg above which atmospheric formation becomes significant. The composition of an atmosphere built during the final stages of planetary accretion would reflect the chemical composition of the final impacting debris (e.g., (2)). The model presented by Benlow and Meadows (6) assumes that the final accretion took a relatively short time (<10^5 yrs). The infalling material is assumed to be very similar in composition to the most primitive materials known today: Cl carbonaceous chondrites and comets (54). The critical feature governing the ambient surface temperature is the magnitude of the greenhouse effect associated with this rapidly evolving atmosphere of CO_2 and H_2O vapour. Despite enhanced levels of CO_2, the authors (6) show that the condensation of surface liquid water is very likely. This is followed by solution of atmospheric CO_2 and finally precipitation of carbonates and bicarbonates. Benlow and Meadows (6) suggest that in the highly acidic oceanic environment this precipitation would be through reaction with alkali metals rather than calcium. The atmospheric mass is stabilized within a short geological time.

The second extreme evolutionary history considers slower atmospheric evolution. The effective planetary temperature, T_e, is calculated from

$$S(1 - A) = f\sigma T_e^4 \tag{1}$$

where S is the solar flux, integrated over all wavelengths at the average distance of the planet from the Sun (i.e., the solar constant in the case of the Earth); A is the Russell-Bond spherical albedo for the planet, and f the planetary flux factor (60). The surface temperature is related to the effective temperature, T_e, via the greenhouse increment. The computation of even average global surface temperature in the case of slow degassing of the atmosphere is complex. The interactions between chemical and physical parameters which characterize the Earth's climate must be considered in detail. The increase in the solar flux is described in Section 2.4. Internal feedbacks, say between water

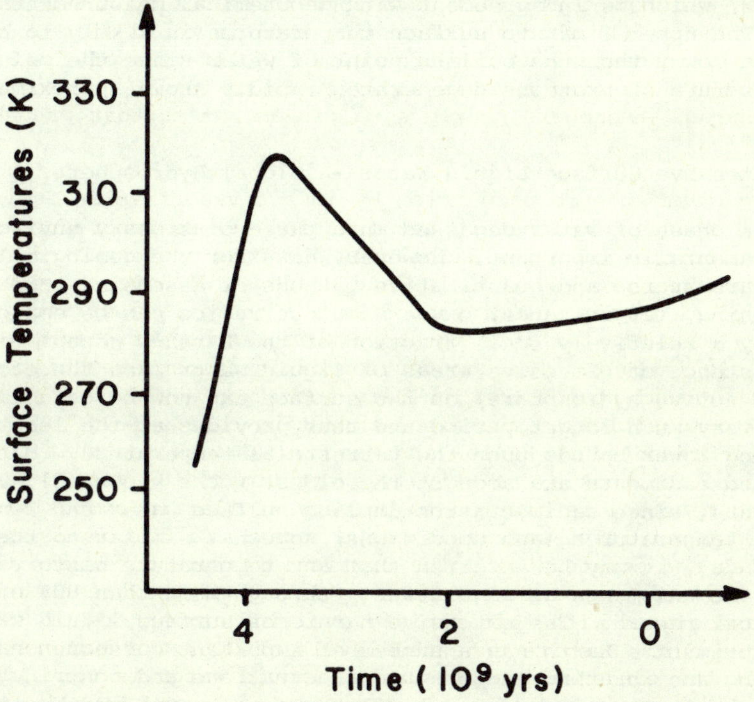

Figure 3. Computed surface temperature evolution of the Earth under the hypothesis of extreme slow degassing (after (42)).

vapour partial pressure, the tropospheric lapse rate and the planetary albedo (which is a sensitive function of global cloud cover) must be considered (see Section 3). Here it seems reasonable to presume that the planetary albedo will increase monotonically over time from a low (≈0.7) initial value as a function of increasing atmospheric mass. The value of the planetary flux factor is still more difficult to establish as it is a function of the rotation rate of the planet and hence, in the case of the Earth, of the gravitational interaction between the moon and the Earth (38, 39). These authors have calculated that the reduced greenhouse effect associated with a tenuous early atmosphere could be compensated for by change in the rotation period upon which f depends and by the lower planetary albedo, A. Figure 3 shows the variation of surface temperatures for the Earth as calculated for the assumption of slow degassing.

Two important results are common to both the slow and rapid atmospheric evolution models. In each case, and therefore by inference in intermediate cases also, the surface temperature of the Earth remains between 275K and 325K, i.e., easily within the freezing and boiling temperatures of water (see Section 2.3 below). Also, each model produces an early peak in temperature (see Figure 3) which is in agreement with geochemical measurements (50). The upper limit to surface temperatures is likely to be somewhat lower than the boiling point of water since the molecular structure of proteins degenerates rapidly above approximately 315K.

2.3 Extensive Surface Liquid Water--A Global Hydrosphere

The phase of water degassed into the evolutionary environment is fundamentally important. However, data for the geological record are sparse and quantitative methods of assessing temperatures using, for instance, oxygen isotope ratios can be employed for only a relatively small fraction of the Earth's recent history. The existence of extensive areas of liquid water (and therefore also an active hydrosphere) on the surface can now be established over a very much longer period and thus provides at the least upper and lower bounds upon the terrestrial temperature. Metamorphosed sediments are amongst the oldest rocks known (61, 75, 76). Thus, since sedimentation implies surface water and hence ambient temperatures (and conditions) somewhat similar to the present day, it may be inferred that any temperature history must permit the existence of liquid water through more than 80% of the geological record. The extensive nature of surface liquid water on the primitive Earth has a number of important consequences, including ameliorating temperature fluctuations and contributing significantly to poleward energy fluxes (e.g., (64) and Section 4) and, apparently, controlling global climatic stability.

2.4 Lowered Solar Luminosity

Newman and Rood (62) state, "a faint young Sun is one of the most unavoidable consequences of stellar structure considerations." Certainly current models of stellar evolution agree that the early Sun emitted appreciably less flux than the present-day Sun. The difference seemed to be sufficiently large to have produced temperatures below the freezing point of water, thus conflicting with known geological data and stimulating a number of early models of the primitive atmosphere (67, 71)--the "faint Sun--enhanced early surface temperature" paradox. The emitted flux has probably increased between 35% and 45% over the lifetime of the solar system (32, 62), approximately a linear increase of between 4-5% over the time span of the evolutionary environment (36). A secondary but important effect for the evolution and stability of the climate will be changes in the wavelength of absorbed and emitted radiation within the atmosphere. Particularly the absorption of ultra-violet radiation, by trace gases within the troposphere (see 2.7) and eventually by the evolving ozone layer in the stratosphere (Section 3), may affect surface and atmospheric temperature profiles.

2.5 Hospitable Average Global Temperatures

Geological evidence suggests the presence of extended sheets of liquid water on the Earth's surface at least 3.85 x 10^9 yrs BP, thus limiting the possible range of surface temperatures. Furthermore, the inference that is drawn from study of the metamorphosed sediments in the Isua rocks is that environmental conditions and particularly the nature of weathering and deposition "were remarkably like those prevailing during the rest of the Archaean" (81). The rest of the Precambrian geological record indicates a continuation of this apparent stability in the depositional environment (45), although fragmentary evidence (27) suggests that climatic variability was at least as pronounced as during more recent gelogical eras. Climatological self-regulatory mechanisms seem to have operated within the Earth-atmosphere-hydrosphere system.

Computation of surface temperatures for the evolutionary environment is only worthwhile for mean global conditions. The surface temperature at any epoch can, in principle, be calculated in two stages. The first involves the computation of the effective temperature of the planet, T_e, from Equation 1. All the factors are straightforward except for f(38) and (39). The second stage of the calculation relates T_e to the surface temperature, T_s, by an equation of the general form

$$T_s = T_e + \Delta T \qquad (2)$$

where ΔT is the greenhouse increment due to the presence of an atmosphere. ΔT is a complex function of the mass, surface pressure and chemical composition of the atmosphere concerned. The solution of Equations 1 and 2 for T_s over a wide range of the relevant input parameters and at any epoch follows (38), incorporating reanalysis of the laboratory data available on infra-red absorption as a function of pressure and was used to produce the surface temperature curve in Figure 3.

These calculated surface temperatures do not conflict with the known geological data, although no attempt has been made in these calculations to include trace gases, which will be discussed in Section 2.7, or to consider climatic feedback mechanisms (Section 3).

2.6 A Convectively Active Atmosphere

The geological data already reviewed suggest mean temperatures $\gtrsim$275-280K and plentiful surface water. Evidence for a fairly substantial atmosphere (Section 2.2) with considerable opacity in the infrared (Section 2.1) also supports these findings. It seems inevitable that all likely atmospheric configurations will support convection since the increased temperatures would lead rapidly to unstable lapse rates in the lower troposphere. The tropospheric lapse rate must have been a function of height in the atmosphere and position on the globe, but for the mean global situation, it is simply defined as the difference between the surface and tropopause temperatures divided by the depth of the troposphere (present day global averages ~285K, 220K and 10 km, respectively).

There are two reasons for requiring evaluation of the lapse rate. They are (i) it permits discussion of convection and hence cloud formation in the atmosphere; and (ii) the tropopause temperature is of vital importance in discussion of loss rates of hydrogen from the evolutionary environment (47). The calculation of mean global surface temperatures in Sections 2.1 and 2.5 assumed radiative equilibrium, i.e., no account was taken of the effect of convective processes. Inclusion of any convective adjustment involves calculation or parameterisation of vertical energy transport. Currently, general circulation models of the Earth's atmosphere (e.g., (57)) assume that an atmospheric lapse rate more unstable than 6.5 K km^{-1} is highly unlikely. Thus it seems reasonable to assume (as indeed we have implicitly throughout) that mixing via convection and cloud condensation can, and does, take place within the atmosphere and therefore unreasonable to assume radiative equilibrium in calcuation of T_s (see Section 3). Cloud formation has the immediate consequence of perturbing the global albedo and hence directly affects the tropopause and surface temperatures. The magnitude of the cloud albedo feedback has been discussed in (34) and (35), but the additional greenhouse contribution means that cloud variations may provide feedback effects in the Earth's atmosphere of either a positive or negative nature (Section 3). The convective adjustment has important consequences, viz. (i) surface temperature is reduced; (ii) tropopause temperature is increased; (iii) the cycle of convective activity and hence cloud formation may become a stabilising influence.

The importance of the tropopause temperatures for troposphere chemistry was established by Hunten (47) when he demonstrated that the rate of hydrogen escape from the present day atmosphere is directly controlled by the water vapour mixing ratio in the stratosphere which in turn is a function of the tropopause or "cold trap temperature" ((87), (89)) rather than the exospheric temperature.

2.7 Trace Gases

Despite the fact that considerable emphasis has been placed here upon demonstrating that the evolutionary atmosphere was dominated by CO_2 and N_2 (see 2.1) with a mixing ratio of water vapour determined by ambient surface temperatures (see 2.6), trace gases are not unimportant. Minor constituents were almost certainly both common and variable in relative and absolute abundances. Attention must be paid to these minor constituents for a number of reasons. For the climate system, the complex interrelationship between surface temperatures and all the other environmental constraints may be perturbed by the presence of any atmospheric gas. The effect of additional neutral gases can probably be neglected for likely surface pressures on Earth. However, any trace constituent possessing infra-red absorption bands may be of importance. The magnitude of the further greenhouse increment is a function of the position and strength of the additional spectral features.

It seems unlikely that free iron existed in the mantle during the time span of the evolutionary environment (86). The closest model atmosphere is therefore that of Holland's ((44), (46)) stage 2. The gases are N_2, H_2O, CO_2, SO_2 with only small amounts of H_2 ($H_2O/H_2 = 60$ and $CO_2/CO = 17$ and trace constituents less than 10^{-4} atm.). However, chemical and photochemical reactions may be of local and even global importance for these trace components and further the extensive abiological and biological reactions associated with the origin and subsequent evolution of life may considerably perturb environmental levels of certain of these components (84). Here the interest lies in the climatological effects that these trace components may have, but it is important to note that constituents such as H_2S and CH_4 also absorb high energy ultra-violet radiation (see (68) and Section 4 below). Ultra violet radiation seems to be favoured as an energy source by a number of researchers in the field of prebiotic synthesis and therefore the likelihood of specific molecular absorption is important. Similarly, the penetration of H_2O into the upper atmosphere may be of critical importance for some reactions (48).

Ammonia has been suggested (e.g., (71) and (31)) as a candidate for enhancement of early surface temperatures because of its considerable infra-red absorption. However, it has been demonstrated that photodestruction of ammonia would have been substantially complete in less than 10^6 yr (26) and Kuhn and Altreya (51) suggest that a mixing ratio $\leq 10^{-4}$ of ammonia would be converted to N_2 in less than 40 yr. The photodestruction of NH_3 is an important problem since most models of chemical evolution require significant concentrations of NH_3 and NH_4^+ in solution of the synthesis of amino acids and other biochemicals (5).

Locally, however, NH_3 levels could be much higher since titanium dioxide acts as a catalyst for the photochemical reduction of N_2 to NH_3 ((9) and (80)). The current global production may be between 6.8×10^9 and 6.8×10^{10} kg per yr, but the rate of nitrogen reduction is obviously a function of the total area of a suitable catalyst and the type of irradiation that it suffers. This previously unrecognized abiological source seems to be of the same order of magnitude as computed destruction rates and thus the global budget of NH_3 could be enhanced. Henderson-Sellers and Schwartz (41) suggest an upper limit of around 10^{13} kg NH_3. They have calculated (Table II) the increase in surface temperature on

TABLE II

The greenhouse effect caused by ammonia in the early Earth's atmosphere for the addition of 10^{13} kg NH_3.

Temperatures in brackets are the recomputed ΔT values when the convective adjustment is made (after (37)).

Time (10^9 yrs BP)	pCO_2 (mb)	10^{13} kg NH_3 ΔT(K)
4.25	310	17.2 (4.2)
3.5	70	15.1 (3.8)
2.5	18	13.7 (3.6)

the Earth. No attempt was made to calculate the effect of adding this amount of NH_3 to $CO_2 + H_2O$ atmospheres more recent than 2.5×10^9 yrs BP since this is outside the time period defined for the evolutionary environment. The carbon dioxide atmosphere is that described in detail in (66).

Climate feedback mechanisms were not considered in (41). In particular, convective adjustments to the surface temperature calculations detailed above seem to be important. Surface temperatures enhanced by up to ~17 K must surely result in considerable tropospheric reorganisation. Specifically, the values T_s have now been recomputed (37) as described above (Section 2.6) and ΔT values are also given in Table II. These recomputations strengthen the conclusions of (41) that this upper limit to the atmospheric level of ammonia is compatible with known geological data (e.g., (94) and (60)) and with locally enhanced surface concentrations of NH_3 (Section 4).

There is some evidence to suggest that either or both methane and carbon monoxide were more abundant in the evolutionary environment than now. For instance, volcanic and mid-ocean emissions can contain significant, but variable, amounts of these gases ((10) and (26)). Also, carbon monoxide is a significant secondary component of the martian atmosphere--e.g., (56). Henderson-Sellers and Meadows (40) have calculated the change in terrestrial surface temperature allowing for variations in carbon monoxide partial pressure from 0.001 to 1 mb with various values of the total atmospheric pressure. Even major changes in carbon monoxide content (1 mb CO in 100 mb CO_2) have a very small effect on surface temperature (T_s raised by $\simeq$0.6K) because the absorption features are both weak and removed from the major atmospheric windows.

The effect of methane upon current surface temperatures has recently been calculated. Hansen et al. (29) used a 1-D radiative-convective model to investigate the sensitivity of surface temperatures to certain perturbations. Their results give a temperature increase of $\simeq$0.4K for a doubling of the present-day level of CH_4 in the atmosphere. Similarly, methane partial pressures ranging from 0.001 to 1 mb with various values of the total CO_2 pressure have been shown (40) to increase the surface temperature more significantly than carbon monoxide, an addition of 1 mb CH_4 leading to a computed temperature increase $\simeq$ 5.8K. Thus, carbon compounds forming trace gases probably cause greater temperature variations in reduced form (CH_4) than the oxidised form (CO).

The amount of free hydrogen in the atmosphere of the evolutionary environment was small (Table I). However, its presence seems to be fundamentally important for the abiological synthesis of simple organics ((83) and (84)). It may further prove an important source of energy in early metabolic evolution (88).

Molecular hydrogen, Walker (88) suggests ~1% or less, will be highly significant for many chemical and photochemical reactions and will also determine the level and evolution of oxygen in the atmosphere. The assertion that photolysis of water vapour provides a source of free oxygen in the prebiological atmosphere is recurrent in the literature (87). Tropospheric photolysis is usually rapidly followed by recombination and only escape to space of hydrogen can result in a net atmospheric source of oxygen. The geological data seem to demand a considerable supply of free oxygen. For instance, Holland's data suggests (87) that a single banded iron formation would require a supply of O_2 more than order of magnitude larger than the present day source provided by the photochemical destruction of water vapour. The appearance of redbeds dates from about 2×10^9 yrs BP ((20) and (76)). The occurrence of this distinctive formation almost certainly indicates an atmospheric oxygen level close to present day (89). However, the banded iron formations require reducing waters if the iron is to be transported in solution, suggesting a complex but quasi-equilibrium situation amongst the various processes controlling environmental level of oxygen and hydrogen. The greenhouse increment caused by, for instance, the addition of abiologically fixed NH_3 as described above could result in greatly enhanced tropospheric temperatures (see Table II). These increased tropopause temperatures could have enhanced the levels of O_2 (see (87)). If abiological processes such as those simulated in laboratory apparatus were global in extent ((84) and (72)), and persistent for a significant time, these might in themselves compose a non-negligible source of free hydrogen in the evolutionary environment. The consideration of possible levels of these trace constituents in the atmosphere is as dependent upon environmental feedbacks as is the level of, for instance, CO_2 discussed above.

The seven characteristics derived here and listed in Table I are vital to any discussion of the evolution of life. It is evident that the evolutionary environment provides hospitable average global surface temperature and zonal energy transport, indicating a climatic regime not too dissimilar from that of the present day. The stability of this water controlled regime is remarkable.

3. THE WATER PLANET: THE CLIMATOLOGY OF AN EVOLVING PLANET CALLED "OCEAN"

In the previous section the basic characteristics of the evolutionary environment have been established (Table I). The existence of surface temperatures which permitted early condensation of water seems to have been fundamentally important and may, as mentioned above, have removed the case of the Earth from

TABLE III

Earth--The Water Planet
(modified from (59))

Orbit:	obliquity considerable eccentricity significant
Interior:	still active
Satellite:	largest in relative mass
Magnetism:	relatively strong, variable
Crust:	non-rigid and non-static
*Hydrosphere:	unique
*Atmosphere:	numerous unique features
Biosphere:	unique
*Climate:	stable (?unique?)

the class of general planetary atmospheric evolution. A.C. Clarke has said, "How inappropriate to call this planet Earth when clearly it is Ocean" (56). Table III, which has been modified from McCrea (59), lists the planetary characteristics of the Earth. The final entry suggesting that the climate has been stable, over geological times, to all perturbations clearly requires some defense.

The range of potential perturbations to the climatic environment is very large. The major astronomical influence upon the evolutionary environment must be the increasing solar luminosity (see Section 2.4). Increased solar luminosity resulting from enhanced accretion during passage of the solar system through the spiral arms of the galaxy has been proposed as a possible cause of ice ages on a time period of about 10^8 yrs (58), although more recently (19) it has been pointed out that the climatic consequences of the passage of the solar system through these interstellar clouds warrants detailed examination. Many other astronomical factors have been investigated as potential triggers of climatic change. The Milankovitch theory of ice ages has been reworked (7) and a number of authors (19) have investigated the effects of other astronomical occurrences, such as supernovae explosions, upon the terrestrial climatic regime.

The most important environmental upheavals on the Earth may have been a direct consequence of the evolution of life. For instance, the evolution of the ability to photosynthesise resulted in release of free oxygen to a previously neutral environment. This must have caused modifications in geophysical processes (chemical weathering and sedimentation changes are illustrated by the discontinuity between banded iron formation and redbeds, e.g., (76)). Climatological effects of a net oxygen source are probably negligible, but the almost simultaneous development of an ozone screen in the stratosphere may have perturbed both surface temperatures and environmental lapse rates. In Figure 4 three temperature profiles computed with a 1-D radiative-convective model illustrate the climatological effects of increasing oxygen and ozone levels in the atmosphere. The curves are for i) present-day $O_2 + O_3$; ii) fifty percent $O_2 + O_3$; and iii) zero $O_2 + O_3$. The

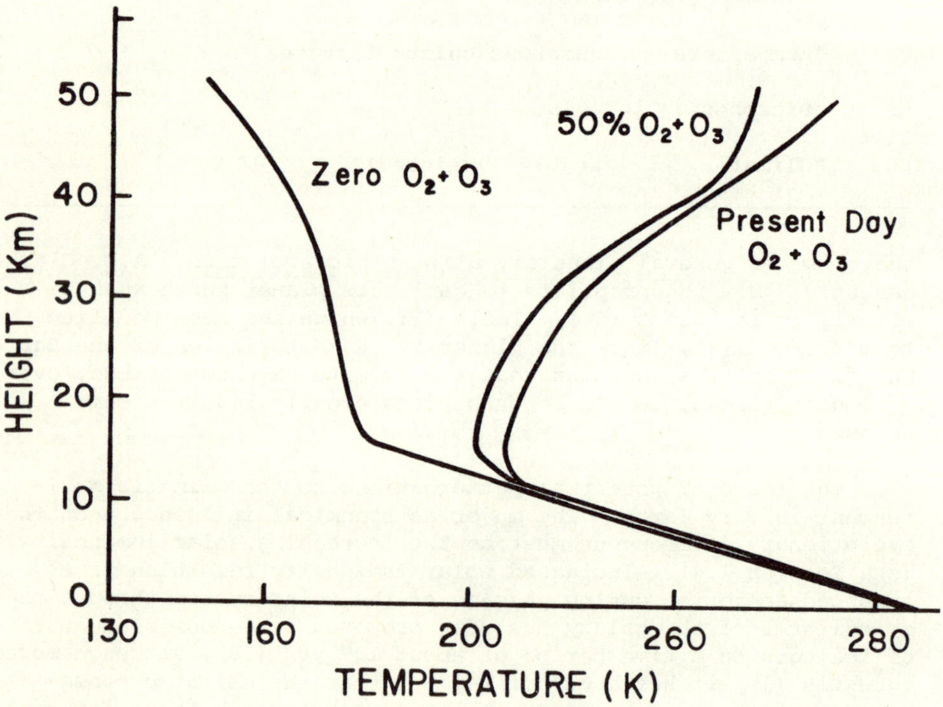

Figure 4. Vertical temperature profiles calculated with a 1-D radiative-convective model for three different O_2+O_3 concentrations: i) present-day; ii) 50% of current; iii) zero $O_2 + O_3$ The model calculations assumed fixed relative humidity and fixed cloud temperature (for description of the model see (29)).

surface temperature decrease caused even by the complete removal of oxygen and ozone is only 2.3K but the stratospheric temperature profile is considerably modified. It is possible that the removal of the temperature inversion that defines the tropopause could cause differences in the upper level clouds. For instance, if this type of vertical temperature profile were linked to higher surface temperatures (e.g., as a result of higher CO_2 levels--see Figure 3), then convective activity would be enhanced from below but less restrained in the upper troposphere. A change in just the height and/or the amount of cirrus cloud could be highly significant for the ambient climate. This is because cirrus clouds, unlike those at lower levels, can increase surface temperatures (e.g., (92) of Section 3.2). There may also have been perturbations caused in the evolutionary environment by prephotosynthetic life. For instance, it has been suggested (88) that the first organisms effectively cleared up the organic molecules which have been abiologically synthesised, whilst later evolutionary steps led to the direct exploitation of the chemical energy available in the atmosphere as a result of the mixture of carbon dioxide and free hydrogen.

Atmospheric evolution was probably a result of both vaporisation of accretional debris and degassing from the Earth's surface (Section 2). Significant bombardment was complete (by definition) before the start of the evolutionary environment, but volcanic activity could have continued throughout this time period. The pattern of tectonic activity was probably a function of the interior planetary temperature and thus a direct result of planetary formation. Evolved gases (as described in Section 2) were predominantly oxidized in nature, but a non-negligible emission of H_2, CO, H_2S and dust particles must be considered if vulcanism was prevalent throughout the evolutionary environment. Volcanic localities have been proposed as possible sites for the genesis and early evolution of life (81). Ejecta may also perturb the environment through, for instance, extensive surface albedo perturbation due to lava flow and dust deposition and insertion into the atmosphere (11). Plate tectonics in general have been suggested as a climate forcing factor for more recent geological epochs through the mechanism of surface albedo perturbation, e.g., (21) and (36). Climatic forcing by configuration may also be a significant trigger under certain circumstances for the evolutionary environment.

Whilst the geological data do indicate environmental fluctuations, there is in each case eventual system recovery. Thus it seems that the evolutionary environment is at least as robust as the post-Precambrian climate system. This apparent long-term climate stability warrants examination.

3.1 Radiati

At the
characterize
assume an ov
(a function
and the emit
for the effe
surface temp
and 2). Cal
assumption o
for any plan
assumption l
Section 2.6)
(e.g., Mars)
equilibrium
atmosphere (
as important
effective em
case of the
considered.
ary state of
less variabl
ating) or Ve
Earth's surf
of the radia
from the sur
climatology
Earth must b
or "runaway"
investigatio
to forcing e
evolution *pe*

Climatic
ture (16,77)
discussion wh
case. These
selected exte
wider investi
I, and the em
This analysis
approach but
model type (6
desired rathe
graph of Figu
between a cur
as a functio

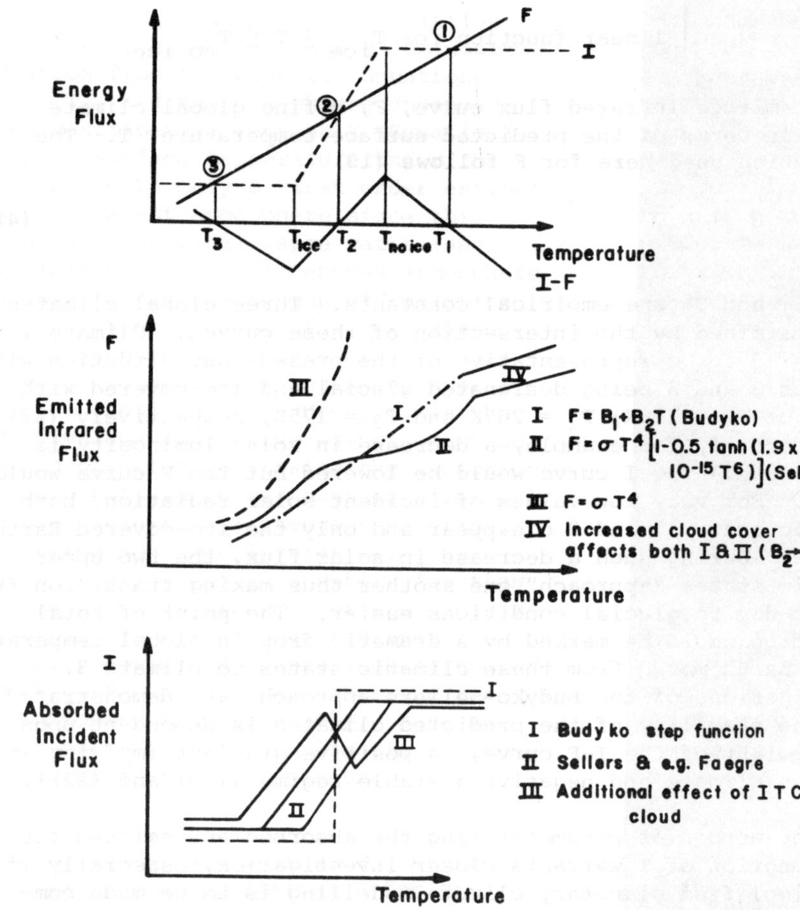

Figure 5. Three graphs illustrate the use of a Budyko-Sellers (13) and (82) climate model to predict possible global climatic states. Mean global temperature is plotted against energy flux. Upper graph shows three climate states at the intersection of the absorbed flux curve, I, and the emitted flux curve, F. The stability of these three regimes, say to a change in solar flux, is given by the slope of the I-F curve (see text for fuller explanation). Centre graph illustrates four methods of parameterising the emitted flux, F. Lower graph illustrates three possible curves for the absorbed flux, I, which varies as a function of the parameterised albedo.

It should be noted that although T is representative of the average surface temperature, it has not been replaced by T_s in these equations for two reasons. Firstly, much of the early observational work and empirical curve fitting used °C rather than K and hence values of constants must be considered carefully. Secondly, the observations referred to above and upon which Equations 4 and 6 are based are made for zonally averaged surface temperatures rather than individual planetary configurations. The latter are clearly difficult to acquire within the confines of the Solar System although the case of Venus should be remarked upon. The emitted thermal fluxes are comparable with those of the Earth, but there is an almost negligible variation with latitutde. Thus a very dense atmosphere would seem to lead, in the limiting case, to an almost constant value of F, i.e., the constant B_2 approaches zero (Figure 5, centre). The effect of clouds on the relationship between F and T is very important (the emitted radiation from clouds is very much less than that from the surface or clear-sky), but difficult to quantify. A possible effect of clouds upon the F curve parameters has been included in Figure 5, but further discussion of cloud-climate feedback is reserved for Section 3.2

The lowest graph in Figure 5 illustrates various curves of absorbed solar radiation, I, as a function of temperature, T. Equation 3 has been reworked (changing gradients and upper and lower albedo values) by a number of authors (see, e.g., (22) and (78)). Such formulae can only include cloud effects in the sense that the overall values assigned to A may be higher than surface albedoes for the ice, no-ice or linear part-ice interpolation. Satellite data now available, e.g., (95), (24) and (14) indicate that in certain temperature regions, especially in the highly baroclinic zone near the snow/ice boundary and in the Inter Tropical Convergence, clouds dominate the planetary albedo. Thus the albedo curve can be redrawn to include these empirical results. This redrawing constitutes a considerable improvement to the overall model formulation since the outgoing infrared radiation curve, F, has always incorporated empirical data ((13) and (82)). This alteration has been made to the planetary albedo and the new absorbed solar radiation curve, I, which results is shown in Figure 5. This modification has a number of consequences. Firstly, as indicated above, the Budyko-Sellers formulation becomes self-consistent. The predicted temperature climates must also change as described below. However, it must be noted that climatic prediction made from even as simple a model as the Budyko-Sellers type radiation balance is hazardous (78). The emitted infrared flux is dependent upon the optical depth and hence gaseous and particulate amounts and the absorbed flux is a function of clear-sky scattering and the surface albedo, but these sources of error are insignificant compared to the potential variability in parameterization of the cloud-climate relationship (see Section 3.2).

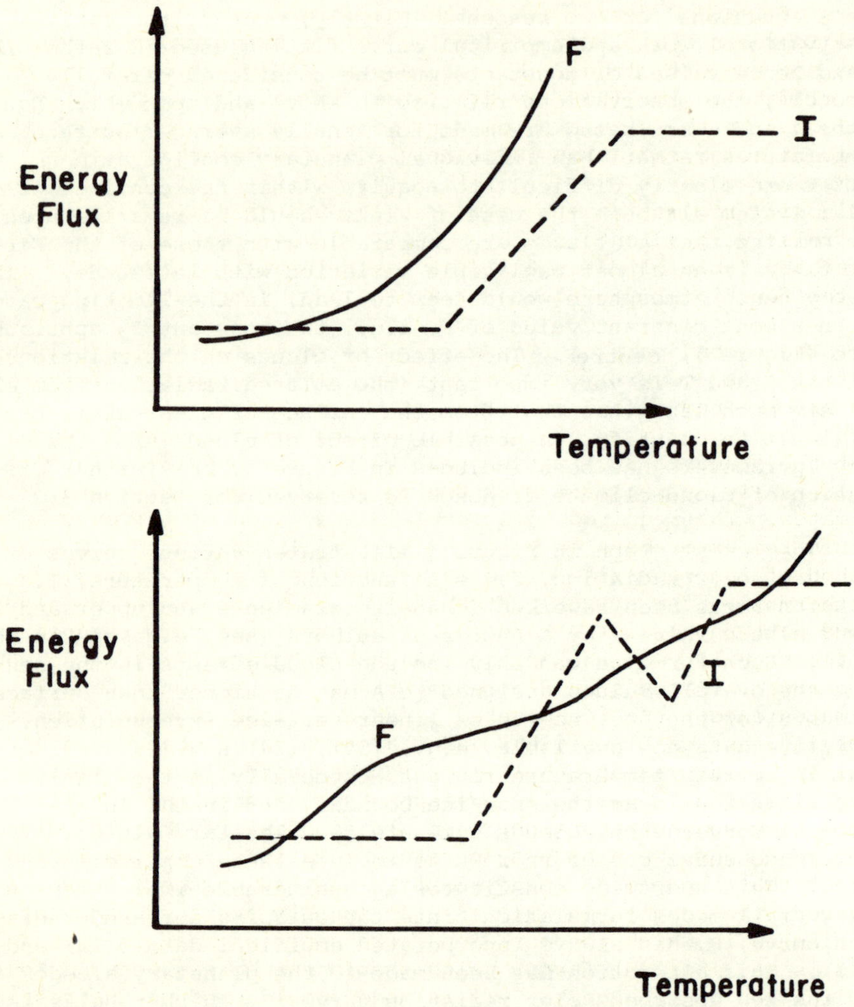

Figure 6. Two planetary configurations are illustrated by different pairs of absorbed flux, I, and emitted flux, F, curves. Upper graph is typical of a planet which does not possess a substantial atmosphere and has no clouds. Only one, deep frozen, climate is possible without a large change in the incident solar energy. Lower graph is representative of a planet possessing an atmosphere and clouds. The cloud cover modifies both the albedo and the emitted flux. The result is to bring the predicted climate states "closer", cf. Figure 5, and to reduce the temperature change expected as a result of external forcing.

Superimposing a pair of I and F curves (as in Figure 5, upper) leads to a "prediction" of climatic states for the Earth. Two pairs of I and F curves are illustrated in Figure 6. The first case represents a planet possessing only a thin atmosphere which does not support clouds. F increases as T^4 (Equation 5) and only one climate regime (totally glaciated) is possible. This is analogous to the situation on Mars today. A large change in the incident solar radiation is clearly required before any other climatic regime is ever possible. The stability of the enhanced luminosity regimes and the temperature predicted will depend upon the relative gradients of the F and I curves; that is, the new, more hospitable climates will, at least initially, be determined by the functional dependence of albedo upon temperature.

The second pair of flux curves are those for a planet with partial cloud cover (e.g., the Earth). The fascinating features now are the number and stability of the predicted climates. The much closer proximity of the absorbed and emitted curves suggests that the standard 3 climates predicted by the Budyko-Sellers models could easily be replaced by a regime of 5, or possibly more, climates. The required additional intersection of the curves is within the error bars on both the system albedoes used and the method of incorporating cloud data into the F and I curves. Furthermore, if a slight perturbation were to occur in the global cloud pattern, e.g. (69) and (92), the predicted temperatures and/ or the stability of the "present-day" and "glacial" climates would be changed. Perhaps even more important is the fact that all climates except the ice-covered Earth are close together (possibly implying ease of transition) and that even large changes in incident solar radiation will not remove these climatic states. This situation is partially analogous to the findings in (91) since they suggest that increased global stability results directly from weakening the albedo-temperature relationship which is the effect that the cloud albedo addition has had on the I curve. However, the results in (91) are for zonally averaged computations of climate in which the meridional transport of heat is included. This is an important factor which should not be neglected. Zonal application of the Budyko-Sellers model would produce interesting results if it could be applied to evolutionary environment, but the method of parameterising both albedo and infrared curves in an environment which differs from the present-day state is unclear.

Any application of the globally averaged Budyko-Sellers model must be viewed with great caution. The parameterisations employed here and indeed the data used are at best representative of zonal fluxes on the present-day Earth. However, it is reasonable to argue that both the infrared and absorbed solar fluxes will be modified by the effects of clouds from original curves that are relatable to physical laws (e.g., Figure 6, upper). The important point to be underlined is the result that clouds and atmospheric water vapour have a moderating influence upon both the 'distance' between predicted regimes and upon the magnitude of perturbations due to external changes.

These conclusions, whilst being highly tentative, are at least in agreement with the working hypothesis that over long time periods the Earth's climate could be stable to most external variations.

3.2 Clouds and Climate

The direction of the cloud-climate feedback is not known. Indeed, even its net effect is questioned by, e.g., (15) and (16), who believe that the effects of perturbed cloud amount upon the albedo and the emitted infrared radiation are mutually compensatory. Certainly the data are, as yet, not adequate to permit conclusive statements about either the sign or the magnitude of feedbacks. Furthermore, it is not only the cloud amount but the cloud type, particularly height, and possibly also the total optical thickness that could control the climatic response. Wang et al. (90) have demonstrated the importance of vertical distribution of clouds using a 1-D radiative-convective climate model. They find that the vertical distribution of cloud is a critical parameter, especially if changes in cirrus cloud percentage and/or optical thickness are anticipated. For instance, they find that different surface temperature changes ($\Delta T = 1.97K$ and $\Delta T = 2.80K$) are caused by perturbations (2% solar increase and CO_2 doubling, respectively) to the present system, in spite of the fact that the total computed cloud cover in each case is the same.

From the discussion in Section 3.1 it is apparent that if, as is suggested in (34), cloud configuration is even a weak function of continent-ocean configuration then over geological time periods this coupling should be incorporated into climate models. However, the preceeding discussion of I and F flux curves underlines the need to include the infrared as well as the shortwave radiation changes. On a local scale these changes need not be compensatory. For instance, a build-up of low level stratus cloud over high latitude ocean areas substantially changes the local albedo, but the emitted flux is unlikely to be greatly modified. (Such a change has been proposed as a possible system response which would dampen if not completely remove the negative ice-albedo feedback anticipated in response to anthropogenically increased CO_2 levels.) However, Wetherald and Manabe (92) suggest that on a global scale cloud changes can become mutually compensatory. It is not clear that results such as these can be extrapolated to evolutionary timescales.

Since cloud-climate responses are so ill-understood in the current situation, and since observational data are still distressingly sparse, methods of parameterising clouds for simpler climate models often result in a wide range of responses. It is hardly surprising that the few attempts that have been made to consider the effect of clouds upon long-term climate trends should produce diverse and even opposing results. Hart's (31) model is extremely sensitive to cloud-climate variations. He assumes that the percentage cloud cover is proportional to the total atmospheric water vapour (which is linked to T_s by the Clausius-Clapeyron equation). Thus his calculated albedoes are a very strong function of T_s, whilst he assumes that the dependence of F upon T_s is similar to that for the present-day. This provides a strong negative feedback through cloud amount on T_s (66). Alternatively (40) suggests that the cloud amount is unlikely to change by a significant amount over geological time periods and thus infers a weak cloud-T_s relationship. Finally, some results from general circulation (GCM) simulations of the current atmosphere (69) have indicated a slight tendency to decrease the overall cloud cover in response to increased surface temperatures.

The pair of curves in Figure 6 (lower) is a first, highly empirical, attempt to represent a parameterization of the cloud-climate relationship in a way that can be used for climate modelling over geological or evolutionary timescales. The cloud-climate problem is potentially more serious for long-term climate modelling than for short-term impact and sensitivity studies since for the latter the parameterisations in GCMs should be adequate. It is not known whether i) the percentage cloud cover is likely to change, ii) whether the cloud type and height will vary, and iii) whether all, some, or none of such changes will be exactly compensated for by opposing modifications in the absorbed solar and emitted infrared fluxes. Clouds could be a cause of long-term climate change or could be purely incidental to it. Figure 6 suggests that clouds are actually an important moderating influence on the climatic regime.

The two topics discussed in this section (i.e., radiation fluxes and clouds) are clearly not independent of one another. On the contrary, the net global radiation balance can be very strongly modified by the nature, extent and vertical position of clouds. If, as the data described in Section 2 seem to suggest, there has been a tendency towards global climatic stability, then it seems probable that clouds could play an important role. Equation 2 can be rewritten in terms of the tropospheric lapse rate (Section 2.6) following (29) as:

$$T_s = T_e + \Gamma H \qquad (7)$$

where Γ is the mean environmental lapse rate and H (km) the globally averaged height of the radiating layer. If the tropospheric lapse rate has not varied greatly, then from Equation 7 it can be seen that surface temperature can be controlled by the movement of this radiating layer. For instance, the height of the Venus cloud tops which radiate to space is ≃70km (cf. the value of H≃6km on the present-day Earth).

Over geological time scales there appear to be effective negative cloud-climate feedback mechanisms which operate to restore climatic stability.

4. CONCLUSIONS--CLIMATIC SENSITIVITY AS A SOURCE OF LIMITS UPON ENVIRONMENTAL CHARACTERISTICS

The evolution of science, like the evolution of life, seems to require a suitable environment and stimulation as well as careful nurturing. It is dangerous but sometimes useful to venture forward despite a dearth of data. It is certainly important to state when scant substantive evidence exists. Table I and the arguments in Sections 2 and 3 are at best 'a stab in the dark' probably closer to a 'feeble groping.' However, if these discussions are to further our understanding, there must be an attempt to extrapolate from such hypotheses towards (hopefully) constructive feedback into the other disciplines concerned with the problems of the origin of life (Figure 1).

I have suggested that the apparently robust nature of the climate system in the face of atmospheric and surface changes of considerable magnitude may be traced back to negative feedback effects which are often under hydrospheric control (see also (37)). The existence of surface liquid water on Earth and the associated global hydrological cycle seem to be the dominant factors in environmental feedback processes. The inference which must be made from Table I and the discussions in Sections 2 and 3 is that the prebiotic and early biotic Earth had a closer resemblance to the present-day state than previously believed. I have incidentally indicated that the origin and evolution of life (if indeed it occurred on the Earth) probably presented the climate-environment system with perturbations larger than any suffered since (including present-day anthropogenic pollution). Certain of these conclusions warrant further elaboration.

4.1 Oceans

The fundamental importance of the existence of extensive oceans on the Earth has been discussed throughout this paper. The availability of water for evaporation and the moderating influence upon diurnal and seasonal variations in temperature are clear, although these effects may have important repercussions. For instance, if the desirability of closely defined water plus dry surface conditions is demonstrated in synthesis experiments (see Section 4.2) these conditions can probably be met in at least some surface location (53). The moderating effect of oceans upon temperatures could be important to theories of climatic change on a number of times scales. For instance (35), the "Milankovitch" mechanism, e.g. (7), is currently opposed by the geographical configuration of ocean, cf. land areas. This complex mechanism had been considerably trivialised by the suggestion that the current distribution of continental masses favours glaciation via the "Milankovitch" orbital perturbation effect because of seasonality of temperatures. Climatological and satellite data confirm that seasonal temperature change is greater in the Northern Hemisphere and that at present the Northern Hemisphere summer is warmer than the same season in the Southern Hemisphere. This is due to the large oceanic areas in the Southern Hemisphere and indicates that the "Milankovitch" mechanism, if shown to be the cause of glacial epochs, operates despite the present global geography which opposes its effects in both hemispheres.

The stratification of the primitive oceans has also been considered (43). This is of interest mainly through its effect upon the atmospheric levels of gases controlled by solution in water (CO_2, O_2, etc.), but may also be important for concentration of simple organics ((81) and (53)).· More recently Broecker (12) has suggested a mechanism to explain the observed increase in CO_2 levels in the atmosphere in interglacial epochs (8). He believes that the atmospheric CO_2 fluctuates in response to decreased oceanic biomass uptake which is itself a result of the flooding of continental shelves and estuaries by the post-glacial ocean transgressions. This flooding effectively removes the element phosphorus from the ocean and hence limits algae growth.

Meridional transport of energy by the oceans (and by the atmosphere) is a significant factor in the general circulation. Poleward heat flux changes are expected to be a further example of negative feedback mechanisms resulting from the nature of the global hydrosphere. Season observations and model results (56) suggest that an increased pole to equator temperature gradient results in a more vigorous meridional transfer resulting in a restoration of the *status quo*.

4.2 Localized Environments for the Origin of Life

The evolutionary environmental characteristics in Table I are apparently in direct opposition to the experimental results of biochemists working on the origin of life. However, the global scale physical and chemical state may not exclude the conditions believed to be necessary for the biochemical building processes. The perceptions of difference scientific groups illustrated in Figure 1 may turn out to be vindicated. For instance, Henderson-Sellers and Schwartz (41) have described the compatibility of different local and global scale conditions. Locally they have fairly high levels of atmospheric and absorbed NH_3, but the global mixing ratio of NH_3 remains below the level at which climatic effects would be seen. The local catalytic mechanism of nitrogen fixation uses titanium dioxide for the photochemical reduction of N_2 to NH_3. A detailed investigation of naturally occurring areas rich in titanium revealed that certain deserts possess an extraordinary activity in the conversion of atmospheric nitrogen to NH_3. It is estimated that ~4,000 m^2 of this type of desert sand generates between 1 and 10 kg of NH_3 in a year. They suggest that this process would provide a significant local abiotic source of ammonia on the primitive Earth. This study provides a clue to the resolution of the apparent conflict between astrogeological theories of neutral planetary atmospheres and the laboratory results of biochemists. The climatic feedbacks described in Section 3 appear to provide a global situation similar to the present-day Earth. Fairly stable global conditions (cf. Mars and Venus) contrast with highly diverse local environments. It is possible that the diversity of local chemical disequilibria in surface, ocean, lakes and even the atmosphere is a requirement for initiation or nurturing of the evolutionary process of life (81). Furthermore, weathering and evaporation processes could be of importance in establishing locally anomolous levels of elements and compounds. For instance, peptides have been successfully built from amino acids in experimental environments possessing considerable temperature and moisture fluctuations (52).

Atmospheric characteristics are clearly important in the provision of hospitable environmental conditions for the genesis and evolution of life; provision of source areas of, e.g., abiologically produced NH_3; as a secondary factor controlling gaseous levels through weathering; and direct weather effects, for instance, solution of atmospheric gases as a result of or enhanced by precipitation processes and the direct formation of trace gases by lightning. The results of (17) suggest that in the evolutionary environment lightning (which we can certainly presume existed) would produce CO, O_2, NO and O. A further, probably minor, effect is the formation and deposition of trace compounds directly as a result of meteoritic and cometory impact ((33) and (54)).

Table IV lists some of the proposed conditions for organic synthesis and evolution together with a comment upon the likelihood of these conditions in the evolutionary environment. It is of the utmost importance that interdisciplinary discussion should focus not only upon the laboratory requirements, but also upon how these relate to the global atmospheric and surface configuration. For instance, it seems reasonable to presume that the negative feedback effects which have stabilized the climate cannot restore equilibrium for all perturbations (see Section 4.5) and thus the global level of disturbance would always be considered.

TABLE IV

A prototype list of "suggested" conditions compared with the appropriate parameter in the established environment as listed in Table I. This list is certainly not exhaustive. It is recommended that all experimental results and inferences from new data should be investigated in such a framework.

Example Requirements	Variable	Relationship to Global Conditions
	1) Hydrogen	
1-10% (Abelson atmosphere)		Sources very small; photolysis of H_2O limited by tropopause temperature; O_2 removal by weathering.
	2) Energy Source	
UV and electrical discharges both available		UV screening by H_2O vapour (height of absorption).
	3) Oceans	
(i) pH 8-9 proposed		(i) Hydrosphere close to present day
(ii) Stratification to reduce effects of dilution of organics		(ii) Conflict with need for "land", also affects O_2, CO_2 levels.
	4) Land Surfaces	
Highly variable! e.g.:		
(i) "evaporating pond"		(i) (?) adequate UV
(ii) Fischer-Troposche 600-700°C + liquid H_2O		(ii) !!!
(iii) Synthesis of nucleotides--diurnal temperature range		(iii) O.K.
(iv) Polymerisation of unactivated amino acids--anhydrising conditions		(iv) O.K. in deserts) localised
(v) Synthesis of amino acids and purines at -20°C		(v) Probably O.K.) localised
	5) Pre-existence of Organics/Life	
Organics certainly in meteorites (? panspermia)		Traumatic bombardment may lead to breakdown; time span strictly less than 10^6 years.
	6) Trace Gases (e.g., NH_3)	
(i) Requirement for many synthesis experiments		(i) "greenhouse" gas and rapidly destroyed.
(ii) TiO_2 a local catalytic source		(ii) global levels remain low.

there is still considerable uncertainty about basic feedback mechanisms such as those relating to the cloud-climate interaction (15, 69, 92) and the appropriateness of convective adjustment schemes and lapse rate choice, e.g. (90). It could be argued that the long-term stability of the Earth may be a reflection of negative feedback mechanisms (Section 3) which are stronger and more effective than has been generally recognized.

4.5 Summary

The early condensation of surface liquid water on the Earth and the early formation of a global hydrological cycle seems to have provided our planet with a remarkably robust climate system. Physically the early Earth seems to provide a most hospitable environment for initiation and especially for the subsequent evolution of life. If, as seems likely, necessary and sufficient chemical conditions are globally incompatible with the atmospheric configuration described in Table I, two possibilities remain. Firstly, it is not impossible that the organic molecules could have effectively "arrived already," say as a result of the final stages of accretion, from cometory impact, or by surviving the transition from an earlier, and for them, more hospitable epoch. The second and more likely alternative rests upon the existence and abundance of local sites providing satisfactory chemical and physical conditions for prebiotic synthesis. Discussions should now follow Table IV in attempting to relate the necessary and sufficient conditions for life to the established global atmospheric chemistry and climate.

ACKNOWLEDGEMENTS

I am grateful to all the participants at the Advanced Study Institute for a most stimulating exchange of ideas. Part of the research described here was undertaken whilst the author was a visiting Research Associate at the Goddard Institute for Space Studies, New York. Discussions with many colleagues and especially the efforts of Ilene Shifrin, Myra Shapiro and Jose Mendoza, as well as financial support from the National Research Council are all gratefully acknowledged.

REFERENCES

(1) Abelson, P.H.: 1966, Proc. Nat. Acad. Sci. U.S.A. 55, p. 1365.
(2) Ahrens, T.J. and O'Keefe, J.D.: 1972, The Moon 4, p. 214.
(3) Arrhenius, G. de B.R. and Alfven, H.: 1974, in E.D. Goldberg (ed.), The Sea vol. 5, "Marine Chemistry." Wiley, New York.
(4) Bacastow, R.B.: 1976, Nature 261, p. 116.
(5) Bada, J.L. and Miller, S.L.: 1968, Science 159, p. 423.
(6) Benlow, A. and Meadows, A.J.: 1977, Astrophys. Space Sci. 46, p. 293.

(7) Berger, A.: 1979, Geophys. Surveys 3, p. 351.
(8) Berner, W.H., Oeschger, H. and Stauffer, B.: 1981.
(9) Bickley, R.I. and Vishwanathan, V.: 1979, Nature 280, p. 306.
(10) Brancazio, P.J. and Cameron, A.G.W. (eds.): 1964, The Origin and Evolution of Atmospheres and Oceans, Wiley, New York.
(11) Bray, J.R.: 1979, Quat. Res. 12, p. 204.
(12) Broecker, W.S.: 1981, in C.I.M.A.S. Symp., E. Kraus (ed.), Univ. of Miami.
(13) Budyko, M.I.: 1969, Tellus 21, p. 611.
(14) Campbell, C.G. and Vonder Haar, T.H.: 1980, Atmos. Sci. Paper 323, Colorado State Univ., 74pp.
(15) Cess, R.D.: 1976, J. Atmos. Sci. 33, p. 1831.
(16) Cess, R.D., Briegleb, B.P. and Lian, M.S.: 1981, J. Atmos. Sci.
(17) Chameides, W.L., Walker, J.C.G., and Nagy, A.F.: 1979, Nature 280, p. 820.
(18) Chang, S., Des Marais, D., Mack, R., Miller, S.L. and Strathearn, G.: 1981.
(19) Clark, D.H., McCrea, W.H. and Stephenson, F.R.: 1977, Nature 265, p. 318.
(20) Cloud, P.E.: 1968, Science 160, p. 729.
(21) Cogley, J.G.: 1979, Nature 279, p. 712.
(22) Crafoord, C. and Kallen, E.: 1978, J. Atmos. Sci. 35, p. 1123.
(23) Craig, H. and Lupton, J.E.: 1979, in The Sea Vol. 7, "The Oceanic Lithosphere." Wiley, New York.
(24) Ellis, J.S.: 1978, "Cloudiness, the Planetary Radiation and Climate," Ph.D. Thesis, Colorado State Univ., Fort Collins.
(25) Enhalt, D.H.: 1974, Tellus 26, p. 58.
(26) Ferris, J.P. and Nicodem, D.E.: 1972, Nature 238, p. 268.
(27) Frakes, L.A.: 1979, Climates Throughout Geologic Time, Elsevier.
(28) Ghil, M.: 1976, J. Atmos. Sci. 33, p. 3.
(29) Hansen, J.E., Johnson, D., Lacis, A., Lebedeff, S., Lee, P., Rind, D. and Russell, G.: 1981, Nature.
(30) Hargrave, R.B.: 1976, Science 193, p. 363.
(31) Hart, M.H.: 1978, Icarus 33, p. 23.
(32) Haselgrove, C.B. and Hoyle, F.: 1959, Monthly Not. Roy. Astron. Soc. 119, p. 112.
(33) Henderson-Sellers, A.: 1977, Atmos. Env. 11, p. 864.
(34) Henderson-Sellers, A.: 1978, J. Geophys. Res. 83, p. 5057.
(35) Henderson-Sellers, A.: 1979, Nature 279, p. 786.
(36) Henderson-Sellers, A.: 1980, S. Afr. J. Sci. 76, p. 373.
(37) Henderson-Sellers, A.: 1981, Geophys. Surveys 4, p. 297.
(38) Henderson-Sellers, A. and Meadows, A.J.: 1976, Planet. Space Sci. 24, p. 41.
(39) Henderson-Sellers, A. and Meadows, A.J.: 1977, Nature 270 p. 589.
(40) Henderson-Sellers, A. and Meadows, A.J.: 1979, Planet. Space Sci. 27, p. 1095.

(41) Henderson-Sellers, A. and Schwartz, A.W.: 1980, Nature 287, p. 526.
(42) Henderson-Sellers, A., Benlow, A. and Meadows, A.J.: 1980, Q. J. R. Astr. Soc. 21, p. 74.
(43) Henderson-Sellers, B. and Henderson-Sellers, A.: 1978, Nature 272, p. 439.
(44) Holland, H.D.: 1962, Petrologic Studies: Geol. Soc. Amer., a volume in honour of A.F. Buddington, p. 447.
(45) Holland, H.D.: 1972, Geochim. Cosmochim. Acta. 36, p. 637.
(46) Holland, H.D.: 1978, The Chemistry of the Atmosphere and Oceans, Wiley, New York, p. 351.
(47) Hunten, D.M.: 1973, J. Atmos. Sci. 30, p. 1481.
(48) Kasting, J.F. and Walker, J.C.G.: 1981, J. Geophys. Res. 86, p. 1147.
(49) Kaula, W.M.: 1980, in The Continental Crust and Its Mineral Deposits, D.W. Strangway (ed.), Geolog. Assoc. Canada. Spec. Pap. 20, p. 26.
(50) Knauth, L.P. and Epstein, S.: 1976, Geochim. Cosmochim. Acta 40, p. 1095.
(51) Kuhn, W.R. and Atreya, S.K.: 1979, Icarus 37, p. 207.
(52) Lahav, N., White, D.H. and Chang, S.: 1978, Science 201,p. 67.
(53) Lahav, N. and White, D.H.: 1980, J. Mol. Evol. 16.
(54) Lazcano-Arujo, A.R. and Oro, J.: 1981, Fifth College Park Colloquium on Chemical Evolution, 'Comets and the Origin of Life', C. Ponnamperuma, ed., D. Reidel, Holland.
(55) Levine, J.S., Hays, P.B. and Walker, J.C.G.: 1979, Icarus 39, p. 295.
(56) Lovelock, J.E.: 1979, Gaia. A New Look at Life on Earth, Oxford Univ. Press, Oxford.
(57) Manabe, S. and Wetherald, R.T.: 1967, J. Atmos. Sci. 24, p. 241.
(58) McCrea, W.H.: 1975, Nature 255, p. 607.
(59) McCrea, W.H.: 1981, Proc. Roy. Soc. London A 375, p. 1.
(60) Miller, S.L. and Orgel, L.E.: 1974, The Origins of Life on the Earth, Prentice-Hall, N.J.
(61) Moorbath, S., O'Nions, R.K., and Pankhurst, R.J.: 1973, Nature 245, p. 138.
(62) Newman, M.J. and Rood, R.T.: 1977, Science 198, p. 1035.
(63) North, G.R.: 1975, J. Atmos. Sci. 32, p. 2033.
(64) Oort, A.H. and Vonder Haar, T.H.: 1976, J. Phys. Oceanogr. 6, p. 781.
(65) Owen, T.: 1978, in Evolution of Planetary Atmospheres and Climatology of the Earth, Centre National d'Etudes Spatiales, Toulouse, France, p. 1.
(66) Owen, T., Cess, R.D. and Ramanathan, V.: 1979, Nature 277, p. 640.
(67) Rasool, S.I. and De Bergh, C.: 1970, Nature 266, p. 1037.
(68) Raulin, F.: 1980, in The Origins and Early Evolution of Life, Council of Europe, Strasbourg, France.
(69) Roads, J.O.: 1978, J. Atmos. Sci. 35, p. 1450.

(70) Rubey, W.W.: 1951, Geol. Soc. Amer. Bull. 62, p. 1111.
(71) Sagan, C. and Mullen, G.: 1972, Science 177, p. 52.
(72) Sanchez, R.A., Ferris, J.P. and Orgel, L.E.: 1966, Science 154, p. 784.
(73) Sanchez, R.A., Ferris, J.P. and Orgel, L.E.: 1967, J. Mol. Biol. 30, p. 223.
(74) Schidlowski, M.: 1979, in B. Ralph, R.A. Trudinger, and M.R. Walter (eds.), Antiquity of Photosynthesis: Possible Constraints from Archaen Carbon Isotope Record, Proc. 4th Int. Symp. Environ., Biogeochem., Canberra.
(75) Schidlowski, M., Appel, P.W.U., Eichman, R. and Junge, C.E.: 1979, Geochim. Cosmochim. Acta. 43, p. 189.
(76) Schidlowski, M.: 1980, "The Atmosphere," in The Handbook of Environmental Chemistry Vol. 1, Part A, O. Hutsinger (ed.) Springer-Verlag, Berlin.
(77) Schneider, S.H. and Mass, C.: 1975, Science 190, p. 741.
(78) Schneider, S.H. and Thompson, S.L.: 1980, Icarus 41, p. 456.
(79) Schrauzer, G.N.: 1978, Chem. Eng. News (Nov. 13, 1978) and personal communication.
(80) Schrauzer, G.N. and Guth, T.D.: 1977, J. Am. Chem. Soc. 99, p. 7188.
(81) Schwartz, A.W.: 1981, in E.K. Duursma and R. Dawson (eds.), Organic Chemistry of Sea Water, Elsevier.
(82) Sellers, W.D.: 1969, J. Appl. Meteorol. 8, p. 392.
(83) Toupance, G., Raulin, F. and Buvet, R.: 1975, Origins of Life 6, p. 83.
(84) Toupance, G., Mourey, D. and Raulin, F.: 1978, in Evolution or Planetary Atmospheres and Climatology of the Earth, Centre National d'Etudes Spatiales, Toulouse, France, p. 31.
(85) Walker, J.C.G.: 1976, in B.F. Windley (ed.), The Early History of the Earth, p. 537, Wiley, New York.
(86) Walker, J.C.G.: 1977, Evolution of the Atmosphere, MacMillan, New York.
(87) Walker, J.C.G.: 1978, Pageoph. 116, p. 222.
(88) Walker, J.C.F.: 1978, "Atmospheric Constraints on the Evolution of Metabolism," presented at 4th College Park Symposium on Chemical Evolution, Univ. of Maryland, Oct. 18-20.
(89) Walker, J.C.G.: 1979, Pageoph, 117, p. 498.
(90) Wang, W.-C., Rossow, W.B., Yao, M.-S. and Wolfson, M.: 1981, J. Atmos. Sci.
(91) Warren, S.G. and Schneider, S.H.: 1979, J. Atmos. Sci. 36, p. 1377.
(92) Wetherald, R.T. and Manabe, S.: 1980, J. Atmos. Sci. 37, p. 1485.
(93) Williams, I.P.: 1975, The Origin of the Planets, Adam Hilger, London.
(94) Windley, B.F. (ed.): 1976, The Early History of the Earth, J. Wiley, London.

(95) Winston, J.S., Gruber, A., Gray, T.I., Varadove, M.S., Earnest, C.L. and Mannello, L.P.: 1979, Earth-Atmosphere Radiation Budget Analyses Derived from NOAA Satellite Data, June 1974-February 1978, Vols. 1 & 2, NOAA/NESS.

THE DATING OF THE EARLIEST SEDIMENTS ON EARTH

Stephen Moorbath

Department of Geology and Mineralogy, University of Oxford, Parks Road, Oxford OX1 3PR, U.K.

Published isotopic age evidence for early Precambrian sedimentary rocks containing biogenic markers is critically reviewed. The age of sedimentation can sometimes be obtained by direct analysis of the sediments themselves, but more frequently by age measurements on stratigraphically related volcanic rocks, or by interpolation between dated older basement rocks and younger intrusive rocks. Provided that stratigraphical/geochronological correlations are correct there is little doubt that sediments containing algal stromatolites were being deposited ∿ 3450 - 3300 m.y. ago, and that true microfossils occur in rocks dated at ∿ 3300 - 3200 m.y. Evidence relating to possible biogenic markers in rocks reliably dated at ∿ 3800 - 3750 m.y. is more difficult to evaluate. Much scope remains for future work.

1. INTRODUCTION

Any convincing report of evidence for life in early Precambrian times must demonstrate that the described structures or molecular markers are unambiguously biogenic and that the geochronological control is accurate. This is easier said than done. Frequently the best evidence for life is found in rocks which are not very old, whilst in the very oldest rocks the evidence for life is not always particularly convincing. Nonetheless, recent work suggests that life began very early indeed in the Earth's history. For example, stromatolites (organo-sedimentary structures resulting from the growth and metabolic activity of micro-organisms) have been reported in rocks from Western Australia and Southern Africa which could be as old as ∿ 3500 - 3400 m.y. (1, 2, 3). In addition, organic-walled

C. Ponnamperuma (ed.), Cosmochemistry and the Origin of Life, 213–233.

Rb-Sr and K-Ar age methods. It is, however, subject to recrystallisation by deep burial, tectonic deformation or metamorphism. Nonetheless, dating of unaltered glauconite from rocks that have not been deeply buried, deformed or metamorphosed may yield the time of its original formation and thus the age of deposition of the sedimentary host rock. K-Ar and Rb-Sr dating of authigenic glauconites has proved immensely valuable for calibrating the younger parts of the Phanerozoic (i.e. post-Precambrian) time-scale (15, 16). Glauconite is rare in Precambrian rocks, but successful K-Ar datings have been reported from independently age-bracketed 1600 m.y. -old sediments from the Northern Territory of Australia (17, 18), whilst reliable K-Ar and Rb-Sr dates of ∿ 1100 m.y. have been reported for the Belt Series of Montana (19). Doubts about the reliability of glauconite dating in more complex situations were expressed early on (20), whilst hydrothermal experiments on glauconite (21) clearly demonstrated that low-temperature metamorphism affects the apparent isotopic age of glauconite grains to an extent which depends upon their chemical composition and the temperature of metamorphism.

In the early 1960's, it was discovered that fine-grained, unmetamorphosed, largely detrital sedimentary rocks, such as shales, may yield Rb-Sr whole rock isochrons whose slopes indicate the time elapsed since isotopic homogenisation of Sr in the rocks. In some cases rather close agreement was obtained for stratigraphically well-defined Palaeozoic sediments between the measured Rb-Sr whole rock isochron age and the previously calibrated age for that part of the Phanerozoic time-scale (22, 23, 24, 25). The method was subsequently tried on various unmetamorphosed Precambrian sedimentary rocks - mostly of Upper Precambrian age - and in some cases the agreement between the measured Rb-Sr isochron age and the independently bracketed age for the sediments was satisfactory (19, 26, 27, 28, 29). The Rb-Sr whole rock isochron age of 1595 ± 24 m.y. for shales from the Gunflint Iron Formation of Ontario, Canada (28) was of particular interest in view of its varied assemblage of well-preserved microfossils in chert horizons.

The favoured explanation proposed for the frequently observed approximate agreement between measured and expected age in many sedimentary rocks is that chemical and mineralogical changes occurring during diagenesis (i.e. compaction and burial metamorphism) of fine-grained, argillaceous (clay) sediments are often sufficient to produce complete homogenisation of $^{87}Sr/^{86}Sr$ ratios, accompanied by localised variations in Rb-Sr ratios, within a single stratigraphic horizon. These are also the formal requirements for Rb-Sr dating of igneous and metamorphic rocks (10). Thus it is possible to date the time of diagenesis which, when analytical error is taken into account, can often be taken as contemporaneous with the time of deposition.

However, it must be emphasised that in many other cases, too numerous to reference here, the Rb-Sr whole rock isochron dates turn out to be much more difficult to interpret. This is because isotopic homogenisation is not necessarily associated only with deposition and penecontemporaneous diagenesis, but may date from much later diagenesis, structural deformation and true metamorphic recrystallisation. For example, it is well known that low-grade metamorphism of the type that transforms shales into slates (greenschist facies) is highly effective in resetting Rb-Sr whole rock dates to the time of metamorphism (30, 31). On the other hand, the presence of detrital minerals derived unchanged from an older source region and resistant to isotopic equilibration (e.g. muscovite) may result in dates that exceed the time of deposition of the sediment (26). These methodological and interpretative difficulties have long been realised (24, 32, 33, 34, 35, 36, 37, 38).

An example relevant to this paper is provided by the Fig Tree Shales from the Swaziland System of southern Africa which have yielded an excellent Rb-Sr whole rock isochron age of 2910 ± 40 m.y. (39), although this most likely dates a mild metamorphic event because cross-cutting, coarse-grained granites yield significantly older Rb-Sr whole rock ages of around 2960 - 3000 m.y. (39). The Fig Tree Shale date is thus a minimum age of sedimentation and for the fossil biota which they have long been recognised to contain (40, 41). The Fig Tree Group, which contains the dated shales, is some 3500 m in thickness, and immediately overlies the Onverwacht Group, whose components are summarised in Table 1.

Subsequently, Hurley et al. (42) reported age determinations on a microcrystalline quartz-mica-calcite-bearing sedimentary rock from the Middle Marker Horizon of the Onverwacht Group (Table 1), and obtained a Rb-Sr whole rock isochron age of 3275 ± 70 m.y., with a low initial $^{87}Sr/^{86}Sr$ ratio of 0.7015 ± 0.0018. (For discussion of the significance of initial isotopic ratios, see ref. 10). This suggested (42) that the effect of metamorphism had been slight and that the measured date was fairly close to the time of deposition of the sediments. Note that the Middle Marker Horizon is more than 8000 m. below the base of the Fig Tree Group, discussed earlier (Table 1). Despite this apparent depth of burial, some parts of the Onverwacht Group have remained essentially undeformed, and show only mild metamorphic effects.

greenschist facies, which imparts a greenish tinge to the more basic rocks because of the formation of chlorite. Such associations of low-grade metamorphosed volcanic and sedimentary rocks are often referred to by field geologists as "greenstone belts", a term which is often incorrectly and confusingly used to cover all Archaean volcano-sedimentary (or "supracrustal") sequences, with a consequent blurring of the very different tectonic and environmental origins of genetically unrelated supracrustal belts. Many of the volcanic or metavolcanic rocks have suitable parent/daughter ratios for whole rock Rb-Sr and Pb/Pb dating. The Rb-Sr method has been the most frequently used until now, although basic volcanics sometimes have impracticably low and uniform Rb/Sr ratios for isochron work. Successful examples of this type of dating have been reported for late Archaean (∿ 2800 - 2600 m.y.) volcanic and metavolcanic rocks from North West Territories, Canada (48), northeast Minnesota, U.S.A. (49), Manitoba, Canada (50), Finland (51), and many other areas.

A common problem with this approach, just as with sediments, is that many fine-grained, acid-to-basic volcanic rocks can easily become open systems to parent and/or daughter nuclides during post-depositional metamorphism, hydrothermal alteration and metasomatism. In such cases, low apparent ages may be obtained, sometimes (not always!) accompanied by excessive scatter of data points on whole rock isochron plots. A rather extreme example is provided by a suite of acid volcanic rocks from near the middle of the Onverwacht Group (Table 1), which yields a Rb-Sr whole rock isochron age of 2560 ± 40 m.y. (52). This result is quite incompatible with Rb-Sr whole rock ages of 3275 ± 70 m.y. and 2910 ± 40 m.y. for the Middle Marker Horizon and Fig Tree Shale discussed previously, or with Rb-Sr whole rock ages of ∿ 3000 m.y. for coarse-grained granites which cut the Swaziland Sequence (52). Felsic (acid) volcanics are particularly prone to loss of radiogenic ^{87}Sr by metamorphism and alteration because, in contrast to most of the commonly dated rocks, they contain little or no Ca-minerals which can act as sinks for migrating Sr atoms, expelled from Rb-bearing minerals. An analogous situation has been reported from the late Archaean Chibougamou greenstone belt of Quebec, Canada, where acid volcanics yield a Rb-Sr whole rock isochron age of 2290 ± 170 m.y., despite the fact that they are cut by a tonalite pluton with a Rb-Sr whole rock isochron age of 2520 ± 160 m.y. (53). In contrast, the isotope systematics of basic volcanic rocks are more commonly subject to post-depositional introduction of mobile elements, such as Rb and U.

Several other techniques have been tried to overcome some of the problems outlined above. Many acid volcanic rocks contain accessory amounts of the comparatively resistant mineral zircon, which can be dated by the U-Pb method. Indeed, in favourable cases, Archaean zircon ages may be analytically precise to less than 10 million years. Thus Nunes and Thurston (54) have dated three acid metavolcanic rocks from the Uchi-Confederation Lakes greenstone belt sequence of northwest Ontario, Canada, which respectively yield U-Pb ages of 2959 ± 2 m.y., ca. 2794 m.y., and 2738 $^{+5}_{-2}$ m.y. In some cases, zircons can yield meaningful U-Pb ages even when the Rb-Sr whole rock systematics of the host acid volcanic rock have been disturbed or reset.

Another important recent development is the application of the Sm-Nd age method to basic volcanic and metavolcanic rocks in Archaean supracrustal sequences (for reviews, see refs. 11, 12). Because Sm and Nd are chemically closely related rare earth elements they may be used in situations where the application of other decay schemes such as Rb-Sr is hampered by post-crystallisation alteration and metamorphism leading to the differential movement of parent and daughter element. Indeed, it appears that Sm and Nd are not significantly separated by crustal processes such as erosion, transport, diagenesis, hydrothermal alteration or metamorphism (55). Several basic metavolcanic sequences from Archaean greenstone belts have been successfully dated by the Sm-Nd method, which in some cases gives an improvement in quality over other age data.

We complete this section by discussing several examples of indirect dating of early Archaean sediments of palaeobiological interest, using stratigraphically related volcanic horizons. Much publicity has been accorded to the discovery of algal stromatolites in the Warrawoona Group of the Eastern Pilbara region of Western Australia (1). The fossils occur in an extensive chert unit (the Strelley Pool chert) near the top of a 10 km section of the Warrawoona Group, which consists large of volcanic rocks showing low greenschist-facies metamorphism interstratified with cherty units generally less than 30 m thick (6). The fossiliferous chert lies 1 - 2 km above acid (dacite) lavas of the Duffer Formation, from which zircons have given a U-Pb age of 3452 ± 16 m.y. (56), interpreted as the date of extrusion of the dacite. Lowe (1) is careful to point out that the "possibility of intervening unconformities cannot be dismissed, although the overall igneous and sedimentological continuity of the Group suggests that major hiatuses are absent. Following deposition of the uppermost part of the Warrawoona Group and the suprajacent Gorge Creek sediments, the rocks were extensively deformed and intruded by granitoid plutons. The main period of deformation and associated plutonism has been dated at ∿ 3100 - 2900 m.y. (57, 58). The possible age range for the stromatolites is, therefore, 3450 - 3100 m.y., but their close association with the older rocks suggests deposition at ∿ 3400 m.y."

Walter et al. (2) have reported stromatolites from a chert-barite unit within the Warrawoona Group at a locality named North

Pole. The unit consists of bedded shallow water metasediments 40 m thick that persist for at least 30 km along strike. The maximum metamorphic grade is lower greenschist facies. The unit is regarded as the approximate stratigraphic equivalent of the Duffer Formation, from which zircons yield a U-Pb age of 3452 ± 16 m.y. (56). Walter et al. (2) also state that "volcanics lower in the Warrawoona Group have recently been dated at 3520 ± 60 m.y. using Sm-Nd (P.J. Hamilton, personal communication.)"

Returning to the Onverwacht Group of Southern Africa, Sinha (59) studied common Pb and U-Pb systematics of five basaltic lavas and obtained a "best" age of 3290 ± 45 m.y. for the Onverwacht Group. For the purpose of palaeobiological correlation the paper leaves something to be desired in that it neither quotes the decay constants used, nor mentions the stratigraphic position of the dated samples within the ∿ 18000 m-thick Onverwacht Group (Table 1). However, the age agrees well with the published Rb-Sr age of 3275 ± 70 m.y. (42) for the Middle Marker Horizon. Jahn and Shih (60) made Rb-Sr analyses on suites of volcanic and sedimentary rocks ranging from the Theespruit and Kromberg Formations (Table 1) and obtained an overall data scatter far outside the analytical uncertainties, except for separated minerals from a basaltic komatiite in the Komati Formation (Table 1), which yielded a Rb-Sr isochron age of 3420 ± 20 m.y., (initial $^{87}Sr/^{86}Sr$ ratio = 0.70048 ± 0.00005), interpreted as the age of eruption. Subsequently, Hamilton et al.(61) reported a Sm-Nd age of 3540 ± 30 m.y. for a suite of basaltic and peridotitic komatiites from the Komati Formation (Table 1). It must be remembered that the top of the Komati Formation is stratigraphically some 8500m below the top of the Swartkoppie Formation, from which the most convincing microorganisms in the Onverwacht Group have been reported to date(4).

In Zimbabwe (formerly Rhodesia), Precambrian stromatolitic limestones have been recognised for many years (62). Some of the finest recently discovered examples come from the Belingwe greenstone belt (63) which is stratigraphically correlated with neighbouring areas containing dated volcanic horizons. The Belingwe sediments and volcanics lie unconformably on granitic crust and indicate a depositional environment which changed from probable beach, to tidal flat, to deeper water conditions. Hawkesworth et al. (64) have reported Rb-Sr whole rock isochrons on volcanics from several greenstone belts in Zimbabwe, but from the Belingwe greenstone belt itself the analytical data yielded too great a scatter for meaningful age interpretation. Stratigraphically broadly equivalent greenstone belt volcanics of the Bulawayo area, of the Que Que area, and from the Maliyami Formation west of Que Que area yielded Rb-Sr isochron ages of 2480 ± 180 m.y., 2470 ± 280 m.y. and 2650 ± 140 m.y. respectively, with correspondingly scattered initial $^{87}Sr/^{86}Sr$ ratios. The scatter on the isochron plots was attributed (64) to locally variable post-depositional redistribution of Rb and/or Sr. In contrast, Jahn and Condie (65) obtained a satisfactory Rb-Sr whole rock isochron of 2690 ± 70 m.y. for eleven samples from the Belingwe belt volcanics. Subsequently, Hamilton et al. (66) reported a pooled Sm-Nd whole rock isochron age of 2640 ± 140 m.y. for 10 basic and ultrabasic lavas from all of the above greenstone belt successions, including four samples from the Belingwe belt itself. At the time the Belingwe stromatolites were reported (63) they were amongst the oldest described until then. Since then, much older ones have been reported from Western Australia (see above), and Zimbabwe (see below).

3.2 Dating of bracketing rock units

This is clearly an extension of the previous approach and is frequently used instead of, or in addition to, the dating of sediments and their intercalated volcanics. Some early Precambrian volcano-sedimentary assemblages are underlain by older basement gneisses, usually consisting of metamorphosed igneous rocks of continental character. In other cases no such basement may be discernible. Even more frequently, the volcano-sedimentary assemblages are intruded by younger igneous plutons. Dating of both older basement and younger intrusives can be used to set limits to the age of deposition. The limits obtained by the bracketing method may be surprisingly narrow, say within ∿ 50 - 100 m.y., or impracticably wide. Where only one bracket can be closed it may be possible to set a close older or younger age limit for deposition. The literature abounds in examples of geochronological bracketing of sedimentary and volcanic rocks of any geological age, and only a few examples relevant to the topic of this paper are quoted here. In some cases, the bracketing method may give more reliable results than the more direct methods. Two cases, from southern Africa and Canada, were already mentioned above where younger, cross-cutting igneous plutons gave older dates than the intruded sediments and volcanics. As explained earlier, this happens because of the higher susceptibility of some fine-grained sediments and volcanics to subsequent alteration and metamorphism than coarse-grained plutonic rocks.

Returning once more to Zimbabwe, Orpen and Wilson (3) have reported the "best examples of ∿ 3500 m.y. biogenic structures (stromatolites) known which add to the growing body of evidence that life was manifest as early as ∿ 1000 m.y. after the formation of the Earth". The stromatolites occur in limestones of the Fort Victoria greenstone belt, assigned to the ∿ 3500 m.y. Sebakwian Group (67, 68, 69). The stromatolitic limestone is regarded as older than the nearby Mushandike granite, dated by the Rb-Sr whole rock isochron method at 3445 ± 260 m.y. (70). The limestone was formerly regarded as part of the ∿ 2600 - 2700 m.y.-old Bulawayan succession. The Sebakwian Group may well turn out to be an important repository for early Archaean biogenic markers. In the Selukwe area, some 90 km northwest of the stromatolite locality, the Sebakwian succession is cut by the Mont d'Or granite which has yielded a Rb-Sr whole rock isochron age of 3340 ± 120 m.y. (71). It is possible, though not yet proven in the field, that the Sebakwian Group is underlain by granitic gneisses which have given Rb-Sr whole rock ages of 3490 ± 400 m.y. (64) and 3480 ± 120 m.y. (65) in the Mashaba and Shabani areas, respectively some 30 km west-northwest and 75 km west-southwest of the stromatolite locality. The dates and their associated errors, as well as the uncertainties in the geological correlation, show that the age of deposition of the Sebakwian Group is not quite as satisfactorily bracketed as might be desired. Further combined field and age work is clearly desirable.

The dating of volcanic rocks from the younger greenstone belts of Zimbabwe and their stromatolitic horizons, described earlier, has been complemented by Rb-Sr whole rock isochron ages of 2560 ± 80 m.y., 2630 ± 140 m.y. and 2590 ± 80 m.y. for the cross-cutting Chilimanzi granite (73), Sesombi and Somabula tonalites (64), which thus provide a younger age limit for the volcano-sedimentary successions. Although it is known that the ∿ 2700 - 2600 m.y. Belingwe greenstone belt, in which convincing stromatolites occur (63), is underlain by gneiss basement, it is not yet clear how much of this basement belongs to older gneisses dated at ∿ 3500 m,y, (see above), or younger gneisses, which dominate this part of Zimbabwe, dated at ∿ 2900 - 2700 m.y. (64, 67, 68, 75).

Another important stromatolite locality occurs in a dolomite unit of the Yellowknife Supergroup in the northern part of the Slave Structural Province, northwest Canada (75). The stromatolites are regarded as having a minimum age of ∿ 2500 m.y. on the basis of age determinations on probably broadly equivalent volcanic rocks of the Yellowknife Group and their intrusive rocks from the southern part of the Slave Province at Yellowknife, some 550 km south of the stromatolite locality (75). Here, the metavolcanics have yielded a Rb-Sr whole rock isochron age of 2570 ± 150 m.y., whilst zircon U-Pb ages of ∿ 2570 m.y. and Rb-Sr whole rock isochron ages of ∿ 2540 m.y. and 2510 m.y. are recorded for three successive suites of intrusive intermediate and acid rocks (48).

Finally, very convincing stromatolites from the ∿ 2500 m-thick Pongola Supergroup of southern Africa have been claimed to be 3000 m.y. old (76). A search through the quoted references (77, 78) yielded only several poorly documented dates and references to unpublished work. It appears that the upper part of the Pongola Supergroup (the Mozaan Group) rests uncomformably on granitic rocks which may be equivalent in age to certain granites in adjacent regions which have been dated at ∿ 3000 - 2900 m.y.. On this basis, Hunter (78) quotes a maximum Rb-Sr age for the Pongola Supergroup of ∿ 2940 m.y., although with revised decay constants this would reduce to ∿ 2870 m.y. A more convincing minimum age is provided by a Rb-Sr whole rock isochron age of 2805 ± 30 m.y. on acid and basic rocks of the Usushwana igneous complex, which cuts the Pongola Supergroup (79).

4. DATING THE ISUA SUPRACRUSTAL BELT, WEST GREENLAND

These rocks are discussed separately because they are the oldest known reliably dated rocks, with an age of nearly 3800 m.y., and because of the debate that currently surrounds the presence or absence of biogenic markers within them. They consist of a complex and varied sequence of metamorphosed supracrustal rocks of igneous and sedimentary origin, exposed in an arc some 15-20 km in diameter around a dome of gneiss of igneous origin (Amîtsoq gneiss) near the margin of the Inland Ice at Isua, some 150 km northeast of Godthaab in West Greenland. The Isua supracrustal belt has a variable width of 1 - 3 km and can be followed as continuous outcrop over a distance of some 40 km. Contacts with the surrounding gneisses are sharp and nearly vertical, and intrusive relationships are preserved at several places along the contact. Brief geological descriptions of the area have been published (80, 81, 82), and the belt is currently being mapped in detail by A. Nutman. Intense deformation and amphibolite facies metamorphism (at a temperature of ∿ 550°C and a pressure of ∿ 5kb, ref. 83) have obliterated many primary igneous textures, but preserved just sufficient primary sedimentary textures to enable the stratigraphy of the sequence to be established (A. Nutman, personal communication). Despite the intense metamorphism and deformation, the original nature of some of the rocks can be clearly discerned, and exhibits many similarities to the much larger greenstone belt assemblages of later Archaean times. The depositional environment of the Isua rocks might be envisaged as one in which island volcanoes, erupting acid, basic and possibly ultrabasic lavas, were surrounded by shallow-water platforms covered by coarse, volcaniclastic deposits, followed seawards by volcanigenic sands, shales and muds, and ultimately by a rich variety of marine chemical sediments, including carbonates, cherts and banded iron-formation which have been the subject of intense inter-disciplinary investigations by palaeobiologists.

So-called "yeast-like microfossils" reported from the cherty layers of an Isua quartzite (84) have been regarded as a misidentification of limonite-stained fluid inclusions (5). Extractable biogenic organic compounds for a sample of Isua banded iron-formation are regarded as contaminants from modern surface-encrusting lichens (85). Current work on the pyrolisis of graphite from Isua metasediments has not yet proved whether the released fragments are of biological or abiogenic origin (86). Isua carbonates are reported by Schidlowski et al. (87) as containing organic and inorganic carbon in approximately modern proportions; they state that "unless this early stabilisation of the terrestrial carbon cycle in terms of a constant partitioning of carbon between the reduced and oxidised species is shown to have been caused by some inorganic geochemical process, a considerably earlier start of chemical evolution and spontaneous generation of life must be considered than is presently accepted." Current debate hinges around the alternative possibility that the carbon isotope characteristics of the Isua rocks could be produced by metamorphic processes alone, without any biogenic contributions. Sulphur isotope measurements on Isua sediments have a $\delta^{34}S$ value close to zero per mil. and show no evidence for bacterial sulphate reducers (88).

Another major controversy centres around the origin of the widely distributed Archaean and early Proterozoic banded iron-formations, of which the one at Isua is a characteristic example. Although the stratigraphic sequences in which banded iron-formations occur are highly variable and indicate several types of depositional environments, their overall similarity in bulk compositions and mineral assemblages imply that the chemical conditions of iron-formation deposition were similar over the period ∿ 3800 - 1800 m.y. (89). Extensive literature is devoted to origins and depositional conditions of banded iron-formation (see refs. 90, 91), and some workers regard the presence of marine photosynthetic organisms as necessary for their formation (see refs. 92, 93, 94). It may be noted that the ∿ 2750 m.y.-old banded iron-formations at Michipicoten and Woman River in the Canadian Superior Province have yielded sulphur and carbon isotope data which provide strong evidence for the existence of autotrophic organisms and reducing bacteria in late Archaean times (95). Evidence for sulphate-reducing bacteria has also been provided (96) by sulphur isotope measurements on sedimentary pyrite from the Deer Lake greenstone sequence in Minnesota, which is probably equivalent to the > 2600 m.y. -old greenstone sequence of the well-known Vermilion District (97).

Published isotopic age determinations on the Isua supracrustals are relatively straightforward. A Pb/Pb isochron age of 3710 ± 60 m.y. for various components of the banded iron-formation (98) and a Rb-Sr whole rock isochron age of 3650 ± 60 m.y. on a suite of acid metavolcanic rocks (99) have been interpreted as the time of deposition and/or metamorphism of the supracrustal sequence. The latter interpretation is the more likely, because the rocks were undoubtedly heated and metamorphosed, and probably isotopically reset, by emplacement of the igneous precursors of the surrounding Amîtsoq gneisses of which two separate suites have yielded Rb-Sr whole rock isochron ages of 3610 ± 140 m.y. and 3690 ± 130 m.y. (100, 101). Zircons from a quartzite and from an acid metavolcanic horizon have yielded U-Pb ages of 3670 and 3770 m.y. respectively (no errors quoted, ref. 102), whilst U-Pb analyses of single zircons from an acid metavolcanic horizon gave an age of $3770 \pm ^{12}_{9}$ m.y. (103). A Sm-Nd isochron age of 3770 ± 42 m.y. has been reported for a combined suite of acid and basic metavolcanics (104). The initial $^{143}Nd/^{144}Nd$ ratio for the isochron is 0.507831 ± 46, which falls on the rather well established growth line of $^{143}Nd/^{144}Nd$ versus age for many Precambrian igneous and metaigneous rocks, with characteristic chondritic Sm/Nd ratio of 0.308 as calculated from the slope of the growth curve (11, 12). The Isua Sm-Nd age, which agrees perfectly with the quoted zircon U-Pb ages, is regarded (104) as providing a precise estimate of the timing of the volcanism which produced the basic and acid igneous rocks from a mantle reservoir with a chondritic Sm/Nd ratio, and therefore also for the timing of intercalated sedimentation.

In addition the Isua supracrustals contain minor sulphide mineralisations, including galena (PbS). One galena sample contains Pb with the least radiogenic $^{206}Pb/^{204}Pb$ and $^{207}Pb/^{204}Pb$ so far reported for any terrestrial material and which yields a single stage model age of 3740 m.y. (105). The isotopic data for this galena fit current Pb isotope mantle evolution models very closely (see refs. 106, 107).

Although the Isua volcanic and sedimentary rocks were obviously laid down on top of older rocks, no evidence has yet been found for the age and nature of this substratum. There is no evidence from initial isotopic ratios that the substratum was of continental, sialic character. The range and variety of rock types at Isua suggest, however, that the conditions of deposition of sediments and volcanics were not greatly different from conditions which may be deduced from the far less heavily metamorphosed and deformed volcano-sedimentary associations of Western Australia, Southern Africa and Zimbabwe mentioned previously (see refs. 6, 7, 8, 68), which post-date the Isua rocks by "only" about 300 - 500 million years.

5. SUMMARY AND CONCLUSIONS

We have tried to provide an objective assessment of the rather limited evidence for the age of deposition of early Precambrian sediments, particularly those of direct interest to palaeobiologists. Taking into account quoted errors on individual age determinations, as well as uncertainties in geochronological/stratigraphic correlations, it now seems likely from the evidence summarised above that stromatolite-bearing rocks were being produced at ∿3450 - 3300 m.y. ago (e.g. Western Australia, Zimbabwe), and that Archaean microfossils showing cell division were in existence at ∿ 3200 - 3000 m.y. ago (Southern Africa). Additional work to confirm and possibly narrow these estimates should be carried out on rocks which are stratigraphically as close as possible to the fossiliferous horizons, or on the latter themselves. Some of the isotopic age methods and dated materials described above are perfectly adequate, whilst there is ample scope for development of promising new variants of existing techniques. We recommend, in particular, further direct Pb/Pb dating of banded iron-formations (so far only reported from Isua, see ref. 98), as well as research into applying different age methods to cherts and limestones. All this will require collaborative, interdisciplinary research between palaeobiologists (sensu lato), geologists and geochronologists.

In view of the intense metamorphism suffered by the Isua supracrustal rocks of West Greenland, and the resulting interpretative problems regarding biogenic markers, it is not yet possible to say with certainty whether life existed at ∿ 3800 m.y. ago. There is nothing in the nature of the palaeoenvironmental evidence, as deduced from the nature of the Isua sediments, to contradict the existence of life, whilst the occurrence elsewhere of stromatolites and cellular micro-organisms some 300 - 500 m.y. later adds an indirect measure of plausibility.

When did life start? A detailed discussion of this question was held at the end of the Symposium on which the present book is based. The concensus view was at ∿ 4000 ± 100 m.y. Some participants thought that life's origin was even older, but nobody seriously considered a younger date to be realistic. The concensus view was based on a charitable interpretation of the evidence from Isua, whilst others were influenced by the general evidence (for review, see ref. 108) for the cessation of extraterrestrial bombardment at around 4000 - 3900 m.y. ago and also the rapidly increasing stabilisation of proto-crust against convective recycling and foundering into the mantle at about the same period, or perhaps somewhat earlier. Intuitively, mechanically stable and cool conditions are prerequisites for biogenic maturation of primordial soup.

In conclusion, we look forward to many exciting developments in this most controversial and problematic field.

ACKNOWLEDGEMENTS: I thank Dr. Paul N. Taylor for critical comments, and Mrs. Joan Morrall for typing the paper.

REFERENCES:

(1) Lowe, D.R.: 1980, Nature 284, pp. 441-443.
(2) Walter, M.R., Buick, R. and Dunlop, J.S.R.: 1980, Nature 284, pp. 443-445.
(3) Orpen, J.L. and Wilson, J.F.: 1981, Nature 291, pp. 218-220.
(4) Knoll, A.H. and Baarghorn, E.S.: 1977, Science 198, pp. 396-398.
(5) Bridgwater, D., Allaart, J.H., Schopf, J.W., Klein, C., Walter, M.R., Baarghorn, E.S., Strother, P., Knoll, A.H. and Gorman, B.E.: 1981, Nature 289, pp. 51-53.
(6) Barley, M.E., Dunlop, J.S.R., Glover, J.E. and Groves, D.I.: 1979, Earth Planet. Sci. Lett. 43, pp. 74-84.
(7) Eriksson, K.A.: 1979, Precambrian Res. 8, pp. 153-182.
(8) Eriksson, K.A.: 1980, Precambrian Res. 12, pp. 141-160.
(9) Dimroth, E., Donaldson, J.A. and Veizer, J. (Editors): 1980, Precambrian Res. 12, pp. 1-470.
(10) Faure, G.: 1977, Principles of Isotope Geology. John Wiley and Sons, New York.
(11) O'Nions, R.K., Carter, S.R., Evensen, N.M. and Hamilton, P.J.: 1979, Ann. Rev. Earth Planet. Sci. 7, pp. 11-38.
(12) De Paolo, D.J.: 1981, EOS 62, pp. 137-140.
(13) Steiger, R.H. and Jäger, E.: 1977, Earth Planet Sci. Lett. 36, pp. 359-362.
(14) Lugmair, G.W. and Marti, K.: 1978, Earth Planet. Sci. Lett. 39, pp. 349-357.
(15) Dodson, M.H., Rex, D.C., Casey, R. and Allen, P.: 1964, Quart. Journ. Geol. Soc. London 120, pp. 145-158.
(16) Owens, J.P. and Sohl, N.F.: 1973, Geol. Soc. Amer. Bull. 84, pp. 2811-2838.
(17) Webb, A.W., McDougall, I. and Cooper, J.A.: 1963, Nature 199, pp. 270-271.
(18) McDougall, I., Dunn, P.R., Compston, W., Webb, A.W., Richards, J.R. and Bofinger, V.M.: 1965, Journ. Geol. Soc. Austral. 12, pp. 67-90.
(19) Obradovich, J.D. and Peterman, Z.E.: 1968, Canad.Journ. Earth Sci. 5, pp. 737-747.
(20) Hurley, P.M., Cormier, R.F., Hower, J., Fairbairn, H.W. and Pinson, W.H.: 1960, Bull. Amer. Assoc. Petrol. Geol. 44, pp. 1793-1808.
(21) Odin, G.S., Velde, B. and Bonhomme, M.: 1977, Earth Planet. Sci. Lett. 37, pp. 154-158.
(22) Compston, W. and Pidgeon, R.T.: 1962, Journ. Geophys. Res. 67, pp. 3493-3502.
(23) Whitney, P.R. and Hurley, P.M.: 1964, Geochim, et Cosmochim. Acta 28, pp. 425-436.
(24) Bofinger, V.M. and Compston, W.: 1967, Geochim. et Cosmochim. Acta 31, pp. 2353-2359.
(25) Bofinger, V.M., Compston, W. and Gulson, B.L.: 1970, Geochim. et Cosmochim. Acta 34, pp. 433-445.

(26) Chaudhuri, S. and Faure, G.: 1967, Econ. Geol. 62, pp. 1011-1033.
(27) Moorbath, S.: 1969, Scott. Journ. Geol. 5, pp. 154-170.
(28) Faure, G. and Kovach, J.: 1969, Geol. Soc. Amer. Bull. 80, pp. 1725-1736.
(29) Pringle, I.R.: 1973, Geol. Mag. 109, pp. 465-472.
(30) Peterman, Z.E.: 1966, Geol. Soc. Bull. Amer. 77, pp. 1031-1044.
(31) Powell, J.L., Skinner, W.R. and Walker, D.: 1969, Geol. Soc. Amer. Bull. 80, pp. 1605-1612.
(32) Dasch, E.J.: 1969, Geochim. et Cosmochim. Acta 33, pp. 1521-1552.
(33) Clauer, N.: 1973, Geochim. et Cosmochim. Acta 37, pp. 2243-2255.
(34) Clauer, N.: 1974, Earth Planet. Sci. Lett. 22, pp. 404-412.
(35) Perry, E.A. and Turekian, K.K.: 1974, Geochim. et Cosmochim. Acta 38, pp. 929-935.
(36) Gebauer, D. and Grünenfelder, M.: 1974, Contrib. Mineral. Petrol. 47, pp. 113-130.
(37) Chaudhuri, S.: 1976, Contrib. Mineral. Petrol. 59, pp. 161-170.
(38) Spanglet, M., Brueckner, H.K. and Senechal, R.G.: 1978, Geol. Soc. Amer. Bull. 89, pp. 783-790.
(39) Allsopp, H.L., Ulrych, T.J. and Nicolaysen, L.O.: 1968, Canad. Journ. Earth Sci. 5, pp. 605-619.
(40) Baarghorn, E.S. and Schopf, J.W.: 1966, Science 152, pp. 758-763.
(41) Schopf, J.W. and Baarghorn, E.S.: 1967, Science 156, pp. 508-512.
(42) Hurley, P.M., Pinson, W.H., Nagy, B. and Teska, T.M.: 1972, Earth Planet. Sci. Lett. 14, pp. 360-366.
(43) Muir, M.D. and Grant, P.R.: 1976, in The Early History of the Earth, B.F. Windley (Editor). John Wiley and Sons, New York. pp. 595-604.
(44) Brooks, J., Muir, M. and Shaw, G.: 1973, Nature 244, pp. 215-217.
(45) Schopf, J.W.: 1976, in The Early History of the Earth, B.F. Windley (Editor). John Wiley and Sons, New York, pp. 589-593.
(46) Clauer, N.: 1976, Mem. Sci. Geol. Strasbourg 45, pp. 1-256.
(47) Clauer, N.: 1979, in Lectures in Isotope Geology, E. Jäger and J.C. Hunziker (Editors). Springer Verlag, Berlin. pp. 30-51.
(48) Green, D.C. and Baadsgaard H.: 1971, Journ. Petrol. 12, pp. 177-217.
(49) Jahn, B.M. and Murthy, V.R.: 1975, Geochim. et Cosmochim. Acta 39, pp. 1679-1689.
(50) Clark, G.S. and Cheung, S.P.: 1980, Canad. Journ. Earth Sci. 17, pp. 560-568.
(51) Vidal, Ph., Blais, B., Jahn, B.M., Capdevila, R. and Tilton, G.R.: 1980, Geochim. et Cosmochim. Acta 44, pp. 2033-2044.

(52) Allsopp, H.L., Viljoen, M.J. and Viljoen, R.P.: 1973, Geol. Rundschau 62, pp. 902-917.
(53) Jones, L.M., Walker, R.L. and Allard, G.O.: 1974. Canad. Journ. Earth Sci. 11, pp. 1550-1561.
(54) Nunes, P.D. and Thurston, P.C.: 1980, Canad. Journ. Earth Sci. 17, pp. 710-721.
(55) McCulloch, M.T. and Wasserburg, G.J.: 1978, Science 200, pp. 1003-1011.
(56) Pidgeon, R.T.: 1978, Earth Planet. Sci. Lett. 37, pp. 421-428.
(57) Oversby, V.M.: 1976, Geochim. et Cosmochim. Acta 40, pp. 817-829.
(58) De Laeter, J.R. and Blockley, J.G.: 1972, Journ. Geol. Soc. Austral. 19, pp. 363-370.
(59) Sinha, A.K.: 1972, Earth Planet. Sci. Lett. 16, pp. 219-227.
(60) Jahn, B.M. and Shih, C.Y.: 1974, Geochim. et Cosmochim. Acta 38, pp. 873-885.
(61) Hamilton, P.J., Evensen, N.M., O'Nions, R.K., Smith, H.S. and Erlank, A.J.: 1979, Nature 279, pp. 298-300.
(62) Schopf, J.W., Oehler, D.Z., Horodyski, R.J. and Kvenvolden, K.A.: 1971, Journ. Palaeont. 45, pp. 477-485.
(63) Bickle, M.J., Martin, A. and Nisbet, E.G.: 1975, Earth Planet. Sci. Lett. 27, pp. 155-162.
(64) Hawkesworth, C.J., Moorbath, S., O'Nions, R.K. and Wilson, J.F.: 1975, Earth Planet. Sci. Lett. 25, 251-262.
(65) Jahn, B.M. and Condie, K.C.: 1976, Contrib. Mineral. Petrol. 57, pp. 317-330.
(66) Hamilton, P.J., O'Nions, R.K. and Evensen, N.M.: 1977, Earth Planet. Sci. Lett. 36, pp. 263-268.
(67) Wilson, J.F., Bickle, M.J. Hawkesworth, C.J., Martin, A., Nisbet, E.G. and Orpen, J.L.: 1978, Nature 271, pp. 23-27.
(68) Wilson, J.F.: 1979, Spec. Publ. Geol. Soc. S. Afr. 5, pp. 1-23.
(69) Wilson, J.F.: 1980, in <u>The Precambrian of the Southern Hemisphere</u>, D.R. Hunter (Editor). Elsevier, Amsterdam.
(70) Hickman, M.H.: 1974, Nature 251, pp. 295-296.
(71) Moorbath, S., Wilson, J.F., and Cotterill, P.: 1976, Nature 264, pp. 536-538.
(72) Moorbath, S., Wilson, J.F., Goodwin R. and Humm, M.: 1977, Precambrian Res. 5, pp. 229-239.
(73) Hickman, M.H.: 1976, Unpublished Ph.D. thesis, University of Leeds.
(74) Hawkesworth, C.J., Bickle, M.J., Gledhill, A.R., Wilson, J.F. and Orpen, J.L.: 1979, Earth Planet. Sci. Lett. 43, pp. 285-297.
(75) Henderson, J.B.: 1975, Canad. Journ. Earth Sci. 12, pp. 1619-1630.
(76) Mason, T.R. and von Brunn, V.: 1977, Nature 266, pp. 47-49.
(77) Burger, A.J. and Coertze, F.J.: 1973, Geol. Surv. S. Afr. 58, pp. 1-46.

(78) Hunter, D.R.: 1974, Precambrian Res. 1, pp. 259-326.
(79) Davies, R.D., Allsopp, H.L., Erlank, A.J. and Manton, W.I.: 1970, Spec. Publ. Geol. Soc. S. Afr. 1, pp. 576-593.
(80) Allaart, J.H.: 1976, in The Early History of the Earth, B.F. Windley (Editor). John Wiley and Sons, New York, pp. 177-189.
(81) Bridgwater, D., Keto, L., McGregor, V.R. and Myers, J.S.: 1976, in Geology of Greenland, A. Escher and W.S. Watt (Editors), Copenhagen, pp. 19-75.
(82) Appel, P.W.U.: 1980, Precambrian Res. 11, pp. 73-87.
(83) Dymek, R.F.: 1981, EOS (Abstracts) V82, p. 420.
(84) Pflug, H.D. and Jaeschke-Boyer, H.: 1979, Nature 280, pp. 483-486.
(85) Nagy, B., Engel, M.H., Zumberge, J.E., Ogino, H. and Chang, S.Y.: 1981, Nature 289, pp. 53-56.
(86) Walters, C., Shimoyama, A. and Ponnamperuma, C.: 1981, in Origin of Life, Y. Wolman (Editor). D. Reidel, Dordrecht, pp. 473-479.
(87) Schidlowski, M., Appel, P.W.U., Eichmann, R. and Junge, C.E.: 1979, Geochim. et Cosmochim. Acta 43, pp. 189-199.
(88) Monster, J., Appel, P.W.U., Thode, H.G., Schidlowski, M., Carmichael, C.M. and Bridgwater, D.: 1979, Geochim. et Cosmochim. Acta 43, pp. 405-413.
(89) Gole, M.J. and Klein, C.: 1981, Journ. Geol. 89, pp. 169-183.
(90) Eugster, H.P. and Chou, I.M.: 1973, Econ. Geol. 68, pp. 1144-1168.
(91) Dimroth, E.: 1975, Geol. Rundschau 64, pp. 751-767.
(92) Cloud, P.: 1973, Econ. Geol. 68, pp. 1135-1143.
(93) La Berge, G.L.: 1973, Econ. Geol. 68, pp. 1098-1109.
(94) Drever, J.I.: 1974, Geol. Soc. Amer. Bull. 85, pp. 1099-1106.
(95) Goodwin, A.M., Monster, J. and Thode, H.G.: 1976, Econ. Geol. 71, pp. 870-891.
(96) Ripley, E.M. and Nicol, D.L.: 1981, Geochim. et Cosmochim. Acta 45, pp. 839-846.
(97) Sims, P.K.: 1972, in Geology of Minnesota: A Centennial Volume, P.K. Sims and G.B. Morey (Editors), Minnesota Geol. Surv., pp. 41-49.
(98) Moorbath, S., O'Nions, R.K. and Pankhurst, R.J.: 1973, Nature 245, pp. 138-139.
(99) Moorbath, S., Allaart, J.H., Bridgwater, D. and McGregor, V.R.: 1977, Nature 270, pp. 43-45.
(100) Moorbath, S., O'Nions, R.K., Pankhurst, R.J., Gale, N.H. and McGregor, V.R.: 1972, Nature Phys. Sci. 240, pp. 78-82.
(101) Moorbath, S., O'Nions, R.K. and Pankhurst, R.J.: 1975, Earth Planet. Sci. Lett. 27, pp. 229-239.
(102) Baadsgaard, H.: 1976, Earth Planet. Sci. Lett. 33, pp. 261-267.
(103) Michard-Vitrac A., Lancelot, J., Allègre, C.J. and Moorbath, S.: 1977, Earth Planet.Sci. Lett. 35, pp. 449-453.

(104) Hamilton, P.J., O'Nions, R.K., Evensen, N.M., Bridgwater, D. and Allaart, J.H.: 1978, Nature 272, pp. 41-43.
(105) Appel, P.W.U., Moorbath, S. and Taylor, P.N.: 1978, Nature 272, pp. 524-526.
(106) Cumming, G.L. and Richards, J.R.: 1975, Earth Planet. Sci. Lett. 28, pp. 155-171.
(107) Stacey, J.S. and Kramers, J.D.: 1975, Earth Planet. Sci. Lett. 26, pp. 207-221.
(108) Smith, J.V.: 1981, Phil. Trans. Roy. Soc. Lond. A 301, pp. 401-422.

INORGANIC CHEMISTRY OF EARLIEST SEDIMENTS:
Bioinorganic Chemical Aspects of the Origin and Evolution of Life

Ei-Ichiro Ochiai

Laboratory of Chemical Evolution, Department of Chemistry,
University of Maryland, College Park, MD 20742, USA

Present Address: Juniata College, Huntingdon, PA 16652

Abstract

A number of inorganic elements are now known to be essential to organisms. Chemical evolutionary processes involving carbon, hydrogen, nitrogen and oxygen have been studied intensively and extensively, but the other essential elements have been rather neglected in the studies of chemical and biological evolution. This article attempts to assess the significance of inorganic chemistry in chemical and biological evolutionary processes on the earth. Emphasis is placed on the catalytic effects of inorganic elements and compounds, and also on possible studies on the earliest sediments, especially banded iron formation and strata-bound copper from the inorganic point of view in the hope of shedding some light on the evolution of the environment and the biological effects on it.

The foregoing two lectures covered the atmosphere, hydrosphere and lithosphere (rocks) during the earlier stages of the earth's history. This lecture, accordingly, attempts to discuss the interactions between the environment (atmosphere/hydrosphere/lithosphere) and the biosphere from the inorganic chemical point of view; in other words, the significance of inorganic chemistry in the study of chemical and biological evolution.

1. INTRODUCTION - WHY INORGANIC CHEMISTRY?

Let us first survey the environments of all the organisms on the present earth. The organisms constitute the biosphere, and the environment is divided into atmosphere, hydrosphere and

C. Ponnamperuma (ed.), Cosmochemistry and the Origin of Life, 235–276.

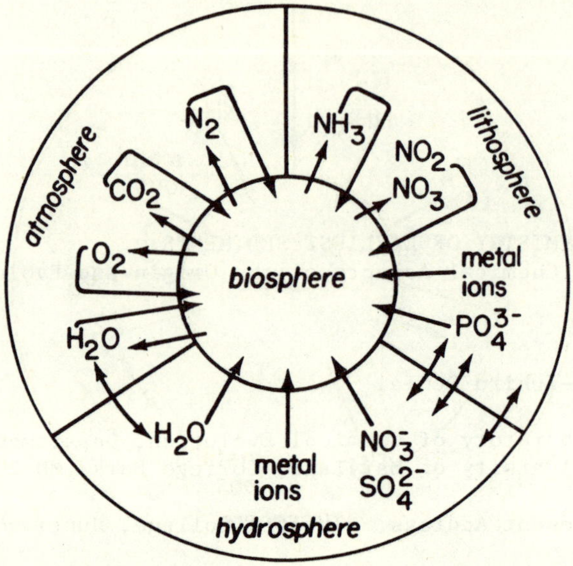

Fig. 1. The interrelationship of the biosphere with its environments [Adapted from E-I Ochiai, "Bioinorganic Chemistry: An Introduction" (Allyn and Bacon, 1977)].

lithosphere. The biosphere maintains an intimate interaction between its environment, atmosphere, hydrosphere and lithosphere, as depicted in Fig. 1 (37). It is noted that all the compounds contained in the environment are inorganic ones. Therefore, it is highly unlikely that the inorganic compounds have no effect on the biosphere. In fact, the inorganic compounds act as nutrients, contaminants or toxicants for the organisms. Accordingly, the organisms have developed metabolic systems to utilize the nutrients (inorganic), and to remove contaminants or toxicants. The interaction between the biosphere and its environment is reciprocative; i.e., the biosphere can affect the environment. Indeed it can be said that a great portion of the present environment is a creation by the biosphere.

Now let us see the situation before the emergence of life. Since the biosphere did not exist yet, we can presume that there was a kind of prebiosphere which consisted of abiotically produced organic compounds. The atmosphere, the hydrosphere and the lithosphere existed. The content of the atmosphere on the primordial earth is believed to be completely different from the present one (see the chapter by Henderson-Sellers). The hydrosphere and the lithosphere also were different from the present ones (see the chapter by S. Moorbath). However, the pre-biosphere, consisting of organic compounds, would have interacted with the atmosphere, hydrosphere and lithosphere which were made entirely

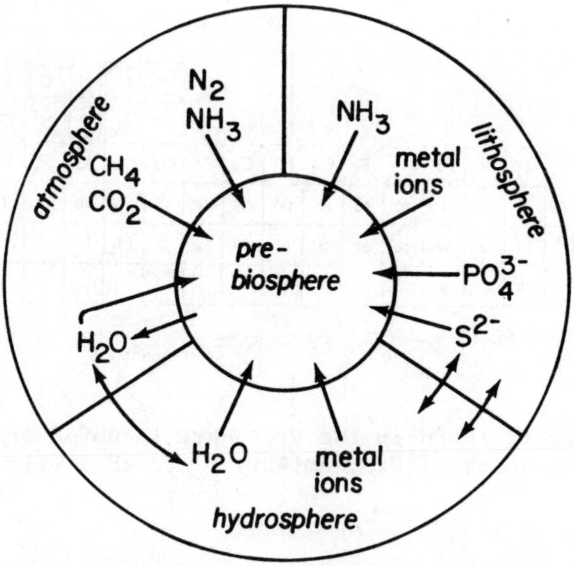

Fig. 2. The interrelationship of the prebiosphere with its environments.

of inorganic compounds (Fig. 2). Here, too, it is likely that the inorganic environment exerted a significant effect on the course of chemical evolution. The organic chemical portion of chemical evolution has been studied extensively. However, little attention has been paid to the inorganic side of the process. The "little warm pond" of primordial soup certainly contained not only the important organic compounds such as amino acids, carbohydrates and bases but also a large number of inorganic elements and compounds. These inorganic compounds must have had a significant effect on the formation of these organic compounds and the subsequent process of chemical evolution.

This article attempts to assess the significance of inorganic chemistry (or bioinorganic chemistry) in the quest of the "origin and evolution of life." The problem is two-fold. One is understanding or enumerating the possible effects of inorganic chemistry on the origin and evolution of life, and the other is seeking clues to the origin and evolution of life in the inorganic chemistry of the geological records.

2. THE EFFECTS OF INORGANIC COMPOUNDS ON THE ORGANISMS

We should first try to learn something from today's organisms. The biological functions of inorganic compounds have been well-documented (37). Here we will give a brief summary of relevant

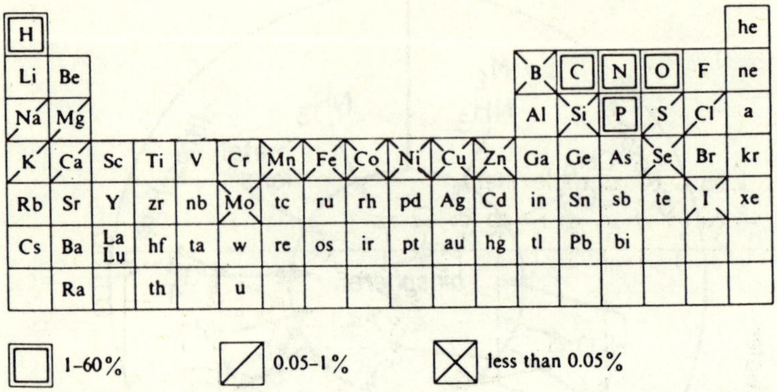

Fig. 3. Elements found in the biosphere [Adapted from R. Mason, "Principles of Geochemistry," 3rd ed., Wiley (1966)].

material.

2.1 Essential Elements

About thirty elements out of a total of ninety-two are known to be essential to organisms (see Fig. 3). Carbon, hydrogen, oxygen and nitrogen constitute the bulk, i.e., organic compounds of the organisms. Sulfur and phosphorus also are the components of the important organic compounds. Potassium, sodium, magnesium and calcium are almost universally required and in significant amounts as well.

The other essential elements are required usually in small quantities; thus, they are called trace elements. It would be expected that the organisms would utilize the more abundant elements rather than the ones in short supply, thus insuring a better chance of survival. Fig. 4 shows the distribution of the elements in the present sea water and the upper crust of the earth. It indicates that generally speaking the more abundant elements are the ones that are utilized by organisms. Exceptions to this general rule include aluminum, titanium and cobalt. The first two are non-essential despite their abundance, and the last one (Co) is essential despite its paucity in the environment. Why this is so is another long story and is beyond the scope of the present article (38).

Organisms developed a variety of mechanisms to deal with the more abundant elements, namely, absorption, metabolism and excretion. These organisms, on the other hand, may not be equipped with mechanisms to deal with elements that are only rarely encountered. These elements may act, therefore, as toxicants for

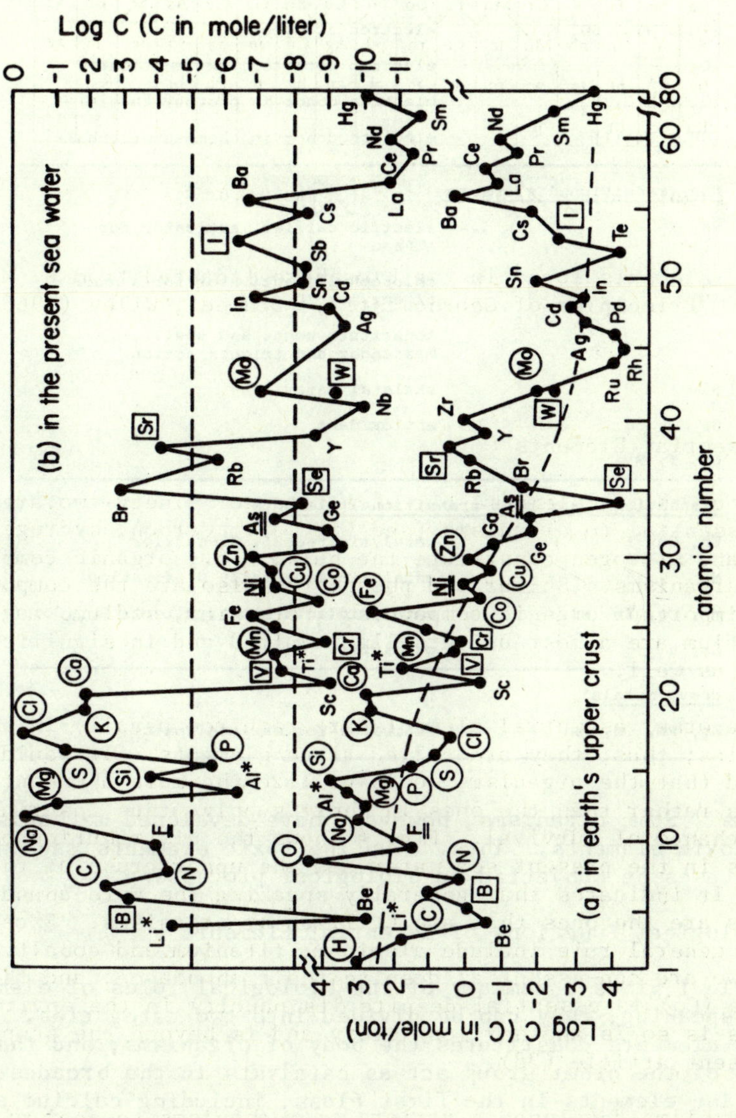

Fig. 4. Distribution of elements in (a) the present sea water and (b) the earth's upper crust.

Table 1 BIOLOGICAL ROLES OF ELEMENTS

Element	Roles
Organic elements	
C, H, N, O, P, S	constitute organic compounds
(CO_2, NO_3^-, SO_4^{2-}	electron acceptor)
(O_2	electron acceptor, oxygenation)
(H_2S, H_2O, H_2	electron donor in photosynthesis)
(NH_3, Fe(II)	electron donor in chemoautotrophs)
Non-transitional elements	
Na, K	electric carrier, activator for ATPase
Mg	chlorophyll, activator for a number of enzymes
Ca	constitute bones and shells, messenger and trigger action
Si	skeletal material
Se	antioxidant
Cl, I, B	
Transitional and post-transitional elements	
Fe, Cu	catalytic(redox), O_2-carrier
Mo	catalytic(especially N_2-fixation)
Mn, Co, Zn, Ni	catalytic(acid-base), vitamin B_{12}(Co)
Cr, V	catalytic
Heavy metals	
Hg, Pb, Cd, As	inhibitor for enzymes(Toxic)

organisms. Some organisms, however, have developed mechanisms to combat toxic elements. Thus, even the toxic elements can in certain cases be subjected to biological processing.

2.2 Biological Functions of Essential Elements

Table 1 gives a summary of the biological roles of elements. Roughly speaking, they can be divided into two categories. One group of elements constitutes the body of organisms, and the elements of the other group act as catalysts in the broadest sense. The elements in the first class, including calcium and silicon, are required in large quantities, whereas those in the second group are required only in minute quantities. As far as quantity is concerned, there is another group of elements whose required quantities would be intermediate between the two groups already mentioned. They are those involved in energy metabolism. Because of the nature of the process (energy metabolism), those elements are utilized in substantial quantities by organisms. They include O (O_2), S (S^2, $^-SO_4{}^{2-}$), N (NH_3, $NO_3{}^-$), Fe ($Fe^{II} \rightarrow Fe^{III}$ by some microbes) and Mn ($Mn^{II} \rightarrow MnO$ by some microbes). Thus, the geochemistry of these elements would be significantly affected by the activities in the biosphere.

2.3 Some Specific Examples

We should like to discuss a few specific topics which have some relevance to chemical evolution.

DNA and RNA polymerases. These enzymes are probably among the oldest since the polymerization of nucleotides is such a basic reaction in organisms.

DNA-dependent DNA polymerases, RNA-dependent DNA polymerases (so-called reverse transcriptase) and DNA-dependent RNA polymerases have been isolated from a number of organisms, ranging from viruses and bacteria to higher organisms. All the enzymes isolated and characterized so far seem to contain zinc as a prosthetic group; zinc [Zn(II)] is the catalytic element. They also require magnesium [Mg(II)] as an activator. The fact that all these enzymes depend on Zn(II) would seem to indicate that such polymerases or their precursors in the first living cells or even in the chemical evolution stage probably utilized Zn(II) as the catalytic element.

In fact it has been shown (45, 46) that some simple metallic ions such as Pb(II) and Zn(II) have some catalytic effect in oligomerizing nucleotides.

Ferredoxin. Ferredoxins, or more generally speaking, iron-sulfur proteins are one of the most widely distributed proteins. It is found even in *Clostridium* species which are considered to be one of the most primitive organisms. Except for a few aromatic amino acid residues, the bulk of the amino acids constituting the ferredoxin molecules are more stable ones which are known to be produced in simulated primordial conditions (see Fig. 5).

Therefore, it is conceivable that ferredoxin or its precursor played an important role even in the chemical evolution stage. Perhaps, iron sulfide, FeS (pyrrhotite) itself may have acted as a precursor of ferredoxin. Alternatively, a composite of FeS and an abiotically formed peptide might have been a better precursor.

This point touches on the controversial problem: which came first, proteins or RNA (DNA)? There is a strong possibility that a ferredoxin-like molecule (a protein) formed spontaneously. This

	1									28
P. ae.	Ala-Tyr-Val-Ile-	-Asn-Asp-Ser	Cys	Ile-Ala	Cys	Gly-Ala	Cys	Lys-Pro-Glu	Cys	Pro-Val-Asn- -Ile-Gln-Gln-Gly- -Ser-
C. bu.	Ala-Phe-Val-Ile-	-Asn-Asp-Ser	Cys	Val-Ser	Cys	Gly-Ala	Cys	Ala-Gly-Glu	Cys	Pro-Val-Ser-Ala-Ile-Thr-Gln-Gly-Asp-Thr-
C. pa.	Ala-Tyr-Lys-Ile-	-Ala-Asp-Ser	Cys	Val-Ser	Cys	Gly-Ala	Cys	Ala-Ser-Glu	Cys	Pro-Val-Asn-Ala-Ile-Ser-Gln-Gly-Asp-Ser-
C. ac.	Ala-Tyr-Val-Ile-	-Asn-Glu-Ala	Cys	Ile-Ser	Cys	Gly-Ala	Cys	Asp-Pro-Glu	Cys	Pro-Val-Asp-Ala-Ile-Ser-Gln-Gly-Asp-Ser-
C. ta.	Ala-His-Ile-Ile-	-Thr-Asp-Glu	Cys	Ile-Ser	Cys	Gly-Ala	Cys	Ala-Ala-Glu	Cys	Pro-Val-Glu-Ala-Ile-His-Glu-Gly-Thr-Gly-
C. th.	Ala-His-Ile-Ile-	-Thr-Asp-Glu	Cys	Ile-Ser	Cys	Gly-Ala	Cys	Ala-Ala-Glu	Cys	Pro-Val-Glu-Ala-Ile-His-Glu-Gly-Thr-Gly-
C. M-E.	Ala-Tyr-Lys-Ile-	-Thr-Asp-Gly	Cys	Ile-Asn	Cys	Gly-Ala	Cys	Glu-Pro-Glu	Cys	Pro-Val-Glu-Ala-Ile-Ser-Glu-Ser-Asp-Ala-
Cl. li. I.	Ala-Leu-Tyr-Ile-	-Thr-Glu-Glu	Cys	Thr-Tyr	Cys	Gly-Ala	Cys	Glu-Pro-Glu	Cys	Pro-Val-Thr-Ala-Ile-Ser-Ala-Gly-Asp-Asp-
Cl. li. II.	Ala-His-Arg-Ile-	-Thr-Glu-Glu	Cys	Thr-Tyr	Cys	Ala-Ala	Cys	Glu-Pro-Glu	Cys	Pro-Val-Asn-Ala-Ile-Ser-Ala-Gly-Asp-Glu-
Chromatium	Ala-Leu-Met-Ile-	-Thr-Asp-Gln	Cys	Ile-Asn	Cys	Asn-Val	Cys	Gln-Pro-Glu	Cys	Pro-Asn-Gly-Ala-Ile-Ser-Gln-Gly-Asp-Glu-
D. gigas	Pro-Ile-Gln-Val-	-Asp-Asn	Cys	Met-Ala	Cys	Gln-Ala	Cys	Ile-Asn-Glu	Cys	Pro-Val-Asp-Val-Phe-Gln-Met-Asp-Glu-Gln-
										56
	29									
P. ae.	Ile-Tyr-Ala-Ile-Asp-Ala-Asp-Ser		Cys	Ile-Asp	Cys	Gly-Ser	Cys	Ala-Ser-Val	Cys	Pro-Val-Gly-Ala-Pro-Asn-Pro-Glu-Asp
C. bu.	Gln-Phe-Val-Ile-Asp-Ala-Asp-Thr		Cys	Ile-Asp	Cys	Gly-Asn	Cys	Ala-Asn-Val	Cys	Pro-Val-Gly-Ala-Pro-Asn-Gln-Glu
C. pa.	Ile-Phe-Val-Ile-Asp-Ala-Asp-Thr		Cys	Ile-Asp	Cys	Gly-Asn	Cys	Ala-Asn-Val	Cys	Pro-Val-Gly-Ala-Pro-Val-Gln-Glu
C. ac.	Arg-Tyr-Val-Ile-Asp-Ala-Asp-Thr		Cys	Ile-Asp	Cys	Gly-Ala	Cys	Ala-Gly-Val	Cys	Pro-Val-Asp-Ala-Pro-Val-Gln-Ala
C. ta.	Lys-Tyr-Gln-Val-Asp-Ala-Asp-Thr		Cys	Ile-Asp	Cys	Gly-Ala	Cys	Gln-Ala-Val	Cys	Pro-Thr-Gly-Ala-Val-Lys-Ala-Glu
C. th.	Lys-Tyr-Glu-Val-Asp-Ala-Asp-Thr		Cys	Ile-Asp	Cys	Gly-Ala	Cys	Glu-Ala-Val	Cys	Pro-Thr-Gly-Ala-Val-Lys-Ala-Glu
C. M-E.	Val-Arg-Val-Ile-Asp-Ala-Asp-Lys		Cys	Ile-Asp	Cys	Gly-Ala	Cys	Ala-Asn-Thr	Cys	Pro-Val-Asp-Ala-Ile-Val-Glu-Gly
Cl. li. I.	Ile-Tyr-Val-Ile-Asp-Ala-Asn-Thr		Cys	Asn-Glu	Cys	Ala Ala Gly Gln Leu Glu Asp	Cys	Val-Ala-Val	Cys	Pro-Ala-Glu-Cys-Ile-Val-Gln-Gly 60
Cl. li. II	Ile-Tyr-Ile-Val-Asp-Glu-Ser-Val		Cys	Thr-Asp	Cys	Glu Ala Gly Pro Tyr Glu Tyr-Asp	Cys	Val-Ala-Val	Cys	Pro-Val-Asp-Cys-Ile-Ile-Lys-Val 61
Chromatium	Thr-Tyr-Val-Ile-Glu-Pro-Ser-Leu		Cys	Thr-Glu	Cys	Val Asp Gly Val His Cys Tyr Gln Glu Ser Thr	Cys	Val-Glu-Val	Cys	Pro-Ile-Lys-Asp-Pro-Ser-His-Glu- Ala-Arg-Leu-Glu-Asp-Glu-Thr-Glu- Lys-Tyr-Glu-Arg-Ile-Thr-Gly-Glu-Gly 81
D. gigas	Gly-Asp-Lys-Ala-Val-Asn-Ile-Pro-Asn-Ser-Asn-Leu-Asp-Asp-Gln-Cys-Val-Glu-Ala-Ile-Gln-Ser-Cys-Pro-Ala-Ala-Ile-Arg-Ser 56									

Fig. 5. Primary structures of bacterial ferredoxins: Pseudomonas aerogenes, Clostridium butyricum, C. pasteurianum, C. acidiurici, C. tartarivorum, C. thermosaccharolyticum, C. M-E, Chlorobium limicola (I and II), Chromatium, Desulfovibrio gigas. (From D. O. Hall, K. K. Rao and R. N. Mullinger, in "Protein Structure and Evolution" (J. L. Fox, Z. Deyl and A. Blazvej, eds.), p. 233-259, Marcel Dekker (1976)).

molecule, if not as effective as the ultimately formed progeny, may have taught the polynucleotide how to build the genetic code. In other words, some spontaneously formed peptides interacted with polynucleotides and cooperated in building up the proper genetic code and the protein synthesis system. The point is that proteins, maybe crude, should have formed spontaneously, which, however, did have properties close enough to the real proteins. It is likely that metallic ions could have had some significant effects on the spontaneous formation of proteins (see section 3.2).

3. INORGANIC CHEMISTRY IN THE PREBIOTIC PROCESS

Having surveyed how important the inorganic elements are in today's organisms, we will now go back to a time when life was in the process of being formed and see what roles the inorganic compounds could have possibly played in those days.

For convenience sake, we will divide the process leading up to the formation of life into four stages; that is:

- Stage I: Abiotic synthesis of essential organic compounds such as amino acids, bases, carbohydrates, lipids and porphyrins, etc.
- Stage II: Abiotic formation of polymeric compounds such as peptides (proteins), nucleic acids and polysaccharides.
- Stage III: Formation of system: Protobiont. Proper combination of material assembled, such as membrane, coacervate, etc.
- Stage IV: Formation of properly interacting system: life such as genetic code, systematization of chemical reactions, specificity and selectivity of enzymes, etc.

Now we will discuss possible effects of inorganic compounds on each of these steps.

3.1 Stage I

Much effort has been expended on the elucidation of abiotic syntheses of essential organic compounds. As a result, virtually all the essential compounds have been shown to form abiogenically. Energy sources of high intensity have been employed in the so-called simulation experiments. It is understandable that high energy sources are required, because fairly stable compounds such as CH_4, $N_2(NH_3)$ and H_2O are to be made to react to each other. These are primary gas-phase reactions.

The subsequent secondary reactions that took place in the aqeous phase could have been affected by the omnipresent inorganic material. The inorganic, particularly metallic ions can coordinate the intermediates or the products of the chemical evolutionary processes or can catalyze such processes.

Beck (6) in his survey of "prebiotic coordination chemistry" emphasized the formation of metallocyano complexes. HCN, being an important intermediate in the chemical evolutionary reactions, is assumed to have been present at a significantly high level in the primordial hydrosphere so that it led to such complexes as $Fe(CN)_6^{4-}$. The latter is supposed to have played some catalytic roles. Beck (6) also reports that an acid hydrolysis of a cyanogen complex $H_2Cu_2(CN)_4C_2N_2$ led to the formation of glycine. These reactions may be possible, but their significance in primordial chemical evolution is rather questionable.

Oro and his coworkers (41) studied the effects of some inorganic compounds on the formation of amino acids from formaldehyde and hydroxylamine. Iron chloride was found to cause a slight decrease of the yield of amino acids, whereas the yield increased exponentially as a function of concentration of molybdate. A similar result was reported earlier for the effect of molybdenum oxide on the photosynthesis of amino acids (5). Although molybdenum is involved in a number of enzymatic reactions of today's organism, it is not participating in any stage of amino acid formation or its reactions. However, it is highly probable that molybdate acted as a general acid catalyst in these reactions.

It is well known that metallic ions enhance the formation of porphyrins. Beck (6) tested this in the reaction of pyrrole and paraformaldehyde. The effect of metal compounds on the yield of porphyrins was found to be: $K_3Fe(CN)_6 > NiSO_4 > K_2Ni(CN)_4 \simeq FeSO_4$. We had a similar experience. A reaction of o-phthalonitrile at an elevated temperature led to an unidentifiable gunk in the absence of a metallic element, while the same led to a ready formation of a metallophthalocyanin in the presence of a metallic ion such as Cu(II), Ni(II) and Fe(II) under a comparable reaction condition. This is certainly due to the stabilization effect by chelation.

Fischer-Tropsch reactions between CO and H_2 produce hydrocarbons, alcohols and aldehydes in the presence of such solid catalysts as iron (Fe). Similar reactions between CO, H_2 and NH_3 have been shown to produce most of the essential organic compounds including amino acids and bases (25, 2, 59). This kind of reaction could probably have occurred on the exposed primordial terrain and in the formation process of carbonaceous chondrites.

It has been shown (49) that radiation by uv light of N_2 adsorbed on wet solid TiO_2 produces NH_3. This reaction was invoked to supply NH_3 to the partially oxidized atmosphere on the primordial earth that is now believed to be more likely (26). A rather simple coordination compound of Mo-Fe-S could have acted as a precursor of nitrogenase, which could have solved the problem of the short supply of NH_3 on the early earth. This leads us to the whole problem of the precursors of metalloenzymes or metal-activated enzymes (38). Undoubtedly, a lot more has to be learned in order to better understand even the chemical evolutionary reactions, if one includes the effects of the inorganic world. And we believe that this (organic reactions in the presence of inorganic compounds) is closer to the reality than the simple Miller-Urey type reactions.

3.2 Stage II

The polymerization of amino acids, nucleotides or saccharides is a condensation reaction, which is usually endothermic. This endothermicity is usually overcome by coupling the condensation reaction with the hydrolysis of ATP or other nucleotides. This is the process employed in today's organisms.

In laboratory experiments simulating the prebiotic polymerization, the dehydration (i.e., condensation) is carried out by using dehydrating agents such as polyphosphate and diamino maleodinitrile (13) or by heating a solution containing mixtures of amino acids to dryness (18, 19, 20). Certainly these reactions produce some peptides. But the question is not just "any polymerization." The polymerizations have to be more specific, in order for the abiotic process to produce meaningful ensembles of polymeric compounds. The key is "specificity." Was the polymerization utterly random or statistical in the chemical evolution stage? Then did a specific polymerization have to wait until the advent of a genetic code system? Or could the polymerization have been specific to a certain extent?

It is to be noted, first of all, that each individual amino acid has a different reactivity, though the difference in reactivity between two amino acids may not be very large. Nature would have taken advantage of this difference in enhancing the specificity in polymerization. The difference in reactivity would be lessened at higher temperatures, because of the entropy effect. However, even the peptides produced at higher temperatures have been shown to exhibit some specificities (21). An analysis of amino acid pairs in peptide sequence has demonstrated that indeed the frequency of different amino acid pairs in abiogenically produced peptides is fairly similar to that in the contemporary biological proteins (52).

It is, therefore, to be expected that the specificity would be more pronounced if the polymerization is effected at lower temperatures. At lower temperatures, the reaction is, of course, slow, maybe too slow to be significant, though it might have been sufficient in view of the geological time scale. Catalysts could speed up reactions.

No metalloenzyme seems to be involved in the contemporary biological protein synthesis system, though Mg(II) is required as a cofactor. This fact does not necessarily preclude an involvement of metallic ions or of metal complexes in the prebiotic polymerization of amino acids. Indeed it is highly likely to be the case. A number of metalloenzymes are known presently, which hydrolyze polypeptides or proteins. Examples include carboxypeptidase A (Zn-enzyme), a dipeptidase (Zn-enzyme) and subtilisin (Zn-proteinase). The hydrolysis of a peptide bond is the reverse of condensation (peptide bond formation). The reaction is reversible and can go either forward or backward depending on the reaction condition, though, of course, the hydrolysis direction is favorable thermodynamically. These enzymes usually catalyze the forward reaction of hydrolysis but have been shown to catalyze its reverse under a certain reaction condition. It is, therefore, probable that metal ions or metal-peptide complexes acted as catalysts in the oligomerization of amino acids in the primordial soup.

It has been reported (58, 10) that amino acid esters oligomerized up to octapeptide in the presence of metal salts such as $CuCl_2$. These reactions were carried out in organic solvents such as methanol and ethanol, while no oligomerization reaction was observed in aqueous media. Therefore, Beck (6) contends that these results are not significant from the prebiotic point of view. The problem here seems to be a thermodynamic one. The peptide formation is a dehydration and would not be favored thermodynamically in an aqueous medium. In organic solvents this problem is non-existent and the forward reaction may be favored also by a fact that the oligomers could precipitate out, thus the forward reaction being carried further. Such a condition, that is, removal of the peptides from a reaction system, might have existed in the primordial soup. Perhaps, for example, different phases created by the formation of coacervates could have performed such a task.

The chelation of peptides or amino acids onto a metallic ion would impose a constraint upon the steric interaction between the reacting centers or upon the conformation of the peptide produced. This would increase the chance of a more specific polymerization. Take ferrodoxin as an example. The structure of a typical bacterial ferredoxin is shown in Fig. 6, and the primary structures (amino acid sequence) of some

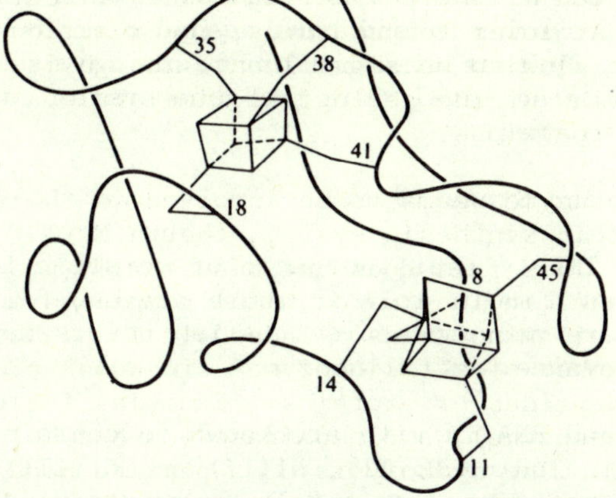

Fig. 6. The structure of Micrococcus aerogenes ferredoxin [From Jensen et al, Biochem. Soc. Trans. 1, 27 (1973)].

ferredoxins are given in Fig. 5. A typical ferredoxin has 56 amino acid residues, and has a two-fold symmetry. Thus it can be regarded to be consisting essentially of 28 amino acid residues. Out of these 28 residues four have to be cysteines, all of which are bound to the iron-sulfur cube Fe_4S_4. The spatial structure about the cube would impose a strong restriction on the amino acid residues to be inserted between the cysteine residues, in terms of both the number of residues and the kind of residues. As is shown in Fig. 5, the number of the intervening residues is strictly conserved (one exception: Chromatium), and the kind of residues, which may not be as critical as the number, is also fairly well preserved. Because of these various constraints imposed upon its structure, a ferredoxin molecule may well have been formed spontaneously without the participation of a m-RNA/translation system. Ferredoxin may well be just an exception, but it is possible that the basic protein units as discussed by Schulz (50) could have been produced spontaneously with some aid of primordial polymerization catalysts. And these proteins could have provided an incentive and basis for the construction of the genetic code.

A possible role of metallic ions in the polymerization of nucleotides was mentioned earlier in connection with DNA or RNA polymerases in today's organisms (section 2.3).

We have no clue as to the polymerization of sugars in terms of our framework. It is very likely that the polymerization of sugars was not essential at least in the first heterotrophic organisms. It might have become important when the organisms invented a photosynthesis and thus needed a means of storing carbohydrates. By that time the organisms have developed to such a sophisticated level that the gene system could produce the necessary enzymes.

3.3 Stage III and Stage IV

It is not likely that inorganic or metallic ions or compounds played a major role in these stages, for the major players here were macromolecules. A few points may, however, be of some relevance.

Both DNA and RNA in vivo are known to contain a number of metallic ions including Mg(II), Ni(II) and Cr(III). These ions are interpreted to play a role in maintaining a specific conformation of the nucleic acids. An x-ray diffraction study showed that Mg(II) ions are indeed strategically located in a t-RNA molecule to maintain its specific structure (29, 43). It is thus highly conceivable that metal ions played some important roles in developing the transcription and translation system of the genetic apparatus.

Ca(II) is known to play a major role in structuring the cell membrane consisting mainly of phospholipids. The phosphate group of a phospholipid binds strongly with Ca(II). Ca(II) thus could have played a decisive role in the formation of coacervates or protocells.

4. A BIOINORGANIC CHEMICAL VIEW OF THE EARLY BIOLOGICAL EVOLUTION

Now we should like to explore how inorganic chemistry could have affected the course of the biological evolution. In order to do it, we need some data from paleogeology and paleoclimatology. Here we will be content with a set of rough data, a more refined discussion of which will be deferred to a later section. The kinds of data we need are i) reduction potentials of relevant elements and ii) the change of the atmospheric oxygen during the course of geological history of the earth. The first set of data is not difficult to find, but the second one is difficult to come by and will be the major topic in the rest of this article (and the chapter by Henderson-Sellers).

The basic argument is as follows. If a kind of organisms requires a specific element and that element is not available to the organism, the organism would not survive, or such an organisms would not appear under such a condition. There are a number of organisms which require specific compounds, especially inorganic ones. Different requirements for different organic compounds are well studied and well documented. Because of the temporal change in the atmospheric oxygen and consequently the availability of inorganic elements, we might be able to say something about the course of biological evolution. An element can be in a number of different oxidation states; e.g., CO_2 (C(IV)), CH_2O (nominally C(0)), C, CH_4(C(-IV)); Fe(III), Fe(II), Fe(0); Cu(II), Cu(I), Cu(0). The distribution of different oxidation states is controlled by the reduction potentials of the element and its compounds. An element can have drastically different properties in different oxidation states, especially a significantly different solubility in aqueous media. This last parameter controls the availability of an element in aqueous media, and consequently to organisms.

4.1 Classification of Organisms

There are a number of ways to do classification but we are not concerned with a rigorous taxonomic system. Table 2 shows a scheme of classification relevant to our discussion. Most of the prokaryotic organisms listed in Table 2 require Na, K, Mg, Ca, Zn, Fe, Mo and Co (some) in addition to the organic elements: C, H, N, O, P, S. Specific compounds are

Table 2 VARIETY OF ORGANISMS

Autotrophic	Photoautotrophic	photosynthetic bacteria
		blue-green algae
		algae, green plants
	Chemoautotrophic	bacteria : oxidize NH_3, Fe(II), etc
Heterotrophic	depend on the others as the source of carbohydrates and other nutrients: bacteria (_Clostridium_, etc), fungi, animals	
Photoautotrophs	source of electrons	
Sulfur bacteria (purple and green)	H_2S, $S_2O_3^{2-}$, etc	
Non-sulfur bacteria	H_2, reduced organic compounds	
Blue-green algae, algae, green plants	H_2O	
Anaerobic	killed by O_2	fermentative
		CO_2 respiration (methane bacteria)
		SO_4^{2-} respiration (_Desulfovibrio_)
Aerobic	depend on O_2	aerobic respiration
Facultative	can live both ways	NO_3^- respiration

required for specific energy metabolisms. For example, sulfur bacteria (purple and green sulfur bacteria) photosynthesize and require H_2S, $S_2O_3^{2-}$ or other reduced forms of sulfur as the electron source. Some non-sulfur bacteria utilize H_2 or partially reduced organic compounds for the same purpose. Some anaerobes utilize CO_2 (methane bacteria) and SO_4^{2-} (sulfur reducer) as the electron acceptor and thus do respirative metabolisms. Chemoautotrophs are all aerobes and obtain energy to produce carbohydrates by oxidizing S^{2-}, NH_3, NO_2^-, Fe(II) or others. The synthesis of NH_3 from N_2 is very important in the global ecosystem and is fairly wide-spread among different classes of microorganisms.

4.2 Basic Assumptions (39)

Having surveyed the organisms, we will now turn to the environment. The basic assumptions we make were mentioned earlier but are reiterated here.

(i) Organisms appeared in sea water and stayed there until the beginning of the Phanerozoic era.

(ii) The temperature and the pH of the sea water were constant at 300°K and 8, respectively.

(iii) The oxidative state of the sea water was controlled by the virtual oxygen pressure $P(O_2)$; thus the oxidative state E_h of the sea water is given by

$$E_h(v) = +0.750 + 0.0148 \log (P(O_2)).$$

The virtual oxygen pressure used here and hereafter is such that it is in thermodynamic equilibrium with the system under consideration. The actual atmospheric oxygen pressure $P^a(O_2)$ may not be the same as the virtual pressure. Oxygen molecules are absorbed by sea water (or other aqueous media) and transported to the target locale in the water to exert its effect which can be an oxidation of metallic ions or otherwise. The oxygen in the immediate vicinity of the reaction system can be assumed to be in equilibrium with the system, but the supply of oxygen is controlled by the kinetic factors among which are $P^a(O_2)$, the diffusion rate of the oxygen molecule and the mechanical mixing of water. This problem of the actual atmospheric oxygen pressure will be discussed in section 8; until then, the oxygen pressure employed should be understood to be the virtual one.

(iv) E_h defined above controlled the oxidation states of the elements present in the environment.

(v) The availability of elements dictated partially the course of evolution.

4.3 The Evolution of the Atmospheric Oxygen and the Dependence of the Oxidation States of Elements and the Biological Evolution on It (39)

Data for the reduction potential are readily available and their application to geochemistry is discussed in detail by Garrels and Christ (1965). We superimpose these data (at 300°K and pH 8) onto the curve of E_h vs. $P(O_2)$ (Fig. 7). This figure gives the relation between the oxidation states of elements and the oxygen pressure.

The difficult task is to find the change of the oxygen pressure with time. A rough estimate of it is given in Fig. 8. A more detailed discussion to justify the curve will be made in Sections 6 and 7. Even though the curve is made of three points and may appear meaningless, it can lead to a number

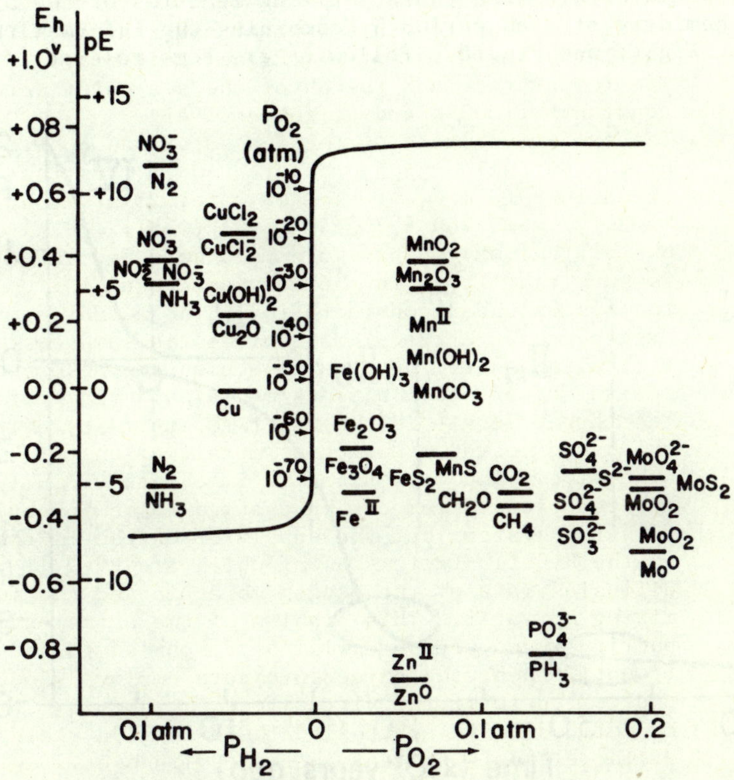

Fig. 7. The dependence of E_h (pE) of the environment upon $P(O_2)$ and $P(H_2)$, and the reduction potentials of relevant compounds at pH 8 and 300°K.

of interesting deductive results. We divide the history of the earth into period I, IIa, IIb, III and IV as shown in Fig. 8. Period I is 4.5 Gy ago to about 3.5 Gy ago (this is equivalent to "Hadean"), and life seems to have started toward the end of this period. Period IIa (3.5 Gy ago - 2.2 Gy ago, approximately the same as "Archean") is the age of non-oxygenic prokaryotes. Period IIb (2.2 Gy ago - 1.5 Gy ago, equivalent to "Proterophytic") may be coined as the age of cyanobacteria. The whole Period II is the age of prokaryote. Period III (1.5 Gy ago - 0.6 Gy ago, equivalent to "Proterozoic") is the age of earlier eukaryotes and Period IV is equivalent to Phanerozoic era.

The $P(O_2)$ was extremely low in Period I. It increased very slowly in Period IIa and then suddenly started to increase rapidly in Period IIb. This was caused by the proliferation of water-splitting and oxygen releasing cyanobacteria.

We would obtain some characterizing features of the bioinorganic chemistry of each period by combining the information given in Fig. 7 and Fig. 8. Following are some relevant points

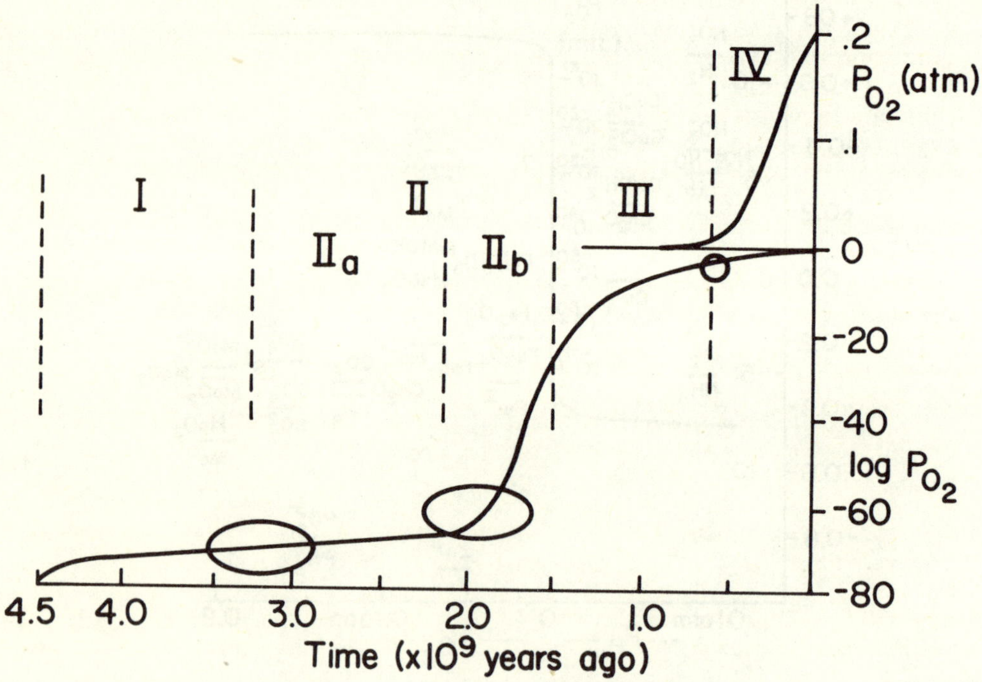

Fig. 8. The evolution of the virtual oxygen pressure $P(O_2)$, as deduced from paleogeological and paleobiological data.

deduced from these figures.

Period I. Because of the low $P(O_2)$, most of the elements were in more reduced states. Although CH_4 and NH_3 may have predominated initially, they seem to have turned largely into CO_2 and N_2 toward the end of this period or by the middle of Period IIa at the latest. The major forms of other elements were S^{2-}, MoO_2 or MoS_2, Fe^{II}, FeS, Mn^{II}, Co^{II} and Cu^0. Of course, most of the other non-transition elements are not affected by the atmospheric condition; these elements include Na, K, Mg, Ca and Zn. Thus, most of the essential elements were available, perhaps with the exception of Mo and Cu. They were sufficient to make chemical evolution proceed.

Period IIa. A little rise in the atmospheric pressure could make a very significant change in the availability of some elements and compounds. It is likely that NH_3 had been depleted by the middle of this period at the latest. About the same time, molybdenum became abundant in the sea water, because it was turned into the very soluble MoO_4^{2-} form. This is a very good coincidence, for the molybdenum which became available was immediately utilized by organisms to build a N_2-fixing enzyme, nitrogenase. Nitrogenase today is widely distributed; the organisms which contain it include the most ancient bacteria such as Clostridium. Since the first organisms that appeared on earth were heterotrophs and they fed on the abiotically produced organic compounds, they would soon deplete the stock of the organic compounds provided by nature. Therefore, autotrophic organisms must have appeared, perhaps in the earlier stage of period IIa. The inorganic compounds required were available to the autotrophs; Mg(II) required for bacteriochlorophyll was certainly available and S^{2-}, the electron donor, was still abundant. It is likely that these autotrophs were sulfur bacteria and non-sulfur bacteria a little later but not oxygen-releasing cyanobacteria.

Carbon was turned into mostly CO_2 by this period, and some organisms which utilize CO_2 probably appeared, such as the methane-forming bacteria.

Period IIb. The main event in this period is the appearance of cyanobacteria. They seem to have evolved from non-sulfur bacteria (Athiorhodaceae). The latter had a photosynthetic apparatus (Photosystem, PSI) and had developed an aerobic respiratory mechanism by this time by utilizing a part of the cyclic photosynthetic apparatus. This organisms could have acquired the second photosystem PSII which has a mechanism of decomposing water (to release free oxygen). One of the important components in PSII is manganese which was amply available.

Table 3 THE EVOLUTION OF THE ENVIRONMENT AND LIFE

Stage	Time[a]	P_{O_2} (atm)	Available compounds[b]	Biological events[c]
I	4.5-3.5	10^{-70}	H_2, CH_4, CH_2O, NH_3, H_2O, H_2S, Fe^{II}, Mn^{II}, MoS_2, Co^{II}, Cu^0	abiogenic synthesis of organic compounds → anaerobic heterotrophs
IIa	3.5-2.2	10^{-70}-10^{-65}	CH_2O, CO_2, N_2, H_2O H_2S, Fe^{II}, Mn^{II}, MoO_4^{2-}, Co^{II}, Cu^0, Fe_3O_4	sulfur photoautotrophs; pro-nonsulfur photoautotrophs; methane bacteria (cobalamin); (N_2-fixation); SO_4^{2-} reducers; nonsulfur photoautotrophs (with respiration)
IIb	2.2-1.5	10^{-65}-10^{-30}	CO_2, N_2, H_2O, SO_4^{2-}, Fe^{II}-Fe^{III}, Mn^{II}, MoO_4^{2-} Co^{II}, Cu^0-Cu_2S	procyanobacteria (no plastocyanin); aerobic heterotrophs
III	1.5-.65	10^{-30}-10^{-3}	CO_2, N_2, H_2O, SO_4^{2-}, Fe^{III}, Mn^{II}-Mn^{III}, Co^{II}, MoO_4^{2-}, Cu^{II}-Cu_2S	cyanobacteria (with plastocyanin); facultative(NO_3^--reducers)
IV	.65-0	10^{-3} -10^{-1}	CO_2, $N_2(NO_3^-)$, H_2O, SO_4^{2-}, Fe^{III}, Mn^{III}, MoO_4^{2-}, Cu^{II}	

[a]units: x10^9 years ago. [b]Other essential elements(Na, K, Mg, Ca, Zn, P) are assumed to have been available more or less readily throughout the whole history. [c]The item in the bracket is a relevant bioinorganic event associated with it.

About this time, the predominant form of sulfur became SO_4^{2-}; this perhaps spurred the development of SO_4^{2-}-reducing bacteria such as Desulfovibrio. This kind of organism, however, might have developed even earlier in association with sulfur autotrophs which produce sulfate as the end product.

Cyanobacteria release oxygen, and thus the atmospheric oxygen content started to rise rapidly hereafter. The virtual oxygen pressure seems to have become about 10^{-30}-10^{-20} atm by the end of this period. This is high enough to mobilize copper in the form Cu(II). Cu(II) thus became available rather late, 1.7 - 1.5 Gy ago.

Period III. A characteristic feature of this period is the emergence of eukaryotic cells, which led eventually to the development of all the higher organisms including ourselves. The continued rapid rise of the atmospheric oxygen encouraged the development of aerobic metabolism. The anaerobes may have been forced to survive in special niches where the prevailing condition was still anoxic.

Toward the end of this period, $NO^-{}_3$ became available. Some aerobes, when faced with lower $P(O_2)$, learned to switch to NO_3^-, whose reduction potential is close to that of oxygen. Consequently, they can grow anaerobically, too, and the facultative organisms developed.

A detailed discussion on the topic: copper and eukaryotes will be given in Section 7.

Summary

The discussion above assumes that the course of the biological evolution could have been dictated at least partially by the availability of elements, which is controlled in turn by the atmospheric oxygen. We may be able to deduce a kind of evolutionary tree from this discussion. This is shown in Table 3. The tree is comparable to those deduced from the sequencing of proteins and RNA (51).

5. BIOINORGANIC CHEMISTRY OF SEDIMENTS - A GENERAL DISCUSSION

Now we can turn to the second question set out at the beginning. That is, what can we learn about the biological activity or the chemical and biological evolution from inorganic chemical studies of the geological records?

Fig. 9 shows a simplified general overview of the geochemical processes. Rocks, igneous or otherwise, would subject to

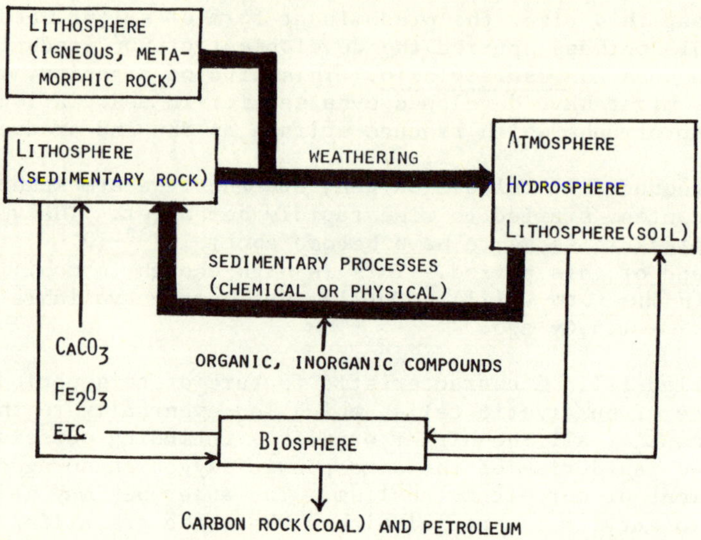

Fig. 9. An outline of geological processes and the effect of the biosphere on them.

weathering; the weathering process can be chemical, physical or biological. Weathered material is then transported, usually by way of the hydrosphere, and deposited, eventually forming sedimentary rocks. Our question here is how and where the biological activities participate in these geological processes.

There are two ways by which biological activities can affect the geological processes. One is "direct" and the other "indirect." The biological activities can produce the geochemical material or affect the geochemical processes directly. An example of the bioproducts that constitute sedimentary rock is calcium carbonate, which comes from the shells of algae and other organisms. Another example is the formation of limonite, $Fe(OH)_3$ by some microorganisms such as <u>Siderocapsa</u> and <u>Gallionella</u>. These bacteria oxidize Fe(II) to Fe(III) and utilize the energy and the electrons in order to produce carbohydrates from carbon dioxide. Some other examples are collected in Table 4. Some chemoautotrophic bacteria (e.g. <u>Thiobacillus</u> species) can leach out metallic elements such as iron, copper and uranium, thus causing a weathering process.

The indirect effect is diverse, but can be summarized as secondary chemical effects by biological products on the geochemical processes. Such products include oxygen, hydrogen sulfide and diverse organic compounds. They affect the geochemical

Table 4 THE DISTRIBUTION OF BIOINORGANIC MINERALS

	MONERA	PROTOCTISTA														FUNGI		ANIMALIA											PLANTAE	
		Sarcodina	Dinoflagellata	Haptophyta	Bacillariophyta	Phaeophyta	Rhodophyta	Chlorophyta	Gamophyta	Rhizopoda	Heliozoa	Actinopoda	Foraminifera	Myxomycota	Ciliophora	Basidiomycota	Deuteromycota	Porifera	Coelenterata	Platyhelminthes	Ectoprocta	Brachiopoda	Annelida	Mollusca	Arthropoda	Sipuncula	Echinodermata	Chordata	Bryophyta	Tracheophyta
Calcium carbonates																														
Calcite	+	+	+	+			+	+					+	+				+	+	+	+	+	+	+	+	+	+	+	+	+
Aragonite	+			?		+	+	+					+					+	+		+		+	+	+	+		+		+
Vaterite							+																	+	+			+		+
Monohydrocalcite	+																							+				+		
Amorph. hydr. carbonate														+						+				+	+			+		
Phosphates																														
Dahllite	+														+		+											+		?
Francolite																						+		+				+		
$Ca_3Mg_3(PO_4)_4$																							+							
Brushite																								+						
Amorph. dahllite precursor																					+			+	+			+		
Amorph. brushite precursor																								+						
Amorph. whitlockite precursor																				+			+	+						
Amorph. hydr. ferric phosphate																							+	+			+			
Sulfates																														
Gypsum																			+										?	+
Celestite												+																		
Barite		+						+		+																				
Silica																														
Opal					+					?	+	+	+					+					+	+	+		+			+
Iron oxides																														
Magnetite	+																							+	+			+		
Maghemite	?																													
Goetnite																								+						
Lepidocrocite																		+						+						
Ferrihydrite	+															+	+						+	+				+	+	+
Amorph. ferrihydrates													+										+					+		
Manganese oxides																														
Burnessite	+																													
Iron sulfides																														
Pyrite	+																													
Hydrotroilite	+																													

Adapted from Margulis(1981)

processes in many different ways. They may affect the transportation, deposition and weathering, mostly through oxidation-reduction and chelation reactions.

The next question is how significant the biological contribution to the geological processes could be. This is highly dependent on the element concerned, and difficult to generalize. Table 5 gives an overall view on the extent of the contribution. It gives the masses (and volumes) of the

Table 5 THE SIZES AND MASSES OF THE COMPONENTS OF EARTH

	Volume (km^3)	Mass (ton)	
Whole earth	1.083×10^{12}	5.976×10^{21}	
Core	1.755×10^{11}	1.881×10^{21}	
Mantle	8.98×10^{11}	4.068×10^{21}	
Crustal (continental and oceanic)	8.89×10^{9}	2.51×10^{19}	Lithosphere
Ocean and sea	1.37×10^{9}	1.41×10^{18}	Hydrosphere
Atmosphere	-	5×10^{15}	Atmosphere
Biota	-	2×10^{13}	Biosphere
Annual production of organic carbon		1.5×10^{11}	

constituents of the earth. Aside from a very long-term effect, we may neglect the masses of the core and the lower mantle.

Obviously the biosphere is insignificant compared to either the upper crust or the hydrosphere in terms of mass; one millionth or ten-millionth, respectively. Therefore, it might be concluded that the biological activities in the biosphere would not leave significant footprints in the geological ground. Besides, the geological activities, i.e., metamorphosis, tend to obliterate any peculiar marks created by the biological activities. This is an overall picture and makes us despair in our attempt to learn anything from the geological records. We are here, of course, excluding the organic compounds found in geological samples (see the chapter by G. Eglinton).

There is a little hope, however, since there are sediments which can only be produced by biological activities. Examples were already mentioned (see Table 4). Although indirect, the change in the atmospheric oxygen which was mostly caused by cyanobacteria is recorded in the geological tablets. The geological activity, being very heterogeneous, may be overridden by the biological activities in limited locales. Especially if a specific inorganic formation is found to be associated with microfossils, that formation is very likely a result of biological activities. This is exemplified by stromatolitic formations (55). When such deposits are found, one can deduce the nature of the biological activities from the characteristics of the inorganic formation. Such an example was found in a 2.3 billion year old South African stromatolite; a micrograph of the sample showed manganese oxide associated with microfossils (35, 36).

In the rest of this article, we will discuss two of the

most important and interesting inorganic elements; they are iron and copper. Iron in the ancient sediments is often found in the so-called "banded iron formation (BIF);" its generation history almost corresponds to period II (Archean and Proterophytic).

Copper, on the contrary, seems to be a marker element of period III (Proterozoic) as far as its biological association is concerned. Thus, these two elements cover the important period of the earth's history, 3.5 - 1.5 and 1.5 - 0.6 Gy ago.

6. BANDED IRON FORMATION (BIF)

It is one of the most conspicuous and yet intriguing geological features. It is found on all continents and constitutes a substantial portion of iron ores accessible in the upper crust. It is also economically very important and, therefore, has elicited extensive research. Useful information can be obtained from Econ. Geology Vol 68, (7) (entire issue, 1973), Eichler (16), James (27), Lundren and Dean (31), and Button, et al. (1981).

6.1 Characteristic features

The iron-rich sedimentary rocks are divided into two classes: iron-stone, which occurs mainly as thin beds in the Phanerozoic age; and iron-formation which is generally chert-banded and mainly of Precambrian age. The former is the so-called "red bed," which consists mainly of hematite (Fe_2O_3), and will not be discussed here.

Three principal types of iron-formation are recognized: Algoma type (A-type), Rapitan type, and Superior and Hamersley type (SH-type). The Rapitan type occurs in several basins of the late Precambrian age and is at least in part associated stratigraphically with glacial deposits. Deposits of significant dimensions of this type are known in the Rapitan group of northwestern Canada and in western Brazil and eastern Bolivia. Unfortunately, these formations are not well dated and rather poorly defined, so we will not discuss them further.

Algoma type BIF's are older than 2.6 Gy, and are closely associated with submarine volcanic stráta and considered to be directly related in origin to volcanic exhalation of iron-containing fluids. These deposits are different from those of Superior and Hamersley type in several respects. These are summarized in Table 6. Superior and Hamersley BIF's represent by far the greatest concentration of iron in the geological record, and occurred in the period between 2.6 Gy ago and

Table 6 CHARACTERISTIC FEATURES OF ALGOMA- AND SUPERIOR-HAMERSLEY-TYPE BANDED IRON-FORMATION

	Algoma-type	Superior-Hamersley-type
Age	pre-2.6 Gy	pre-1.8 Gy
Extent	commonly lenticular bodies of a few km	extensive, over some 100 km to more than 1,000 km
Thickness	0.1 m to some 10 m	several m to more than 100m(1,000 m)
Volcanics	close association to volcanism in time and space	no direct association with contemporaneous volcanism
Sedimentary facies	oxide-facies predominant; carbonate- and sulfide-facies thin and discontinuous; silicate facies.	oxide-facies most abundant; silicate- and carbonate-facies frequently intergradational.
	sulfide- and carbonate-facies near the center of volcanism; oxide-facies on the margins.	sulfide-facies insignificant or absent.
	heterogeneous lithological assemblages with fine grained clastic beds.	more homogeneous (especially oxide-facies); little or no detritus.

Adapted from J. Eichler, in "Handbook of Strat-bound and Stratiform Ore Deposits"(K. H. Wolf, ed.), Vol. 7, p157-201, Elsevier, Amsterdam (1976)

1.8 Gy ago. It is depicted as a huge band (2.6 - 1.8 Gy ago) in Fig. 10, whereas the first two humps belong to the Algoma type. This type of BIF has never occurred since. Such a deposit can extend more than 1,000 Km aerially and can be as thick as 1,000 m.

As Table 5 indicates, A-type seems to be volcanogenic in origin, whereas the SH-type is sedimentary. Although oxide-facies usually predominate, carbonate and sulfide (and separately silicate) facies are closely associated in BIF's of the A-type. In contrast, oxide facies are most abundant and sulfide-facies are insignificant in SH-type BIF's. One of the most conspicuous features of the BIF is the structure of alternating layers of chert and iron-rich minerals.

The oxide facies are usually predominantly magnetite (Fe_3O_4). A-type BIF's thus consist of alternating layers of iron-mineral and chert, and the iron minerals are made of $Fe_3O_4 \geqslant FeCO_3 \simeq FeS$ ($\simeq Fe_2SiO_4$). The development of the layered structure here is considered to be related to periodic tectonic events (12), rather than to the biological processes which seem to be responsible for the SH-type BIF.

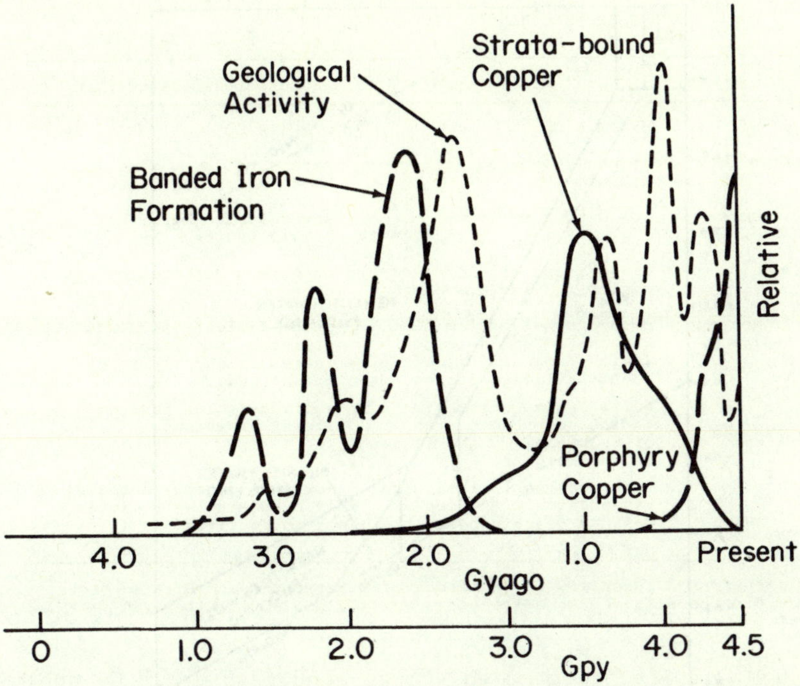

Fig. 10. The age profile of (a) geological (magmatic) activities, (b) BIF formation and (c) stratabound and porphyritic copper ore (produced in 1974). [Adapted from (a) and (b): J. Eichler, in "Handbook of Stratabound and Stratiform Ore Deposits," (K.H. Wolfe ed.) Vol. 7, pp. 157-201, Elsevier (1976); (c) R. Bowen and A. Gunatilaka, "Copper: Its Geology and Economics," Halstead Press (1977)].

6.2 Prevailing Atmosphere

It is to be noted that siderite $FeCO_3$, pyrite FeS_2 or pyrrhotite FeS and iron-silicate Fe_2SiO_4 contained in A-type BIF's all have iron in the reduced state, Fe(II), whereas magnetite Fe_3O_4 contains Fe(II) and Fe(III) in one-to-two ratio. This indicates that the predominating atmospheric condition at the time of deposition should be very reducing. Figs. 11 and 12 give E_h vs. pH diagrams for these minerals. If the ambient hydrosphere is assumed to have pH = 8, E_h at which Fe_3O_4, $FeCO_3$ and FeS_2 (and FeS) exist is below approximately -0.2 v, which corresponds approximately to $P(O_2) = 10^{-64}$ atm (see Fig. 7). Therefore, the ambient E_h at the time of the

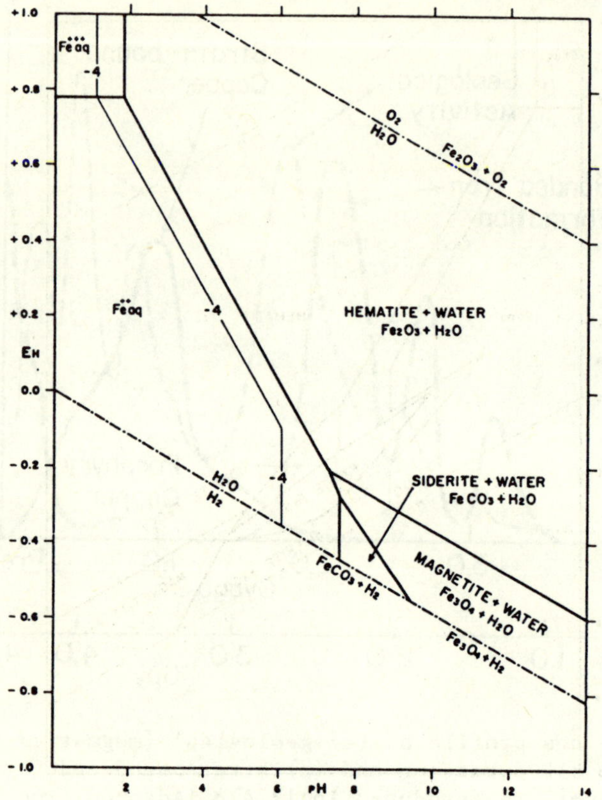

Fig. 11. Stability of hematite, magnetite and siderite in aqueous solution containing total dissolved carbonate species of 10^{-2}m at 25°C and 1 atmosphere total pressure [From R.W. Garrels and C.L. Christ, "Solutions, Minerals and Equilibria," Freeman, Cooper and Co. (1965)].

deposition of these BIF's must have been $E_h < -0.2$ v. This condition must have continued to prevail through the age of BIF, 3.5 - 1.8 Gy ago, because the predominant mineral even in the SH-type BIF is still magnetite (Fe_3O_4).

The distribution of rare earth elements (REE), La ~ Lu, associated with the BIF and other iron-formations has been studied extensively (23). The REE distribution patterns (normalized against that in chondrites) in about thirty samples are given in Figs. 13 and 14. It is to be noted that there are two different patterns. One pattern has no or a small negative

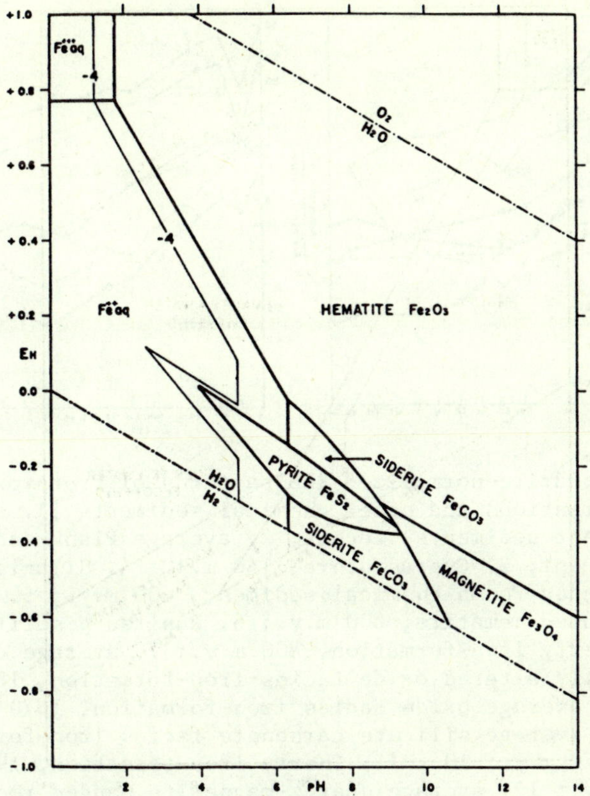

Fig. 12. Stability relations of iron oxides, sulfides and carbonate in water at 25°C and 1 atmosphere total pressure (total dissolved sulfur = 10^{-6}m and total dissolved carbonate - 10°m).

Eu (europium) - deviation; this pattern is found in samples 1 through 16. The other pattern exhibits a distinct positive deviation at Eu. Interestingly, the ages of samples 1 through 16 range from recent to up to about 2.2 Gy ago, whereas samples 17 through 30 are 2.5 - 2.9 Gy old. That is, the samples older than about 2.3 Gy contain abnormally high concentrations of Eu. The interpretation is as follows. Europium, unlike other REE's, can take both Eu(II) and Eu(III). The other REE's can take only M(III) state (exception: Ce can take both Ce(IV) and Ce(III)). Most REE M(III)'s precipitate out as hydroxide about pH 7, whereas Eu(II) is very much more soluble around the neutral pH. Therefore, if the prevailing E_h was very low and thus the significant portion of Eu was present in

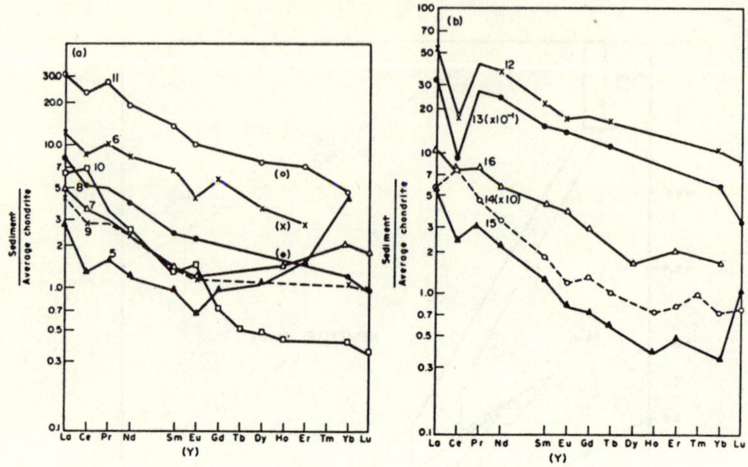

Fig. 13. Chondrite-normalized REE patterns of Proterozoic iron-formations and other chemical sediments [1. average crest sediments, recent; 2. average Plank sediments, recent; 3. Cyprus ochre, 140 m.y.; 4. Silurian banded Fe-Mn chemical sediment, 400 m.y.; 5. Rapitan banded hematite, 800 m.y.; 6. Rapitan pisolitic cherty iron-formation, 800 m.y.; 7. average diagenetically altered oxide facies iron-formation, 1870 m.y.; 8. average oxide facies iron-formation, 1870 m.y.; 9. average silicate-carbonate facies iron-formation, 1870 m.y.; 10., 11. Cherty iron-formation, 1900-2700 m.y.; 12. average quartz-magnetite banded rock, 2075 m.y.; 13. apatite-rich iron-formation, 2075 m.y.; 14. Transvaal chert, 2000 m.y.; 15. Red creek marble, 2300 m.y.; 16. average iron-rich zone, Krivoy Rog, 2000-2200 m.y.] [From B.J. Fryer, Geochim. Geophys. Acta, 41, 361-367 (1977)].

Eu(II) state, the hydrosphere should have contained an abnormally high concentration of Eu (compared to the other REE's). Thus the sediment which deposited from this medium would show a positive Eu-deviation. E_h (Eu(III)/Eu(II)) is -0.43 v. Therefore, it might be concluded that the E_h of the environment before 2.3 Gy ago was lower than -0.43 v.

6.3 Hamersley BIF

This BIF has been studied in detail by Trendall (53). It has been dated at 1.8 to 2.3 Gy old, and is typically about 2,400 m thick. Some 1,000 m consists of iron-formation in five stratigraphic units, separated by shale, dolomite, lava and tuff.

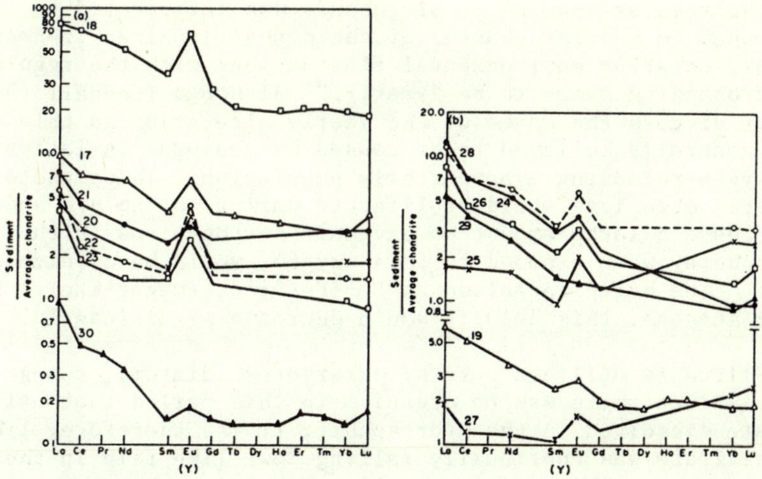

Fig. 14. Chondrite-normalized REE patterns of Archean iron-formations and other chemical sediments [17. quartz-magnetite-amphibole banded rock, 2740 m.y.; 18. Soudan carbonaceous pyritic slate, 2600 m.y.; 19. banded quartz-rich iron-formation, 2700 m.y.; 20. sulfide-carbonate facies iron-formation, 2700-2750 m.y.; 21. carbonate facies iron-formation, 2700-2750 m.y.; 22., 23. Jasper-magnetite BIF, >2490 m.y.; 24.-27. massive magnetite, Mary River, Baffin Island, 2600 m.y.; 28. banded magnetite-hematite iron-formation, Mary River, 2600 m.y.; 29. high grade banded quartz-magnetite iron-formation, Mary River, 2600 m.y.; 30. Bulawayan limestone, 2900 m.y.] [From B.J. Fryer, Geochim. Geophys. Acta, 41, 361-367 (1977)].

The fourth iron-formation from the base, called Weeli Wolli Formation, is about 450 m thick, and was reported in detail. Macrobanding consists of alternations of shale and BIF, typically at a scale of several meters. Within each macroband of iron-formation is a conspicuous alternation of sharply defined iron-rich and iron-poor layers, referred to as mesobands, typically between 0.1 and several centimeters. The main mesoband types are chert, chert-matrix and magnetite. Chert mesobands normally have an internal lamination (microband) and are defined by thin and regularly spaced layers of some iron-bearing mineral within the general framework of fine-grained quartz. Microbands have a usual spacing of between 0.2 and 2.0 mm. The number of microbands per stripes within a mesoband was counted to be 23.3 on the average, ranging from 21 to 28.

The regular spacing of microbands was interpreted to correspond to regular changes of the depositional environment. The only existing environmental fluctuations with the regularity of microbanding seems to be "yearly." Although Trendall (53) did not discuss the cause of the yearly alteration in this paper, it is generally believed to be caused by seasonal variation of the oxygen-releasing cyanobacteria population. These photosynthetic organisms would proliferate during spring and summer, and release a large amount of oxygen. Ferrous ions in the hydrosphere, when exposed to this oxygen, would be oxidized to ferric, thus being deposited as $Fe^{II}Fe_2{}^{III}O_4$ (magnetite). During leaner seasons, this deposit would decrease significantly.

Silica is utilized only by eukaryotes (diatoms, sponges, etc.). Since there was no organism in this period that utilized silicate dissolved in the hydrosphere, it is, therefore, likely that silicate was continually falling down like rain in the ancient ocean. This process would account for the chert part of the BIF.

At any rate, even at this time (1.8 - 2.3 Gy ago), the upper limit of E_h of the environment seems to be that of magnetite (Fe_3O_4). The oxygen required for the formation of magnetite appears to be the same one produced the cyanobacteria. As soon as oxygen was released from cyanobacteria, it was consumed by Fe(II); thus the presence of Fe(II) seems to have kept E_h rather low.

6.4 Isua Iron Rock

It appears fairly conclusive that the Hamersley BIF was associated with oxygen-releasing cyanobacteria. How about the other BIF's? Can we reach a similar conclusion with regard to the old BIF's? How far can we go?

Since magnetite is the major component in the BIF, oxygen must be present, oxidizing Fe(II) to Fe(III). What was the source of oxygen? A likely source of free oxygen, other than a biological one, is the photolysis of water, which forms free hydrogen and oxygen. Hydrogen, being light, can escape into the space, leaving oxygen behind. However, the rate of oxygen production by photolysis of water in the primordial atmosphere has been estimated (54) to be inadequate to explain the extensive oxidation of iron (II). There is no way of knowing how accurate this estimate is. If we suppose that the amount of oxygen from this source was adequate, then can one explain the periodicity? There are a few suggestions for it. One is to assume periodic tectonic activities, as mentioned earlier (12). Another possible mechanism is "Liesegang banding" that developed within a gelatinous colloidal precipitate after deposition (53).

There are other physical, chemical mechanisms possible.

Or was oxygen really necessary? The following reaction was suggested for the formation of carbonaceous material embedded in iron-formation(42).

$$6\ FeCO_3 \rightarrow 2\ Fe_3O_4 + C + 5\ CO_2$$

This seems to be feasible at least for the older type (A-type) of BIF, which often contains graphitic carbon associated with the iron-rich layer. The ΔG for the reaction is positive at the ambient temperature, but can become negative at temperatures higher than about 330°K.

Isua rocks, dated 3.73 Gy old, are a good example, and may fit this kind of mechanism. It is known to have undergone a number of metamorphoses, as high as to an amphibolitic one. Therefore, the original sediment of siderite $FeCO_3$ could have been converted into magnetite and carbon according to the equation above. The entire Isua rock formation is intimately associated with volcanic rocks, as is usual in the cases of Archean sediments. Thus, the original deposit could have been predominantly siderite, pyrite-pyrrhotite and some iron-silicates, and at least part of the magnetite found today may have been derived from siderite. In accord with this volcanogenic hydrothermal mechanism, Isua rocks contain relatively higher concentrations of base metals (their sulfides) than the BIF of a later period.

Interestingly, the REE patterns of some Isua rocks show negative deviation at Eu (3) contrary to the prevailing positive deviation as discussed above (23, 56, 57). This result is not inconsistent with the volcanogenic mechanism of formation, rather than the chemical sedimentation of the later BIF (SH-type).

This whole argument would indicate that biological activities, more specifically those of the oxygen-releasing photosynthetic cyanobacteria, may not necessarily be responsible for the formation of the BIF of the Algoma type (of the Archean age).

7. COPPER: AN INDICATOR FOR OXYGENIC ACTIVITIES (40)

Up until about 1.8 - 1.9 Gy ago, the E_h value of the environment was about -0.2 v or below. Cyanobacteria seem to have appeared around 2.2 - 2.5 Gy ago from the stromatolitic (4) and other evidence including the BIF discussed above (53).

The appearance of cyanobacteria would soon release enough oxygen to make the atmosphere oxygenic. Is there any indication when this happened? Copper seems to be a good indicator for this event (39, 40).

7.1 Biochemical Data

First let us take a look at Table 7, which gives the distribution of copper proteins and enzymes in various organisms.

The conspicuous features found in this listing are (1) most copper enzymes and proteins are found only in eukaryotes, and (2) a few copper proteins, azurin and plastocyanin are found in prokaryotes which are aerobic. The most interesting cases are plastocyanin and superoxide dismutase.

Plastocyanin is a component involved in the electron

Table 7 DISTRIBUTION OF COPPER ENZYMES AND PROTEINS

Protein or Enzyme	Distribution	
	Prokaryotes	Eukaryotes
Stellacyanin		Rhus vernicifera, horse-radish, Mung bean, etc
Plastocyanin	cyanobacteria: Anabaena variabilis, Phormidium ludium (not isolated from Spirulina maxima, Synechococcus 6312)	green algae, red algae, plants (not found in Euglena gracilis)
Azurin	Pseudomonas aeruginosa, Ps. fluorescens, Ps. denitrificans, Bordetella pertussis	
Hemocyanin		invertebrates
Superoxide dismutase	found only in Photobacterium leiognathi	animals (not found in Pleurotus olearius (fungus), Tetrahymea pyriformis, paramecia)
Luciferase		Pholus dactylus
Amine oxidase (pyridoxal dependent)		fungi, plants
D-Galactose oxidase		Dactilium dendroides, Polyporus circinatus
Ascorbate oxidase		plants, etc
Tyrosinase		animals
Luccase		Polyporus versicolor, Rhus vernicifera
Ceruloplasmin		animals

transfer system connecting the photosystem I and II in chloroplast. It is found in probably all plants, most green algae, red algae (these are eukaryotes) and some cyanobacteria (prokaryotes). It is not found in Euglene gracilis (eukaryote) (17). With cyanobacteria, it is found in some filamentous species such as Anabasna variabilis (30) and Phormidium ludium (8), but has not been isolated from others including Spirulina masima and Synechococcus 6312 (1).

There are three distinct superoxide dimutase (SOD); one contains manganese as the catalytic element, the second one, iron, and third, copper. Mn-SOD and Fe-SOD are found in most prokaryotes and mitochondria of the eukaryotic cells including those of human beings, whereas Cu-SOD is found in the cytoplasmic fraction of the eukaryotic cells (22). It has been pointed out (38) that copper, iron and manganese are suitable for the catalytic function for the superoxide dismutase reaction. Organisms, having faced the toxic superoxide, devised SOD. However, the organisms that had developed before perhaps about 1.7 Gy ago had to utilize manganese or iron which was available in that period, rather than the more efficient copper which was unavailable (38). Cu-SOD has been found in a prokaryote, Photobacterium leiognathi; this seems to be the only bacteria which contains a Cu-SOD (34). The major SOD in P. leiognathi is, however, a Fe-SOD. This marine bioluminescent bacterium is symbiotic with pony fish, and thus it could have been transferred to the bacterium by a kind of natural genetic engineering. Therefore, the Cu-SOD may not be original in this prokaryote.

Evidence is mounting to indicate that eukaryotic cells first appeared about 1.4 - 1.5 Gy ago (47, 48). This is based mainly on the microfossil records. Eukaryotic cells contain not only a nucleus but also organelles such as mitochondria and chloroplasts. It has been suggested (32, 33) that both mitochondria and chloroplasts may be evolutionary derivatives of what were once free-living microorganisms. Mitochondria, for example, may have been derived from aerobic bacteria that were engulfed by pro-eukaryotic cells and that later established a symbiotic relationship with them. It seems that copper had become readily available by the time the pro-eukaryotes came into being as discussed above (Section 4.3) and detailed shortly, and that the aerobic bacteria which had developed before that time contained SOD relying either Mn and/or Fe. Therefore, it could be deduced that SOD in the cytoplasm of eukaryotic cells would contain Cu but SOD in mitochondria should be dependent on Mn and/or Fe, if the theory of symbiotic origin (32, 33) is correct. It turned out that this is indeed the case, as discussed earlier. This fact is consistent with the theory (32,33). Everything, thus, seems to fit together nicely, if copper did become available about this time, as postulated. Is there any geological data to support this?

7.2 Geological Data

There is no direct data available to indicate the change in the oxidation state of the geological copper. Figure 10 shows the world copper production (in 1974) from the two major sources, porphyry and stratabound deposits. The age profile indicates that the so-called stratabound deposits became abundant, starting about 1.7 Gy ago. The origin of the stratabound deposits is still debated, but is believed (9) to be of a chemical sediment. The porphyritic copper which is the major copper source of more recent times is of an igneous origin.

The question is then how the stratabound copper formed; the source of copper and the condition of the sedimentary deposition. The major ore minerals in the stratabound sediment are chalcocite Cu_2S, chalcopyrite $CuFeS_2$ and bornite Cu_5FeS_4 and are usually underlain by pyrite FeS_2. In any copper ore, the sulfides are the major one; this is because the sulfides are much less soluble than other copper compounds. Other notable copper minerals such as azurite $2CuCO_3Cu(OH)_2$ are secondary products (by oxidation of copper(I) and sulfide ores). One of the controversial points regarding the stratabound copper is the source of copper and the form of transportation (9). It is not certainly volcanogenic, though the ultimate source might be. Somehow, the copper became quite transportable and chemically sedimentable during 1.7 - 0.3 Gy ago. A prevailing theory is that Cu(I) was transported as $CuCl_4^{3-}$ in a brine medium. This requires a source of a very concentrated brine, which may have existed only under a very special circumstance.

In our frame of concept, the oxygen pressure became high enough ($E_h \simeq + 0.2$ v) to oxidize Cu(I) to Cu(II) about this time of the earth's history. Cu(II) is usually much more soluble than Cu(I) compounds. However, this condition could have been met only in shallow water. In a little deeper level, an anoxic condition still prevailed. Therefore, copper was mobilized by atmospheric oxidation (perhaps with the aid of bacterial activities) and transported as Cu(II) of soluble forms in shallower water, but then was reduced in the anoxic area and deposited as Cu(I) sulfides where sulfide (H_2S) was produced by SO_4^{2-}-reducing bacteria. There seems to be an agreement among investigators as to the latter part of the process, i.e., the sulfide formation (9).

Renfro (1974) argued for a coastal sabkah type of mechanism in which copper is mobilized (he did not specify the oxidation state) in a terrestrial water having low pH and a high E_h value, and then reduced and deposited when the terrestrial water goes through a hydrogen sulfide-charged algal mat. This is essentially the same as the mechanism described here. This condition did not last very long, because the steady and rapid rise of the atmospheric oxygen level would have soon destroyed the appropriate condition for this kind of deposition process. The curve for the stratabound copper in Fig. 10 looks as though it lasts until the present time. This is because the curve was redrawn from a histogram. The fact is that no stratabound copper ore younger than 0.25 Gy is known. Today, the deposition of copper-sulfide is found only in limited locations, such as the bottom of the Black Sea.

This could explain why the stratabound deposit of copper formed only in that limited period of time (1.7 - 0.3 Gy ago), and solve the problem of transportation of copper at the same time.

7.3 Concluding Remarks

As indicated above, oxygen-releasing cyanobacteria first appeared about 2.3 Gy ago. One of the components in their chloroplast is plastocyanin, a copper-containing protein. If the argument developed here including the date of copper-mobilization is correct, the cyanobacteria that had developed before the copper-mobilization (which started about 1.7 Gy ago) should not contain plastocyanin. Indeed this seems to be the case, as discussed above. That is, cyanobacteria of the earlier type, mostly unicellular, do not seem to contain plastocyanin, whereas some of the more developed filamentous cyanobacteria contain it.

The arguments developed here, especially in sections 6 and 7, appear to support the curve of the virtual atmospheric oxygen development as depicted in Fig. 8.

8. "ACTUAL" VS. "VIRTUAL" ATMOSPHERIC OXYGEN PRESSURE

The discussion above was based on "virtual" oxygen pressure, $P(O_2)$, which is a thermodynamic parameter and may not necessarily represent the "actual" prevailing atmospheric oxygen pressure, $P^a(O_2)$.

If a chemical reaction was taking place on a locale exposed to the atmosphere, $P^a(O_2)$ can be assumed to be the same as $P(O_2)$. However, it is more likely that most of the chemical reactions of our interest, including the BIF formation and mobilization of copper, took place in water. $P(O_2)$ represents, then, the oxygen pressure in the immediate vicinity of, e.g., a BIF which is forming in water. The oxygen for this formation of BIF would be supplied from the atmosphere whose $P^a(O_2)$ could be much higher than $P(O_2)$. If the relation between $P(O_2)$ and $P^a(O_2)$ was known, one can easily estimate $P^a(O_2)$ values from $P(O_2)$, and thus can deduce the evolution of $P^a(O_2)$, which is more meaningful than that of $P(O_2)$.

No general relationship can be found between $P^a(O_2)$ and $P(O_2)$. The oxygen content of the ocean depends not only on $P^a(O_2)$, but also many other factors including the depth, the diffusion rate of oxygen, and mixing of water by convection or other means. Some of the latter factors unfortunately cannot be specified for the ancient ocean which is of our interest.

Kasting and Walker (28) attempted to estimate $P^a(O_2)$ based on the rate of photolysis of H_2O in the atmosphere and the volcanic production rate of hydrogen or the consumption of O_2 by iron (Fe^{II}) in sea water. Using the latter assumption (steady consumption of O_2 by iron), they estimated $P^a(O_2)$ to be 4×10^{-8} of the present air level (PAL), that is, about 8×10^{-9} atm. The former hypothesis (capturing oxygen by volcanic hydrogen gas) yielded a $P^a(O_2)$ value of 10^{-13} PAL, i.e., 2×10^{-14} atm. Other estimates of $P^a(O_2)$ range from 10^{-3} - 10^{-4} PAL (7) to 0.27 PAL (11). These values are far larger than the virtual $P(O_2)$ values as estimated in the foregoing sections. In the calculation of $P^a(O_2)$ by Kasting and Walker (28), the sea water was implicitly assumed to be still, that is, no mixing occurred. If there was a significant mixing, a much lower value of $P^a(O_2)$ would suffice. Their value, 2×10^{-14} or 8×10^{-9} atm, therefore, may be regarded to be the upper limit of $P^a(O_2)$. Thus $P^a(O_2)$ was anywhere between 10^{-70} and 10^{-8} atm in the Archean era. It could have been much higher than 10^{-70} atm, but how close it was to the upper limit cannot be estimated. Hematite (Fe_2O_3) has been reported to occur occasionally among BIF's. The equilibrium $P(O_2)$ value for hematite is higher than 10^{-64} atm (at 300°K and pH = 8). This indicates that $P^a(O_2)$ should have been higher than 10^{-64} atm most of the time when the BIF was formed. This would allow for an occasional formation of hematite in a shallower water or an exposed area, whereas only magnetite (and siderite, pyrite as well) was formed in a deeper water.

In the Proterozoic era when copper was mobilized, $P(O_2)$ was about 10^{-30} atm. The corresponding $P^a(O_2)$ could have been as high as 10^{-4} atm, which is about 0.05% of PAL.

The curve of the $P^a(O_2)$ evolution may be similar in its shape to the one shown in Fig. 8, but the scale on the $P(O_2)$ axis must be changed to range from 10^{-8} to 10^0 atm instead of 10^{-70} to 10^{-0} atm, if one assumes the upper limiting case.

REFERENCES

1. Aitken, A.: 1975, Biochem. J. 149, pp. 675-83.

2. Anders, E., Hayatsu, R. and Studier, M.H.: 1973, Science 182, pp. 781-90.

3. Appel, P.W.U.: 1980, Precambrian Res. 11, pp. 73-87.

4. Awramik, S.M., Margulis, L. and Barghoorn, E.S.: 1976, in "Stromatolites" (ed. M.R. Walter), Elsevier, pp. 149-162.

5. Bahadur, K., Ranganayaki, S. and Sanatamaria, L.: 1958, Nature 182, p. 1668.

6. Beck, M.T.: 1978, in "Metal Ions in Biological System" (ed. H. Sigel), Vol. 1, Marcel Dekker. pp. 1-28.

7. Berkner, L.V. and Marshall, L.L.: 1966, J. Atmos. Sci. 23, pp. 133-43.

8. Biggins, J.: 1967, Plant Physiol. 42, pp. 1447-1456.

9. Bowen, R. and Gunatilaka, A.: 1977, "Copper: Its Geology and Economics," Halsted Press.

10. Brack, S., Louembe, D. and Spach, G.: 1974, Origins of Life 6, p. 407.

11. Brinkman, R.T.: 1969, J. Geophys. Res. 74, pp. 5355-68.

12. Button, A., Brock, T.D., Cook. P.J., Eugster, H.P., Goodwin, A.M., James, H.L., Margulis, L., Nealson, K.H., Nriagn, J.O., Trendall, A.F. and Walter, M.R.: 1981, in "Mineral Deposit and Evolution of the Biosphere" (ed. H.D. Holland and M. Schidlowski), Springer-Verlag, pp. 259-73.

13. Calvin, M.: 1969, "Chemical Evolution," Oxford University Press.

14. Cloud, P.: 1973, Econ. Geol. 68, pp. 1135-43.

15. Econ. Geol. 68, No. 7 (1973), pp. 913-1179.

16. Eichler, J.: 1976, in "Handbook of Strata-bound and Stratiform Ore Deposits" (ed. K.H. Wolf), Vol. 7, Elsevier, pp. 157-201.

17. Ellefson, W.L., Ulrich, E.A. and Krogman, D.W.: 1980, "Methods in Enzymology," Vol. 69 (ed. A. San Pietro), Academic Press, p. 223.

18. Fox, S.W. and Harada, K.: 1958, Science 128, p. 1214.

19. Fox, S.W., Harada, K. and Vegotsky, A.: 1959, Experientia 15, pp. 81-4.

20. Fox, S.W. and Harada, K.: 1960, J. Am. Chem. Soc. 82, pp. 3745-51.

21. Fox, S.W. and Yuyama, S.: 1963, Am. N.Y. Acad. Sci. 108, pp. 487-94.

22. Fridovich, I.: 1975, Ann Rev. Biochem. 44, pp. 147-59.

23. Fryer, B.J.: 1977, Geochim. Cosmochim. Acta 41, pp. 361-67.

24. Garrels, R.M. and Christ, C.L.: 1965, "Solutions, Minerals and Equilibrium," Harper and Row.

25. Hayatsu, R.: 1964, Science 146, pp. 1291-93.

26. Henderson-Sellers, A. and Schwartz, A.W.: 1980, Nature 287, pp. 526-28.

27. James, H.L.: 1966, U.S. Geol. Surv. Prof. paper, 440(W), 66 pp.

28. Kasting, J.F. and Walker, J.C.G.: 1981, J. Geophys. Res. 86, pp. 1147-58.

29. Kim, S.H., Quigley, G.J., Suddath, F.L., McPherson, A., Sneden, D., Kim, J.J., Weinzel, J. and Rich, A.: 1973, Science 179, p. 285.

30. Lightbody, J.J. and Krogman, D.W.: 1967, Biochim. Biophys. Acta 131, pp. 508-15.

31. Lundgren, D.G. and Dean, W.: 1979, in "Biogeochemical Cycling of Mineral Forming Elements" (ed. P.A. Trudinger and D.J. Swaine), Elsevier, pp. 211-51.

32. Margulis, L: 1970, "Origins of Eukaryotic Cells," Yale University Press.

33. Margulis, L.: 1981, "Symbiosis in Cell Evolution," W.H. Freeman and Co.

34. Michelson, A.M.: 1978, in "Frontiers in Physicochemical Biology" (ed. B. Pullman), Academic Press, pp. 309-55.

35. Nagy, L.: 1978, Abst. G.S.A. Ann. Meeting, Toronto, p. 461.

36. Nagy, L.: 1980, 26 Int. Geol. Congr., Paris, Abst. Vol., p. 179.

37. Ochiai, E.-I.: 1977, "Bioinorganic Chemistry: An Introduction," Allyn and Bacon.

38. Ochiai, E.-I.: 1978a, Biosystems 10, pp. 329-37; J. Chem. Ed. 55, pp. 631-33.

39. Ochiai, E.-I.: 1978b, Origins of Life 9, pp. 81-91.

40. Ochiai, E.-I.: 1981, Biosystems, submitted.

41. Oro, J., Kimball, A., Fritz, R., and Master, F.: 1959, Arch. Biochem. Biophys. 85, pp. 115-30.

42. Perry, E.C. and Ahmad, S.N.: 1977, Earth Planet. Sci. Letters 36, pp. 280-84.

43. Quigley, G.J., Teeter, M.M. and Rich, A.: 1978, Proc. Nat. Acad. Sci. U.S.A. 75, pp. 64-68.

44. Renfro, A.R.: 1974, Econ. Geol. 69, pp. 33-45.

45. Sawai, H. and Orgel, L.E.: 1975, J. Am. Chem. Soc. 97, pp. 3532-33.

46. Sawai, H.: 1981, J. Mol. Evol. 17, pp. 48-51.

47. Schopf, J.W.: 1975, Ann. Rev. Earth and Planet. Sci. 3, pp. 212-49.

48. Schopf, J.W.: 1978, Sci. Am. 239(9), pp. 111-38.

49. Schrauzer, G.N. and Guth, T.P.: 1977, J. Am. Chem. Soc. 99, pp. 7189-93.

50. Schulz, G.E.: 1981, Angew. Chem. Internat. Ed. Engl. 20, pp. 143-51.

51. Schwartz, R.M. and Dayhoff, M.O.: 1978, Science 199, pp. 395-403.

52. Steinman, G. and Cole, M.N.: 1967, Proc. Nat. Acad. Sci. U.S.A. 58, pp. 735-42.

53. Trendall, A.F.: 1973, Econ. Geol. 68, pp. 1089-97.

54. Walker, J.C.G.: 1978, Pageoph. 116, pp. 222-31.

55. Walter, M.R.: 1967, ed., "Stromatolites," Elsevier.

56. Wildeman, T.R. and Condie, K.C.: 1973, Geochim. Cosmochim. Acta 37, pp. 439-53.

57. Wildeman, T.R. and Haskin, L.A.: 1973, Geochim. Cosmochim. Acta 37, 419-38.

58. Yamada, S., Wagatsuma, M., Takeuchi, Y. and Takeshima, S.: 1971, Bull. Chem. Pharm. 19, p. 238.

59. Yang, C.L. and Oro, J.: 1971, in "Chemical Evolution and the Origin of Life" (ed. R. Buvet and C. Ponnamperuma), North Holland, p. 155.

BIOLOGICALLY MEDIATED ISOTOPE FRACTIONATIONS: BIOCHEMISTRY, GEOCHEMICAL SIGNIFICANCE AND PRESERVATION IN THE EARTH'S OLDEST SEDIMENTS

Manfred Schidlowski

Max-Planck-Institut f. Chemie, Saarstr. 23,
D-6500 Mainz, W. Germany

Abstract

Preferential metabolization of isotopically light carbon and sulfur has resulted in a fractionation of the stable isotopes of these elements on a global scale, with the light species (^{12}C, ^{32}S) markedly concentrated in biogenic materials. Since the biological effects are basically retained when carbon and sulfur are incorporated in sediments, the respective fractionations are propagated into the rock section of the geochemical cycle, this having consequently caused a characteristic bipartition of both elements between "light" and "heavy" crustal reservoirs. Preservation of the biological isotope effects in sedimentary rocks permits us to trace the underlying biochemical processes back over most of the geological record. According to the available evidence, biological (autotrophic) carbon fixation arose prior to 3.5 (if not 3.8) billion years ago, while the emergence of dissimilatory sulfate reduction antedates the appearance of the oldest presumably bacteriogenic sulfur isotope patterns in rocks between 2.7 and 2.8 billion years old. Hence, biological control of the terrestrial carbon and sulfur cycles has been established very early in the Earth's history.

INTRODUCTION

Since the pioneering studies by Nier and Gulbransen (60), Murphey and Nief (58) and Thode et al. (97, 98), it has become firmly established that incorporation of carbon and sulfur into living systems, or their biological processing in the broadest sense, entails sizable isotope effects. These result from both thermodynamic and kinetic fractionations which are imposed on

C. Ponnamperuma (ed.), Cosmochemistry and the Origin of Life, 277–322.

the principal assimilatory and metabolic pathways. As most biochemical processes are enzymatically controlled, and since living systems represent stationary states undergoing rapid cycles of anabolism and catabolism, it is generally believed that the differentiations observed are mostly due to kinetic rather than to equilibrium effects. On the other hand, claims have been voiced that the inter- and intramolecular distribution of carbon isotopes in biological systems often displays regularities resembling those of thermodynamically ordered states (30).

Of the six elements that usually account for >99% of the elemental composition of organic matter (C, O, H, S, N, P), all but phosphorus are mixtures of isotopes and, accordingly, subject to biologically mediated isotope fractionations. In the case of carbon and sulfur dealt with in this paper, the mean terrestrial abundances of the stable isotopes come close to the following percentages:

^{12}C = 98.894%	^{32}S = 94.941%
^{13}C = 1.106%	^{33}S = 0.769%
	^{34}S = 4.273%
	^{36}S = 0.017%

These abundances and the resulting ratios $^{12}C/^{13}C \approx 89.42$ and $^{32}S/^{34}S \approx 22.22$ are those of geochemically undifferentiated or "primitive" carbon and sulfur from the Earth's upper mantle believed to be inherited from late-stage chondritic condensates of the parent solar nebula. The percentages for ^{12}C and ^{13}C listed above reflect the average isotopic composition of the principal forms of deep-seated carbon (carbonatites, diamonds), while the respective figures for sulfur are those of troilite (FeS) from meteorites (1, 51) thought to be representative of primordial mantle sulfur.

Ever since the appearance of life on Earth, both primordial carbon and sulfur have been continuously processed by the terrestrial biota. This has given rise to a marked enrichment of the light isotopes (^{12}C, ^{32}S) in biosynthesized materials and a corresponding retention of the heavy species in the residual inorganic reservoirs. As the biological effects were preserved in sediments with but minor alteration, they were propagated into the rock section of the respective element cycles. As a result, biological processes have left their isotopic signature on the sedimentary inventories of carbon and sulfur which can be traced back to almost the beginning of the geologic record. The geochemical cycles of carbon and sulfur are, therefore, outstanding examples of element cycles governed by the terrestrial biosphere. Biological processing through the ages of these two elements has ultimately resulted in their large-scale isotopic disproportionation, with consequent formation of crustal twin reservoirs comprising an isotopically light biogenic and an isotopically heavy inorganic compartment in each case (Fig. 1).

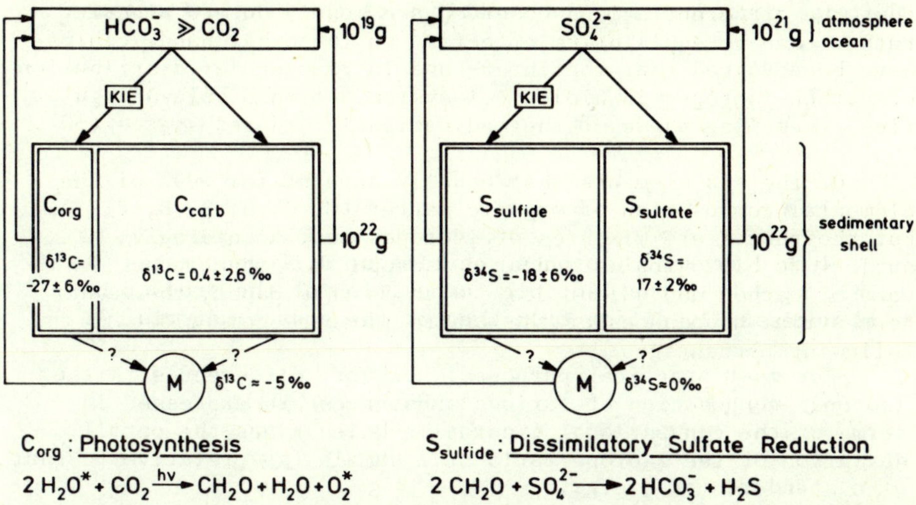

Figure 1. Synopsis of the global carbon and sulfur cycles showing partitioning of mantle(M)-derived carbon and sulfur between biogenic and non-biogenic crustal reservoirs. The kinetic isotope effects (KIE) inherent in the processes of photosynthetic carbon fixation and bacterial (dissimilatory) sulfate reduction have been propagated from the exogenic exchange reservoir (atmosphere, ocean) into the crust. This is ultimately responsible for the isotopic disproportionation of both elements into "light" reduced and "heavy" oxidized phases. While the biogenic species (organic carbon, bacteriogenic sulfide) preferentially concentrate the light isotopes, the oxidized phases (carbonate, sulfate) are enriched in the heavy ones.

EXPRESSION OF QUANTITATIVE CHANGES IN ISOTOPIC COMPOSITION

Isotopic fractionation between two substances A and B may be conveniently expressed by a fractionation factor, K, e.g.,

$$K = \frac{(^{12}C/^{13}C)_B}{(^{12}C/^{13}C)_A}$$

In a thermodynamically controlled equilibrium reaction, K represents the (temperature-dependent) equilibrium constant while, in a uni-directional reaction, it reflects the magnitude of the kinetic isotope effect that equals the ratio k_1/k_2 of the rate constants for the reactions of the single isotope species,

$$^{12}C_A \xrightarrow{k_1} {}^{12}C_B$$

$$^{13}C_A \xrightarrow{k_2} {}^{13}C_B.$$

With different isotopes reacting at different rates (i.e., $k_1 \neq k_2$), the isotopic composition of the product B will necessarily differ from that of the source material A.

For most practical purposes, however, differences in the isotopic composition of various substances are expressed in terms of the conventional δ-notation which gives the permil deviation of the isotope ratio of a sample (sa) relative to that of a standard (st), e.g.,

$$\delta^{13}C_{sa} = \left[\frac{(^{13}C/^{12}C)_{sa}}{(^{13}C/^{12}C)_{st}} - 1\right] \times 10^3 \qquad [‰, \text{PDB}]$$

and

$$\delta^{34}S_{sa} = \left[\frac{(^{34}S/^{32}S)_{sa}}{(^{34}S/^{32}S)_{st}} - 1\right] \times 10^3 \qquad [‰, \text{CDT}].$$

Positive values of $\delta^{13}C$ or $\delta^{34}S$ would, accordingly, indicate an enrichment of the sample in heavy carbon or sulfur relative to the standard, while negative values stand for a depletion in ^{13}C and ^{34}S, respectively.

The rationale for preferring the δ-notation to other means of expressing isotope changes lies in the fact that the precision of the measurement of an isotope ratio is greater than the precision obtained in a determination of the absolute amounts of individual isotopes in a sample. Hence, it seems convenient to relate measured isotope ratios to the ratio of a standard on which a high-precision determination of the constituent isotopes has been carried out. For carbon, this standard normally is Peedee belemnite (PDB) with $^{12}C/^{13}C = 88.99$; for sulfur, it is

Canyon Diablo troilite (CDT) with $^{32}S/^{34}S = 22.22$. The $\delta^{13}C$ and $\delta^{34}S$ values of these standards are 0.0‰ by definition.

CARBON

Not surprisingly, carbon as the very key element of life has served as the first object for studying biological isotope fractionations (20, 58, 60). Subsequent work has confirmed that the most important fractionations in the terrestrial carbon cycle are linked to processes of photoautotrophic carbon fixation. These fractionations mostly derive from the bias in favor of light carbon (^{12}C) exercised by the ribulose-1,5-bisphosphate (RuBP) carboxylase reaction of the Calvin cycle which brings about the bulk of the carbon transfer from the inorganic to the organic reservoir.

Isotope Fractionations in Biological Carbon Fixation

Biological carbon fixation is primarily the assimilation of carbon dioxide (CO_2) and bicarbonate ion (HCO_3^-) by autotrophic microorganisms and plants. By a limited number of biochemical pathways, all cellular components of primary producers are synthesized from these simple inorganic precursor molecules (see Table 1). Assimilation of carbon monoxide (CO) seems to be negligible, while the rate of methane (CH_4) utilization in methanotrophic pathways probably does not exceed a few percent of the total rate of biological carbon fixation which amounts to some 10^{16} g organic carbon per year (114). Hence, biological carbon fixation is largely synonymous with assimilation of carbon dioxide since the bulk of primary production in the present carbon cycle is clearly due to either chemically or solar-powered conversion of CO_2 to organic matter.

All pathways of autotrophic carbon assimilation hitherto investigated entail sizable fractionations of the stable carbon isotopes. These fractionations are primarily due to kinetic isotope effects inherent in the principal carbon-fixing enzymatic reactions all of which favor, though to varying degree, the preferential uptake and metabolization of the light isotope (^{12}C). The difference between the isotopic composition of cells and substrate,

$$\Delta\delta_c = \delta^{13}C_{cells} - \delta^{13}C_{substrate}$$

is a convenient measure of the magnitude of ^{13}C-discrimination exercised by a specific organism.

Data on carbon isotope fractionation are available for

(1) CO_2 + ribulose-1,5-bisphosphate → phosphoglycerate

Operated by: Green plants [*C3 plants rely on (1) exclusively], *algae, *cyanobacteria, *purple photosynthetic bacteria, chemoautotrophic bacteria

(2) CO_2/HCO_3^- + phosphoenolpyruvate/pyruvate → oxaloacetate

Operated by: Green plants [*C4 and *CAM species combine this reaction with (1)], anaerobic and facultatively aerobic bacteria

(3) CO_2 + CO_2 → acetyl coenzyme A/acetate

Operated by: *Green photosynthetic bacteria [primary CO_2 fixation via succinyl CoA and α-ketoglutarate]; anaerobic bacteria (Acetobacterium woodii, Clostridium acidiurici), *methanogens (?) [primary CO_2 fixation via C1 acceptors]

(4) CO_2 + acetyl coenzyme A → pyruvate/phosphoenolpyruvate

Operated by: *Green photosynthetic bacteria [combined with (3) and (2)], Clostridium kluyveri, autotrophic forms of sulfate reducing bacteria, *methanogenic bacteria

Table I. Synopsis of geochemically relevant autotrophic pathways of biological carbon fixation 10, 70, 95, 108). Reduction of CO_2 primarily gives rise to C2 (acetate, acetyl CoA), C3 (phosphoglycerate, pyruvate) and C4 (oxaloacetate) compounds. The quantitatively most important pathway is the RuBP carboxylase reaction of the Calvin cycle ("C3 photosynthesis," No. 1). The details of the assimilatory pathway of methanogens are not known, but the presence of both α-ketogluterate and pyruvate synthases suggests the involvement of reactions (3) and (4). Groups of organisms for which data on carbon isotope fractionations are available are marked by an asterisk.

several major groups of organisms that utilize the principal CO_2 and bicarbonate-fixing pathways responsible for the bulk of the carbon transfer from the inorganic to the organic reservoir. (Table 1). The quantitatively most important of these pathways is linked to the operation of the reductive pentose phosphate (or Calvin-Benson) cycle which constitutes the main contrivance for effecting a large-scale reduction of inorganic (oxidized) carbon to the carbohydrate level.

The biochemical background of carbon isotope fractionation by plants and autotrophic microorganisms has been recently summarized by O'Leary (61). As a result of numerous detailed studies there is widespread consensus that this process is largely governed by enzymatically controlled (and other) kinetic fractionations rather than by equilibrium effects [for dissent see (30)]. The salient isotope-discriminating steps are associated with

(i) the diffusion of CO_2 into, and out of, the photosynthetically active plant tissue;

(ii) the attainment of isotopic equilibrium between CO_2 and HCO_3^- [the latter serving as "active" carbon species in selected pathways (C4, CAM), see below];

(iii) the first CO_2-fixing carboxylation reaction by which inorganic carbon is incorporated into the carboxyl group of an organic acid; and

(iv) decarboxylation and related dissimilatory processes.

In this scheme, step (ii) represents a noteworthy instance where equilibrium chemistry may come into play in the assimilatory pathway.

Fig. 2 gives a graphic synopsis of these reactions which differ in both rate-constants and magnitudes of fractionation. Total fractionation over the pathway will, therefore, depend on an interplay of these processes. Any of the component reactions that becomes dominant or rate-controlling in the specific instance is bound to impose its intrinsic isotope discrimination on the pathway as a whole.

According to O'Leary (61), we may assume that the kinetic fractionation effect inherent in the initial diffusion step (Fig. 2, k_1 and k_2) amounts to about -4‰ (-4.4‰ is the value for CO_2 diffusion in air which figure is, within broad limits, independent of temperature, pressure and CO_2 concentration). Although this value may be modulated by additional effects caused by dissolution, liquid diffusion and hydration of CO_2 (notably in the case of aquatic plants), there is widespread agreement that the overall fractionation linked to the diffusion of external CO_2 into the plant tissue is relatively small.

In contrast, fractionations in the first enzymatic carboxylation reaction (Fig. 2, k_3) are considerably larger though highly variable in detail. Quantitatively, the most important

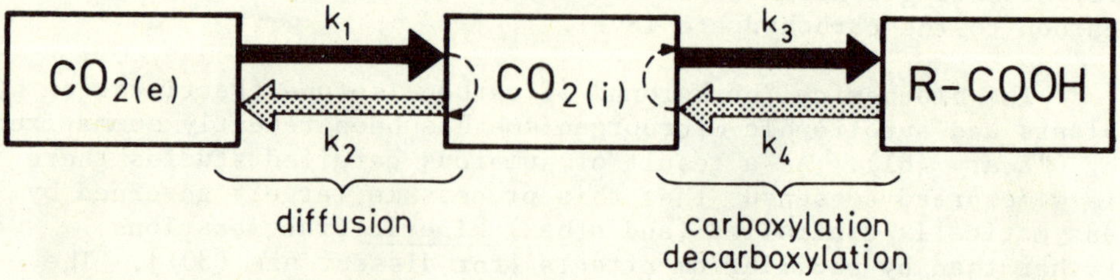

Figure 2. Summary of the principal isotope discriminating steps responsible for the gross isotopic composition of plants and autotrophic microorganisms (black: assimilatory reactions; stippled: dissimilatory and other reverse reactions). $CO_{2(e)}$ and $CO_{2(i)}$ stand for external and internal CO_2, respectively, and R-COOH represents the product of the initial CO_2-fixing carboxylation; k_{1-4} are the rate constants assigned to the individual reactions. The overall result of this reaction sequence is a sizable enrichment of light carbon (^{12}C) in organic matter as compared to the parent substrate (inorganic carbon dioxide).

carboxylation reaction is that catalyzed by the enzyme ribulose-1,5-bisphosphate (RuBP) carboxylase by which CO_2 enters the reductive pentose phosphate ("Calvin") cycle, yielding a C3 compound (phosphoglycerate, PGA) as initial product of carbon fixation (Table 1, No. 1). In fact, autotrophy has often been biochemically defined by the presence of this key enzyme of the Calvin cycle (apart from algae and the bulk of photosynthetic and chemosynthetic bacteria, this cycle is operated by all green plants; those relying on it entirely are termed C3 plants). Values reported for ^{13}C discrimination in the RuBP carboxylase reaction range from -17 ‰ (64) to about -42 ‰ (27), a large number of determinations having yielded fractionations within the range -25 to -35 ‰ (61, 111). The wide range observed seems to be due to the fact that fractionations in enzymatic reactions are apt to vary with pH, temperature, metal ion availability, and other variables (61).

On the other hand, carboxylation by phosphoenolpyruvate (PEP) carboxylase yielding a C4 dicarboxylic acid (oxaloacetate, OAA) as first product of carbon fixation (Table 1, No. 2), results in a fractionation of -2 to -3 ‰ only (73) relative to bicarbonate ion which figures as the "active" carbon species in this reaction (19). This reaction is the crucial step in another carbon fixing pathway that is a relatively late achievement in angiosperm evolution based on the so-called Kranz-type leaf anatomy (Fig. 3). Here, the reactions of the assimilatory sequence proceed in

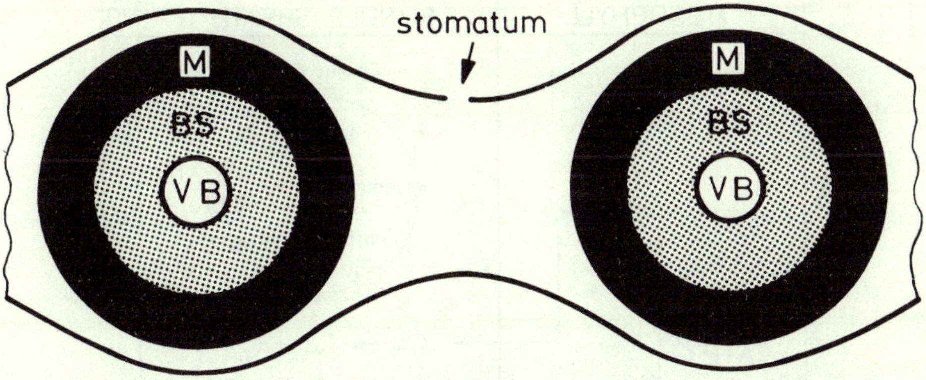

Figure 3. "Kranz" anatomy of C4 plants as exhibited in the cross-section of a leaf. Carbon dioxide enters via stomatal apertures situated in the depressions between the ridges or "veins." These indicate the position of the "Kranz" structures centered by the vascular bundle (VB) and surrounded by two rings of photosynthetic tissue: the outer mesophyll (M) and the inner bundle sheath (BS). The assimilatory pathways of CO_2 in these compartments is depicted in Figure 4.

different compartments of the leaf structure (Fig. 4). A basic advantage of such compartmentation of the assimilatory pathway is an increase in CO_2 pressure in the inner compartment (bundle sheath) which facilitates the operation of the Calvin cycle working most effectively "behind closed doors." After the first product of carbon fixation, this pathway has been termed C4 dicarboxylic acid pathway; higher plants operating this sequence are called C4 plants (cf. 6, 108).

The small gross fractionation typical of C4 plants (Fig. 6) has been usually ascribed to the poor isotope selection properties of PEP carboxylase. However, since this enzyme works on bicarbonate instead of CO_2, the equilibrium fractionation between CO_2 and HCO_3^- of 7 - 8 ‰ (57) is necessarily superimposed on this pathway. Assuming diffusion were fast and the CO_2-HCO_3^- interchange kept close to equilibrium, the $\delta^{13}C$ value of external CO_2 (-7 ‰) would change to about +1 ‰ in the internal bicarbonate substrate whose subsequent carboxylation would yield -2 ‰ for the resulting assimilate (OAA). The fact that the $\delta^{13}C$ values of C4 plants do not come close to this value, but rather approach averages between -12 and -14 ‰ (cf. Fig. 6) indicates that the basic control of isotope fractionation in C4 photosynthesis is exercised by the initial diffusion step (Fig. 2, k_1).

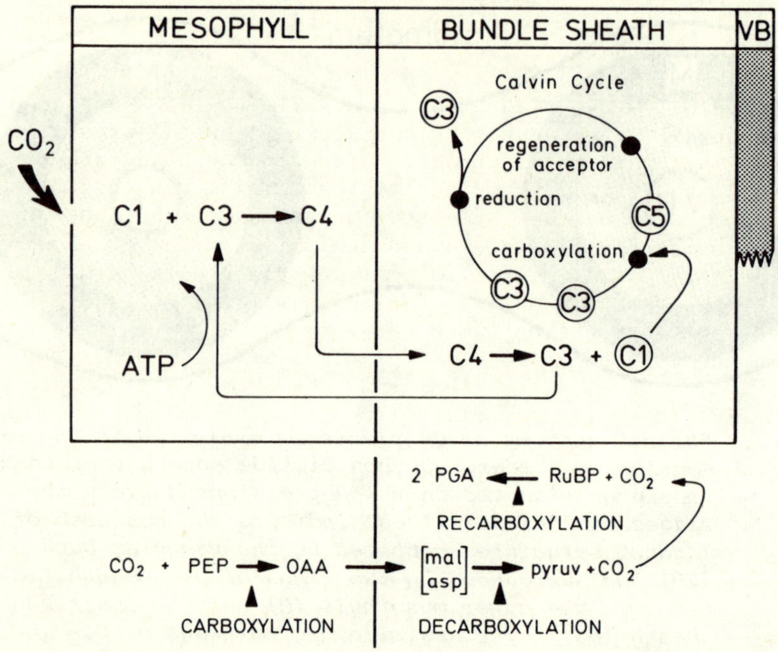

Figure 4. Compartmented biochemistry (VB = vascular bundle) of the C4 dicarboxylic acid pathway (adapted from 108). The sequence of actual assimilatory reactions shown below (comprising two carboxylation steps separated by an interceeding decarboxylation) is retraced within the compartments as reactions of the carbon skeletons of the respective compounds. After entering the mesophyll, external CO_2 (a C1 compound) is carboxylated by phosphoenolpyruvate (a C3 compound) to give a C3 compound (pyruvate) and CO_2. While later is subsequently converted into related C4 compounds (malate, aspartate) and transferred to the bundle sheath. Here, these C4 compounds are decarboxylated to give a C3 compound (pyruvate) and CO_2. While pyruvate is returned to the mesophyll (restoring the original CO_2-acceptor phosphoenolpyruvate), the CO_2 released is immediately recarboxylated by RuBP (a C5 compound) in the Calvin cycle operating under closed system conditions in the bundle sheath chloroplasts. This recarboxylation yields two molecules of phosphoglycerate (C3) for each CO_2 fixed (C1 + C5 → 2 C3). Abbreviations: PEP = phosphoenolpyruvate, PGA = phosphoglycerate, OAA = oxaloacetate, mal = malate, asp = aspartate, pyruv = pyruvate, RuBP = ribulose-1,5-bisphosphate.

The fractionation of -4 ‰ linked to this step would yield a value of -11 ‰ which closely approximates the average actually observed in C4 plants.

Accordingly, the minor isotope discrimination associated with the uptake of external CO_2 turns out to be the crucial contribution toward total fractionation in C4 photosynthesis while this step (Fig 2, k_1) is almost negligible in the C3 pathway dominated by the large fractionations of the RuBP carboxylase reaction. Since subsequent biochemical processing of the primary C4 assimilate (OAA) takes place within the closed system of the "Kranz" anatomy, the sizable fractionation of the second carboxylation step (when decarboxylated CO_2 is recarboxylated in the Calvin cycle, cf. Fig. 4) is largely suppressed.

The intrinsic fractionations of the assimilatory reactions (3) and (4) listed in Table 1 are not well known. However, isotope data are available for groups of organisms that have these reactions integrated in either a cyclic or sequential assimilatory pathway. In green photosynthetic bacteria (Chlorobiaceae), reaction (3) proceeds via the reductive carboxylic acid cycle and is subsequently followed by reaction (4) (Fig. 5). Preliminary studies conducted on _Chlorobium_ sp. suggest that this pathway discriminates against ^{13}C less effectively than the Calvin cycle, namely, between -12 and -14 ‰ (2, 90) or even less (-2.5 to -5.2 ‰; cf. 69). Therefore, the isotope selection properties of neither of these carbon fixing reactions is likely to exceed the maximum fractionation observed in the sequence as a whole, with -14 ‰ as a probable upper limit. Recent investigations have shown that reactions (3) and (4) also form part of a non-cyclic (sequential) assimilatory pathway believed to be utilized by methanogenic bacteria (95), but their specific fractionations appear to be completely masked by other steps of the assimilatory sequence.

Apart from fractionations accompanying carbon assimilation, enzymatic decarboxylations and related dissimilatory processes (Fig. 2, k_4) may also contribute to the overall isotopic composition of autotrophs. Although there are differing reports regarding the magnitude and even the direction of the isotope effect in such processes, most biological decarboxylations apparently release CO_2 which is isotopically lighter than that of the organic material (61), thus running counter to fractionations in the assimilatory pathway. In fact, discrepancies between fractionations predicted from the isotopic selection properties of specific enzymes and those actually observed may be ascribed to a possible contribution by respiratory processes. While, for instance, fractionations caused by RuBP carboxylase mostly average between -25 and -35 ‰, gross fractionations displayed by C3 plants lie close to a mean of -20 ‰ [i.e., at -27 ‰ when

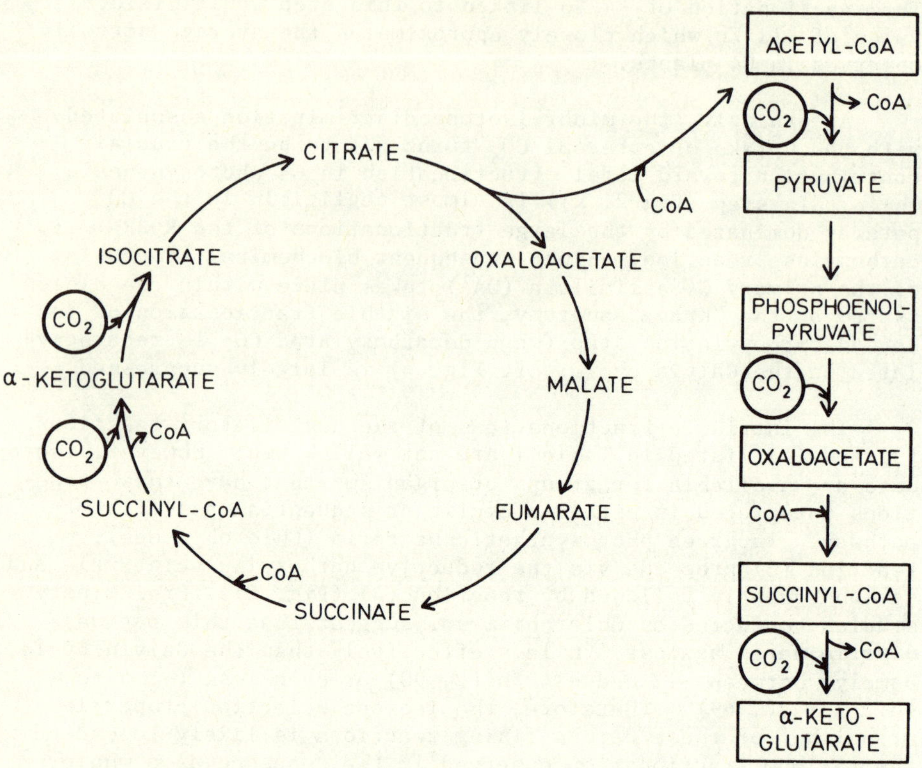

Figure 5. Synopsis of carbon fixing reactions in reductive carboxylic acid cycle, a pathway unique to photosynthetic prokaryotes. In one turn, this cycle produces one molecule of acetyl coenzyme A from two molecules of CO_2 (primary CO_2 acceptors are syccinyl CoA and α-ketoglutarate). Superimposed on the cycle proper are a number of subsequent CO_2 fixing steps (right column) required for the synthesis of different metabolites (amino acids, carbohydrates). Adapted from (10).

atmospheric carbon dioxide ($\delta^{13}C$ = -7 ‰) serves as inorganic substrate, cf. Fig. 6]. It is known that photorespiration is particularly pronounced in C3 species due to the "oxygenase" activity of RuBP carboxylase, this also placing constraints on the productivity of C3 plants (108). Hence, respiratory processes are likely to make substantial contributions to the overall isotope economy of plants, but the small number of detailed investigations hitherto reported leaves important details of the issue open (61).

Isotopic Composition of Plants and Autotrophic Microorganisms.

Depending on which of the various isotope discriminating steps summarized in Fig. 2 and/or which of the carbon fixing enzymatic reactions listed in Table 1 become dominant or rate-controlling in the specific instance, the isotopic composition of autotrophic organisms may vary over a considerable range. Fig. 6 gives a summzry of these variations, listing observed $\delta^{13}C$ ranges for some major groups of extant higher plants, algae, and autotrophic prokaryotes as compared to the terrestrial feeder pools of atmospheric CO_2 and marine HCO_3^-. Since the isotopic composition of atmospheric CO_2 ($\delta^{13}C$ = -7 ‰) constitutes the base-line for fractionations in the natural environment, data obtained from culture experiments relying on different carbon sources have been normalized to atmospheric CO_2.

<u>C3 Plants</u>. Higher plants operating the C3 pathway range from about -23 to -34 ‰ with an average close to -27 ‰ (5, 91). The most important single contribution toward total fractionation in this group is provided by the RuBP carboxylase reaction of the Calvin cycle. Since discrimination against ^{13}C in this reaction is usually larger (-20 to -40 ‰) than the average observed for the group as a whole (-20 ‰ vs. atmospheric CO_2, see Fig. 6), we must conclude that either diffusion (Fig. 2, k_1) becomes partially rate-limiting, or that fractionations in the carboxylation step are largely offset by a reverse effect inherent in photorespiration which is particularly pronounced in C3 plants.

<u>C4 Plants</u>. $\delta^{13}C$ values for C4 plants range from -6 to about -23 ‰ with an average between -12 and -14 ‰ (5, 91). This average suggests that carbon assimilation in the C4 pathway must be mainly diffusion-controlled because the fractionation of approximately -4 ‰ linked with this step would yield values closely approaching the mean obtained for the group as a whole. However, the wider range of values observed indicates that, in the specific instance, neither diffusion nor carboxylation is entirely limiting. It is also safe to say that the sizable fractionation inherent in the second carboxylation step of this pathway (by RuBP carboxylase) is largely suppressed as a result of the

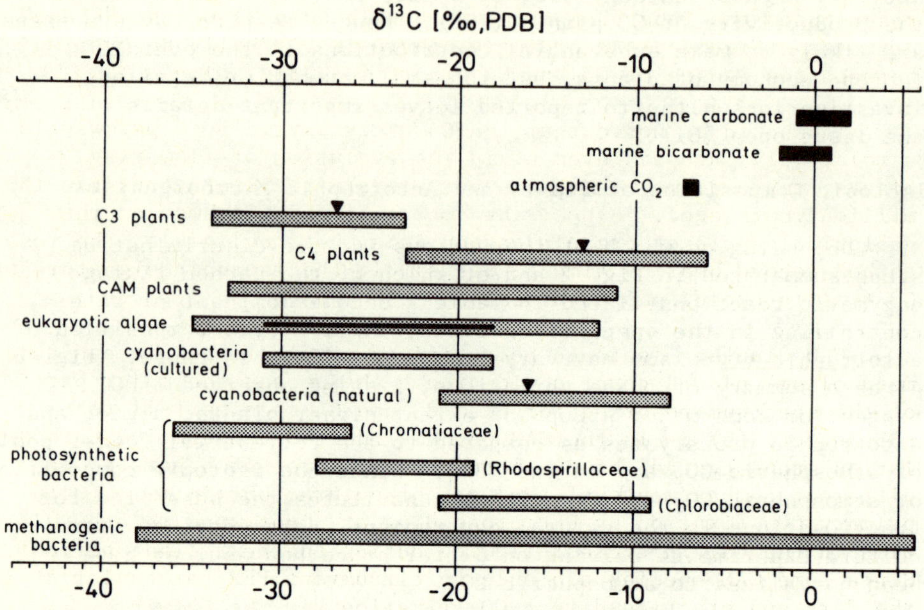

Figure 6. *Isotopic composition of higher plants, algae, photosynthetic bacteria and chemosynthetic archaebacteria (methanogens) as compared to the inorganic carbon pools of atmospheric carbon dioxide (CO_2) and marine bicarbonate (HCO_3^-). Fractionations obtained in culture experiments have been normalized to atmospheric CO_2 ($\delta^{13}C = -7$ ‰) as carbon source; estimated means for some groups are indicated by black triangles. It is obvious that the average biomass is enriched in ^{12}C by about 20 - 30 ‰ relative to oceanic HCO_3^-, the most abundant carbon species in the environment (the reservoir of atmospheric CO_2 is smaller by almost two orders of magnitude). Note that there is virtually no overlap between C3 and C4 plants; black line inserted in the bar for eukaryotic algae indicates the principal isotope spread of marine plankton. The higher fractionations for cultured cyanobacteria (formerly "blue-green algae") as compared to natural communities were obtained at partial pressures of CO_2 > 0.5 % (63). The positive part of the spectrum for methanogens has resulted under conditions of extremely low P_{CO_2}. Here, preferential conversion of CO_2 into CH_4 (with a strong bias in favor of ^{12}C) is believed to give rise to heavy residual CO_2 that is subsequently incorporated into the cell substance (29).*

"closed system" conditions imposed by the confinement of this reaction to the inner compartment (bundle sheath) of the "Kranz" structure (Fig. 4). Accordingly, there is virtually no overlap between the ranges of C3 and C4 plants (cf. Fig. 6).

It should be noted that the C4 dicarboxylic acid pathway is a relatively late achievement in the evolution of flowering plants (angiosperms) that does not predate the Cretaceous (70-130 million years ago). Hence, the impact of this pathway on the geological cycle of organic carbon is as yet negligible, the fossil carbon record bearing almost exclusively the isotopic signature of C3 (Calvin cycle) photosynthesis.

CAM Plants. Isotopic compositions of CAM (crassulacean acid metabolism) plants cover a wide range that spans the major part of the spectrum reported for both C3 and C4 species [-11 to -33 ‰, cf. (5)]. In the CAM pathway the two consecutive carboxylation steps typical of C4 photosynthesis are not separated in different compartments of the leaf structure, but in time (with primary CO_2 fixation by PEP proceeding at night, and recarboxylation in the Calvin cycle following during the next light period). Since some CAM species are able to circumvent PEP carboxylation in the dark and practice direct CO_2 fixation via the Calvin cycle during the day, the sizable fractionation of the RuBP carboxylase reaction is sometimes expressed in CAM plants while it is largely suppressed in diffusion-limited C4 photosynthesis. The considerable spread in $\delta^{13}C$ values found in CAM species appears primarily due to the capability of these plants to preferentially entertain one carboxylation option at the expense of the other (cf. 61).

Eukaryotic Algae. $\delta^{13}C$ values of marine and freshwater algae (exclusive of blue-green algae, see below) are reported to range from -12 to -35 ‰ (89, 91, 112). Eukaryotic algae also form the bulk of marine phytoplankton which, either directly or via consumption, will ultimately impart their isotopic composition to the total plankton population. Accordingly, the principal isotopic range of marine plankton [-18 to -31 ‰, (77)] is likely to hold, with fair approximation, for the bulk of marine algae. Since the positive part of the spectrum shown in Fig. 6 ($\delta^{13}C > -20$ ‰) is not consistent with fractionations commonly obtained in Calvin cycle photosynthesis, either partial CO_2 fixation by the PEP carboxylase reaction (112) or diffusion-limited CO_2 supply in the aquatic environment (61) has been invoked for the suppression of the expected magnitudes of C3 fractionation.

Cyanobacteria. Naturally occurring mat communities of prokaryotic cyanobacteria (formerly termed blue-green algae or "cyanophytes") have yielded $\delta^{13}C$ values between -8 and -21 ‰, with an average of -16 ‰ (2, 4, 13, 91). Fractionations reported for cultured taxa were invariably in excess of those observed in the natural environment (63), with maximum values approaching -31 ‰ (normalized to atmospheric CO_2 with $\delta^{13}C$ = -7 ‰ as carbon source).

Photosynthetic Bacteria. Photosynthetic sulfur bacteria that presumably operate the Calvin cycle have furnished fractionations consistent with isotope discriminations in the RuBP carboxylase reaction. $\delta^{13}C$ values of autotrophically grown cultures of *Chromatium* sp. were found to range between -30 and -36 ‰ (113), while single values reported by Sirevåg et al. (90) and Quandt et al. (69) for this purple sulfur bacterium were -29.5 and -26.6 ‰, respectively (normalized to atmospheric CO_2). Corresponding figures obtained for purple nonsulfur bacteria (*Rhodospirillum rubrum*) were -27.5 ‰ (90) and -19.4 ‰ (69). On the other hand, green photosynthetic bacteria (Chlorobiaceae) utilizing the reductive carboxylic acid cycle have yielded consistently lower fractionations (cf. Fig. 6). Sirevåg et al. (90) and Barghoorn et al. (2) have reported values of -19.2 and -20.8 ‰ for *Chlorobium thiosulfatophilum*, while Quandt et al. (69) give a range of -9.5 to -12.2 for various species of this bacterium (all values recalculated for atmospheric CO_2 as carbon source).

Chemoautotrophic Bacteria. Isotope data for chemoautotrophic bacteria are hitherto restricted to methanogens only, deriving mostly from work on *Methanogenium thermoautotrophicum*, a strictly anaerobic species growing on CO_2 and H_2O as only carbon and energy sources. Recalculation of fractionations reported by Fuchs et al. (29) for atmospheric CO_2 as substrate gives a range from +6 to -31 ‰. Similar recalculation of figures obtained by DeNiro et al. (21) for *M. thermoautotrophicum*, *M. barkeri* and M. strain *ivanovii* yields values between -26.8 and -37.8 ‰, with an average of -30.6 ‰.

As is obvious from the graphic summary of biological carbon isotope fractionations presented in Fig. 6, the terrestrial biomass is enriched, on the average, by about 20 to 30 ‰ in light carbon (^{12}C) as compared to oceanic bicarbonate, the most abundant inorganic carbon species in the environment. It is worth noting that this conspicuous enrichment of ^{12}C in living matter derives, for the most part, from the isotope selection properties of the key enzyme of the Calvin cycle, RuBP carboxylase (cf. Table 1, No. 1).

Impact of Biological Carbon Fixation on the Isotope Geochemistry of Carbon

Preferential accumulation of light carbon in the biosphere has necessarily exercised a profound impact on the geochemical distribution of the stable carbon isotopes. As from the advent of autotrophic life on Earth, any input of CO_2 to the exogenic exchange reservoir (atmosphere and ocean) was eventually subject to an isotopic disproportionation into a light and a heavy fraction due to the sizable fractionation inherent in biological carbon fixation, notably by the RuBP carboxylase reaction. Enrichment of organic substances in ^{12}C has, conversely, increased the ^{13}C content of the residual reservoir, shifting the $\delta^{13}C$ values of inorganic carbon in positive direction in direct proportion to the fraction of total carbon ending up as organic matter. The bulk of the residual CO_2 was usually left to remain in the environment in hydrated form (as marine bicarbonate, HCO_3^-) which presently exceeds the gaseous (atmospheric) species by almost two orders of magnitude.

Quantitatively, the interrelationship between the relative proportions and the isotopic compositions of the geochemical reservoirs of organic (C_{org}) and inorganic (bicarbonate and carbonate) carbon (C_{carb}) can be expressed by an isotope mass balance. For instance, a primary carbon influx with $\delta^{13}C_{prim} \approx -5$ ‰ that is partitioned between organic and inorganic carbon with an average biogenic fractionation of -25 ‰ and eventually yields $\delta^{13}C_{carb}$ values close to zero permil would indicate a ratio $C_{org}/C_{carb} \approx 20/80$ in the exogenic compartment. This follows from the constraints of the mass balance equation

$$\delta^{13}C_{prim} = R\delta^{13}C_{org} + (1 - R)\delta^{13}C_{carb} \qquad (1),$$

rendering $R = C_{org}/(C_{org} + C_{carb}) \approx 0.2$ for the above numerical parameters (implicitly, $\delta^{13}C_{org} = \delta^{13}C_{carb} - 25$ ‰). If the exogenic carbon reservoir is being flushed at a relatively rapid rate as it is at present (the residence time of bicarbonate in the oceans is currently $\sim 10^5$ yr, cf. 39, p. 156), a stationary state complying with Eq. (1) will become established within a geologically short time interval.

Accordingly, any large-scale biological fractionation of the stable carbon isotopes will be monitored by the $\delta^{13}C$ values of both terrestrial carbon species, i.e., organic <u>and</u> carbonate carbon. Considering the wide range in biological fractionations and the resulting spread in $\delta^{13}C_{org}$ values of biological materials (cf. Fig. 6), the state of the system is probably best reflected by the mean for marine bicarbonate. Because of the rapid mixing of the oceans that store more than three quarters of all carbon residing in the exogenic compartment in the form of HCO_3^-, the bicarbonate mean is likely to integrate most faithfully over the environmental exchange reservoir as a whole.

Fate of Organic Carbon and Carbonate Carbon During Storage in Sediments

Minor fractions of both carbon species are continuously transferred from the surficial environment to newly-formed sediments. The flux of C_{org} amounts, for instance, to some 10^{14} g C/yr which corresponds to between 10^{-2} to 10^{-3} of the global rate of primary production of organic matter (39, 114). Since C_{org} and C_{carb} enter the Earth's crust with almost the same isotopic composition as imparted to them in terrestrial near-surface environments, the kinetic isotope effect involved in biological carbon fixation is propagated into the rock section of the carbon cycle (cf. Fig. 1).

Sedimentary <u>carbonates</u> usually preserve the isotopic composition of their parent carbonate muds within ± 1 ‰ of the original value which, in turn, is inherited from a bicarbonate precursor with a minor shift of about + 1 ‰ due to the equilibrium fractionation between HCO_3^- and CO_3^{2-} (107) (see also Fig. 6). Isotope changes during burial and postdepositional (diagenetic) alteration of sedimentary organics are more pronounced. These shifts may span the range from a few to several permil over the maturation pathway of organic matter in the sediment. Ultimately, this maturation results in the formation of <u>kerogen</u>, the polycondensed, acid-insoluble end product of the diagenetic reconstitution of the primary organic materials. The concomitant isotope changes are linked to various factors, notably

(i) preferential loss of isotopically light lipids and hydrocarbons during dehydrogenation of kerogen,

(ii) preferential loss of isotopically distinctive functional groups (e.g., "heavy" carboxyl groups) during microbiological and/or thermal degradation, and

(iii) selective scission of ^{12}C-^{12}C bonds in response to thermal stress, with subsequent removal of the crack products (cf. 30, 66, 78).

In sum, these and related processes tend to make the residual kerogen fraction slightly heavier with regard to its progenitor, while accumulating the light complement in the form of oil and gas (methane). With the magnitude of these secondary effects usually small, the isotopic signature of the primary biological materials is, as a rule, never seriously obscured.

On the other hand, both C_{org} and C_{carb} may undergo marked changes in their isotopic composition in response to rock metamorphism. Except for occurrences of pure marbles or major coal measures which retain the $^{12}C/^{13}C$ ratios of their precursor materials virtually unchanged through successive metamorphic regimes, isotope effects are usually pronounced in the common forms of sedimentary carbon, notably impure carbonates and finely dispersed kerogenous materials. Siliceous (quartz-bearing) carbonates are invariably subjected to metamorphic decarbonation reactions during which the original carbonate minerals (calcite, dolomite) react with silica to give Ca-Mg-silicates (tremolite, actinolite, and others), with a concomitant liberation of CO_2, e.g.,

$$5 \text{ dolomite} + 8 \text{ qtz} + 1 \, H_2O \rightarrow 1 \text{ tremolite} + 3 \text{ calcite} + 7 \, CO_2.$$

Decarbonation processes of this type start in the greenschist facies (between 300 - 450 °C) and increase with metamorphic grade. They give rise to residual carbonates enriched in ^{12}C since equilibrium fractionation between CO_3^{2-} and gaseous CO_2 at metamorphic temperatures favors the accumulation of heavy carbon in the volatile phase (7). As a result, the $\delta^{13}C$ values of metamorphosed carbonates may be shifted by 2 - 5 ‰ in negative direction relative to their original values around zero permil.

Metamorphic alteration of sedimentary organic matter has been reported to shift $\delta^{13}C_{org}$ values from primary means between -25 and -27 ‰ to -10 ‰ and less (26, 38, 54). In part, this can be ascribed to excessive dehydrogenation of kerogen and concomitant preferential cracking of ^{12}C-^{12}C bonds in response to increasing thermocatalytic stress. This may lead to the formation, and subsequent removal, of an isotopically light hydrocarbon fraction. Dehydrogenation of the residual kerogen may ultimately give rise to graphite as the hydrogen-free end member of the alteration series.

There is, however, reason to assume that this process cannot account for the extreme depletion in ^{12}C observed in some metamorphosed kerogens. The most important agent responsible for the formation of ^{13}C-enriched sedimentary organic matter seems to be isotopic reequilibration between kerogen and coexisting carbonates. According to Valley and O'Neil (103), isotope exchange between kerogen and isotopically heavy CO_2 released by decarbonation reactions commences already in the lower greenschist facies (300 - 400 °C) and increases with increasing temperature. As a result, the magnitude of fractionation between C_{org} and C_{carb} decreases, with thermodynamic equilibrium usually attained in the granulite facies ($\gtrsim$650 °C, see Fig. 7). These findings are crucial for a conclusive interpretation of carbon isotope data from metamorphic terranes that store the oldest ($t > 3.5 \times 10^9$ yr) segment of the carbon record.

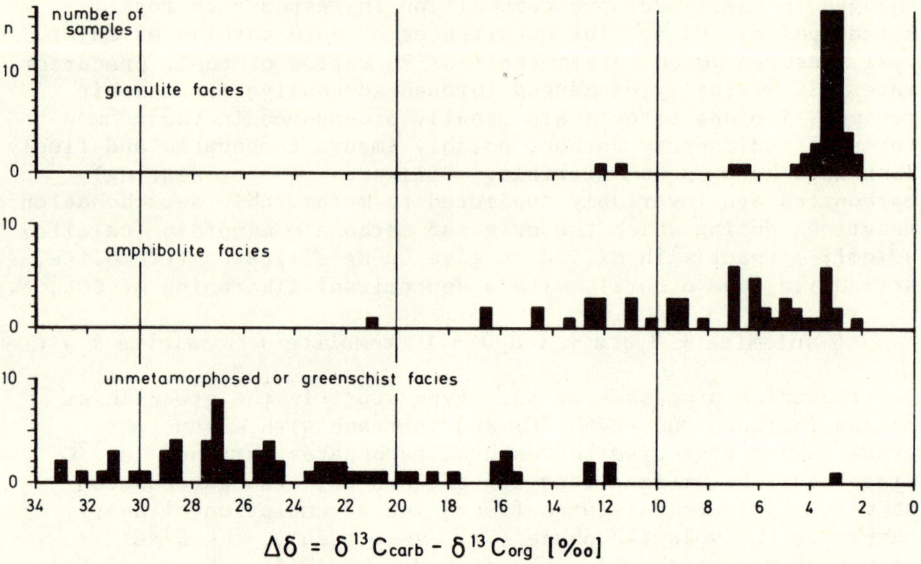

Figure 7. Decrease of isotopic fractionation between coexisting sedimentary carbonate (calcite) and organic carbon in response to increasing metamorphic grade (greenschist facies: 300 - 450 °C; amphibolite facies: 450 - 650 °C; granulite facies: ≳650 °C). Changes observed result from high-temperature reequilibration between the two carbon phases, with ^{13}C exchange promoted by the release of isotopically heavy CO_2 during metamorphic decarbonation reactions. The narrow spread of values in the granulite facies suggests that thermodynamic equilibrium is just about attained at this grade. Adapted from (103).

We may conclude from this overview that both diagenetic and metamorphic processes are capable of imposing measurable overprints on the primary isotopic compositions of both sedimentary carbon species. However, these secondary changes usually just blur, and never seriously obliterate, the isotopic signature of the primary carbon source. This even holds for the extremely positive $\delta^{13}C_{org}$ values of some graphitized kerogens from high-grade metasediments which, due to unavoidable ^{13}C exchange with carbonate, cannot be taken as evidence of an abiogenic origin of these organics.

Biological Carbon Isotope Fractionation Through Geologic Time

With carbonate (C_{carb}) and organic carbon (C_{org}) preserved

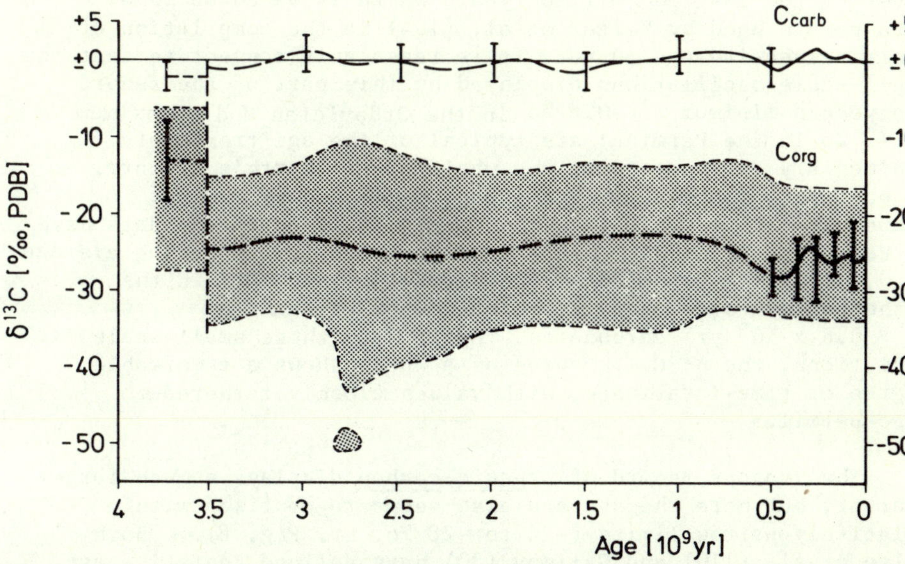

Figure 8. Isotope age functions for sedimentary carbons and organic carbon over the last 3.8×10^9 yr. The large number of measurements available for Phanerozoic kerogens allows the definition of fairly reliable means and standard deviations (vertical bars) for organic carbon during the last 600 million years. The scatter shown for the total spread of $\delta^{13}C_{org}$ over the Earth's history covers the data base presented in detail by Schidlowski et al. (84). The mainstream of this band can be best explained as geochemical manifestation of the activity of the key enzyme of the Calvin cycle (RuBP carboxylase). The discontinuity at $t \simeq 3.5 \times 10^9$ yr is almost certainly due to the amphibolite-grade reconstitution of the Isua suite with concomitant isotopic reequilibration of both carbon species (cf. Fig. 7).

in sediments, and the isotope effects linked to the formation, deposition and postdepositional alteration of both carbon species fairly well understood, we may attempt to decode the record retained in ancient sediments by tracing biologically mediated isotope effects back into the geologic past. Fig. 8 summarizes the age functions of both sedimentary carbon species based on some 8,000 measurements which span the time interval from Recent to about 3.8×10^9 yr ago.

The best-documented segment of the carbonate record is that covering the last 800 million years which is based on 3016 data points used by Veizer et al. (104) in the compilation of their respective age curve. It is reasonable conjecture that the small-scale oscillations displayed by this part of the record (between a minimum of -0.8 ‰ in the Ordovician and a maximum of +2 ‰ in the Permian) are typical of the age trends of sedimentary carbonate over the whole of the Earth's history. In any case, the mean of 0.5 ± 2.6 ‰ yielded by 2.916 Phanerozoic ($t < 0.6 \times 10^9$ yr) samples included in the data base of Veizer et al. (104) compares well with a corresponding average of 0.4 ± 2.7 ‰ reported by Schidlowski et al. (83) in the hitherto largest single investigation of Precambrian ($t > 0.6 \times 10^9$ yr) carbonates. Apart from these small-scale variations, the $\delta^{13}C_{carb}$ curve as a whole shows a remarkable degree or time-invariance, with values closely tethered to zero permil.

The isotope record of organic carbon displays a much larger scatter, but here the average also seems to be fixed within relatively narrow limits (-24 to -28 ‰, cf. Fig. 8). Both Welte et al. (110) and Galimov (30) have defined tentative age curves for Phanerozoic kerogens that show a marked covariance with the respective carbonate curve (with a minimum close to -29 ‰ during Ordovician-Silurian and a maximum around -24 ‰ in the Carboniferous). Incidentally, this sympathetic relationship would be a necessary consequence of a coupling of the isotopic compositions of both carbon species to their relative amounts in the environmental exchange reservoir in terms of an isotope mass balance [cf. Eq. (1)], with the average fractionation between C_{org} and C_{carb} ($\Delta\delta_c = \delta^{13}C_{org} - \delta^{13}C_{carb} = -25 \pm 1$ ‰) entering the relationship as a biological constant. Accordingly, the oscillations of the isotope curves almost certainly reflect moderate changes in the relative proportions of C_{org} and C_{carb} through geologic time. This is not surprising in view of the considerable chemical inertia of the atmosphere-ocean-biosphere-crust system [see also (85)].

While the data base presently available for the Phanerozoic allows the definition of reasonable $\delta^{13}C_{org}$ averages for several geological formations (cf. Fig. 8), this is not possible, with a comparable confidence level, for Precambrian formations. It is, however, obvious from previous (26) as well as recent investigations (84) that both the $\delta^{13}C_{org}$ mean and the total isotope spread of Precambrian kerogens do not differ very much from those of younger sediments. As a whole, the envelope for the Precambrian data is slightly inflated in both positive and negative directions as compared to the Phanerozoic which, on the other hand, does not seem to affect the trend-line for the average to any significant extent (cf. Fig. 8).

As pointed out before, positive excursions of $\delta^{13}C_{org}$ are typical of "high-rank" or metamorphosed kerogens, and notably those that have suffered isotope exchange with "heavy" carbonate which reequilibration starts already at temperatures slightly above 300 °C (103). Hence there is little doubt that the bulk of isotopically heavy organics encountered in Precambrian sediments have been subjected to varying degrees of metamorphism, the probability of which proportionately increases with increasing geological age of the rocks. More difficult to explain than these positive extremes are negative spikes imposed on the mainstream of the record, particularly between 2.5 and 3.0 x 10^9 yr ago (Fig. 8). Kerogens from the 2.7 x 10^9 yr old Australian Fortescue Group and the nearly coeval South African Ventersdorp Supergroup were found to range between -35 and -52 ‰ (36, 84), while Schoell and Wellmer (87) have reported lower extremes of -44 ‰ for Canadian samples of similar age. Since these isotopically anomalous kerogens are coeval with a host of "normal" kerogen provinces (cf. 84, Fig. 3), these anomalies must reflect changes in the local (rather than global) environment in which the organic precursor materials were formed.

Since negative $\delta^{13}C$ values of this range have hitherto only been reported in connection with methane, it seems appropriate to view the bacteria of the methane cycle (methanogens and methanotrophs) as most likely agents responsible for the formation of such "light" kerogens. With considerably lower oxygen levels assumed during the Archaean and Early Proterozoic, methane as end-product of the anaerobic degradation of organic matter may have played a quantitatively more important role in those times as a transient in the global carbon cycle than it does today. Increased availability of CH_4 would, in turn, be conducive to a proliferation of microorganisms relying on methane-based assimilatory pathways (methanotrophic bacteria). In this way, the isotopically light methane would be incorporated into the cell substance of methanotrophs and, consequently, into sedimentary organics. A crucial requirement of this pathway is, however, the presence of an oxidant promoting the oxidation sequence

$$CH_4 \rightarrow CH_3OH \rightarrow HCHO \rightarrow HCOOH \rightarrow CO_2$$

of which the HCHO (formaldehyde) intermediate is metabolized by the bacteria. The necessity for such an oxidant certainly poses problems for the operation of this pathway in the Precambrian. However, recent investigations have shown that the biologically mediated oxidation of methane can also be achieved by "mild" oxidants such as sulfate (SO_4^{2-}) instead of free oxygen (62, 71). Since the widespread occurrence of Precambrian banded iron-formation is believed to indicate at least transient oxygen levels in the ancient seas (18), it would not seem impossible that some of the oxygen flux into the iron oxides was deflected for sustaining the activities of methane-oxidizing bacteria in suitable environments. This interpretation of the isotopically light kerogens seems to be the most plausible way to account for the extremes at the lower fringe of the C_{org} envelope shown in Fig. 8. It should be noted that extremely light nonvolatile organics (with minimum values around -80 ‰) have also been reported from geologically younger sediments (43).

The only conspicuous discontinuity in the sedimentary $\delta^{13}C_{org}$ record is the break between the 3.8 x 10^9 yr old graphitic kerogen constituents of the Isua meta-sedimentary suite and post-Isua sediments. As has been set out in full elsewhere (80, 81, 82, 84), the most convincing way to explain the isotope shifts noted for both C_{org} and C_{carb} in the Isua rocks is to invoke the effects of a metamorphic overprint. Recorded changes in $\delta^{13}C_{org}$ and $\delta^{13}C_{carb}$ are most decidedly consistent with the amphibolite-grade metamorphism experienced by the suite as a whole, with both carbon species responding in a predictable manner, i.e., in keeping with presently available thermodynamic data and with observations from younger metamorphic terranes. Notably these latter have provided ample evidence that initial sedimentary $\delta^{13}C_{carb}$ and $\delta^{13}C_{org}$ values around zero and -26 ‰ are readily shifted to the ranges observed in Isua ($\delta^{13}C_{carb}$ = -2.3 ± 2.2 ‰; $\delta^{13}C_{org}$ = -13.0 ± 4.9 ‰) in response to increasing metamorphism.

SULFUR

Sulfur is another element whose geochemistry is largely governed by the activity of the Earth's biosphere. The decisive impact is exercised by the process of dissimilatory (bacterial) sulfate reduction which is responsible for the basic configuration of the terrestrial sulfur cycle (Fig. 1) and for the isotopic composition of the principal sulfur reservoirs. As was the case with carbon, the biogenic sulfur phases preferentially concentrate the light isotope (^{32}S).

Biological Fractionation of Sulfur Isotopes

Biological utilization of sulfur proceeds, for the most part, as assimilatory or dissimilatory reduction of sulfate (SO_4^{2-}). Sulfate is the thermodynamically stable phase of the element at or near the Earth's surface, the large marine SO_4^{2-} reservoir storing almost all sulfur residing in the exogenic compartment (cf. Fig. 1). Although both pathways of sulfate reduction show striking similarities in their biochemistries (Fig. 9), their isotope selection properties are rather different. While the assimilatory pathway (responsible for the primary synthesis of sulfur-containing cell constituents) discriminates

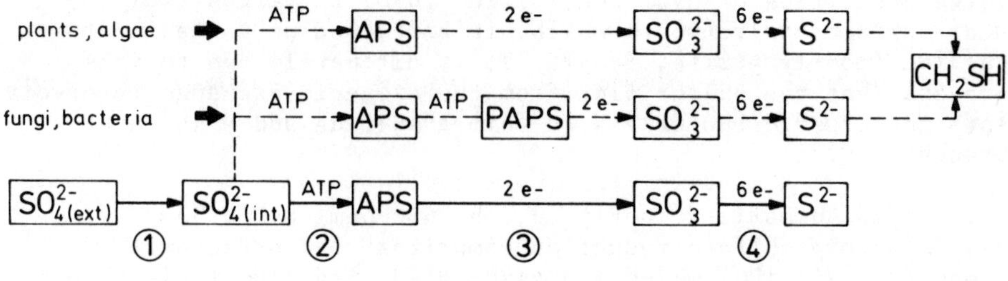

Figure 9. Summary of biochemical pathways of sulfate reduction (solid lines: dissimilatory reaction; dashed: assimilatory pathways). After uptake by organisms, SO_4^{2-} is phosphorylated to give adenylyl sulfate ("adenosine-5'-phosphosulfate," APS) and, in part, phosphoadenylyl sulfate (PAPS) which "activated" species are subsequently reduced to sulfite (SO_3^{2-}) and sulfide (S^{2-}). In both assimilatory pathways, sulfur ends up in the sulfhydryl (CH_2SH) group of cysteine; in dissimilatory reduction it is released to the environment as H_2S. Note that the initiation of these reactions depends on an expenditure of energy (ATP), and that the oxidation state of sulfur changes from S^{6+} to S^{2-} over the whole pathway. From (84).

against heavy sulfur ineffectively (-4.5 ‰ as a maximum; see 42, 44, 55), fractionations in the dissimilatory reaction are formidable, amounting to as much as -60 ‰, with averages between -35 and -45 ‰ (15, 32, 84, and others).

Dissimilatory sulfate reduction is practiced by few genera of strictly anaerobic bacteria (viz., Desulfovibrio, Desulfotomaculum, Desulfomonas) which couple this reduction to the oxidation of organic substrates (e.g., lactate, $CH_3CHOHCOOH$) in an energy-yielding reaction:

$$2\ CH_3CHOHCOOH + SO_4^{2-} \rightarrow 2\ CH_3COOH + 2\ H_2O + 2\ CO_2 + S^{2-}$$
$$(\Delta G_0' = -88\ kJ).$$

Since SO_4^{2-} is serving here as ultimate electron acceptor instead of oxygen, this process may be conceived as a form of anaerobic respiration. Sulfate reducing bacteria as a group (68, 101, 102) are of paramount importance as the principal agents for catalyzing the reduction of sulfate to sulfide (S^{2-}) at low temperatures. This conversion constitutes a pivotal link in the global sulfur cycle, but has never been demonstrated to proceed at t < 200°C unless biologically mediated. It releases large quantities of hydrogen sulfide (H_2S) to terrestrial near-surface environments which, in part, end up as sedimentary sulfide (mostly pyrite, FeS_2). It is ultimately due to this process that the sulfur flux from the exogenic exchange reservoir into the crust dichotomizes to form a sulfide and a sulfate branch (Fig. 1).

As is summarized in Fig. 9, the biochemical pathway of dissimilatory sulfate reduction comprises four principal steps (72, 75, 100) which are associated with kinetic isotope effects of variable magnitude. Fractionations over the total pathway have been shown to add up to -46 ‰ in culture experiments and to about -60 ‰ in newly-formed sediments, the natural habitats of sulfate reducers (15, 84). Accordingly, the $\delta^{34}S$ values of bacteriogenic sulfide are distinctly displaced in negative direction as compared to the parent sulfate pool (Fig. 10). Since total fractionation over the biochemical

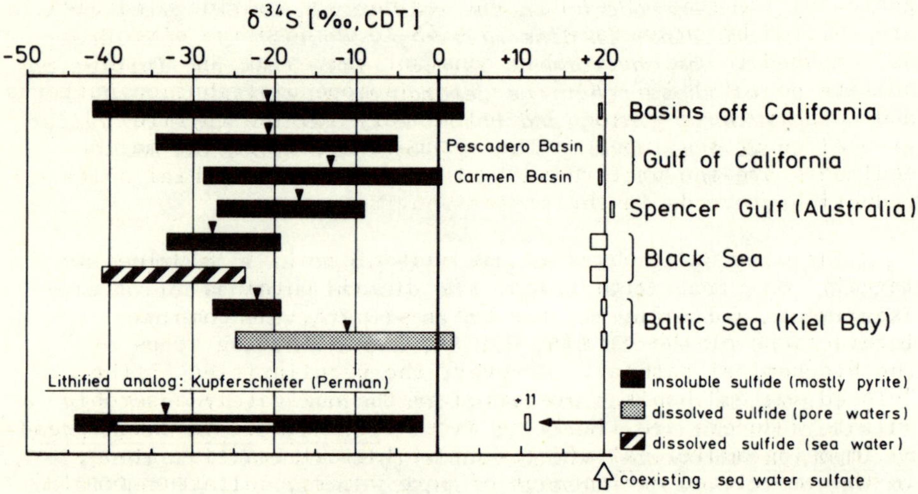

Figure 10. *Isotope fractionation between sulfate and bacteriogenic sulfide in contemporary marine environments. Approximate means for sulfide are indicated by triangles and seem to scatter around the -20 permil line (corresponding to an average fractionation* $\Delta\delta_S = \delta^{34}S_{sulfide} - \delta^{34}S_{sulfate} \simeq -40$ *‰). The sulfide pattern shown for the Permian Kupferschiefer exemplifies a lithified facies (main range of values indicated by white bar;* $\delta^{34}S$ *for Permian seawater sulfate was approximately +11 ‰). From (84); sources of original data: (14, 33, 35, 42, 53, 94, 105).*

pathway is a composite of the fractionations associated with the individual reactions, its magnitude may vary depending on which of steps (1) to (4) indicated in Fig. 9 becomes rate-controlling under specific conditions (15, 44, 45, 72). There is as yet no general consent on the magnitude of the kinetic effects associated with the single steps. It is, however, widely assumed that the effects linked to reactions (1) and (2) amount to a few permil only, while values proposed for step (3) are -25 ‰ (72) and -10 ‰ to -15 ‰ (15). Fractionations inherent in reaction (4) were found to be extremely variable (0 - 33 ‰, cf. 15). A generalization that mostly holds is that fractionations over the total pathway increase with decreasing rates of sulfate reduction (44, 45, and others).

Fig. 10 gives a graphic synopsis of sulfur isotope fractionations observed in surficial sediments from various parts of the contemporary ocean, along with the values obtained for an ancient lithified analog (Kupferschiefer). It is obvious that the $\delta^{34}S$ values of the two principal forms of bacteriogenic sulfide (dissolved H_2S and sedimentary sulfide minerals) are shifted by an average of 35 - 45 ‰ in negative direction as compared to the remarkably constant values of the oceanic sulfate pool ($\delta^{34}S \simeq +20$ ‰). Such isotope distribution patterns, and notably those yielded by sedimentary sulfide (pyrite), give eloquent testimony to the assumption that recent marine sediments are the preferential sites at which microbial sulfate reduction proceeds in the present environment.

The wide range observed for bacteriogenic $\delta^{34}S$ values is largely consistent with respective spreads obtained in culture experiments and reflects, for the most part, the complex interactions of the various isotope discriminating steps of the biochemical pathway. However, the peculiarities of the "closed system" habitat preferred by the obligately anaerobic sulfate reducers are likely to further contribute to this spread by imposing additional effects on biological fractionations. In the restricted environment of pore waters, sulfate reduction is usually coupled with sulfate depletion. Because of preferential utilization of light sulfur by the bacteria, the $\delta^{34}S$ values of residual sulfate and, consequently, of late-formed sulfide can be expected to increase with a progressive exhaustion of the initial sulfate pool. Applying a fractional distillation (Rayleigh) model to these relationships, it is possible to calculate a theoretical or "apparent" fractionation factor

$$K = \frac{(^{32}S/^{34}S)_{sulfide}}{(^{32}S/^{34}S)_{sulfate}}$$

from the isotope distribution pattern observed in a specific sedimentary environment (32). This and related approaches (cf. 15) have yielded average fractionation factors in the range 1.015 to 1.045 for sulfate reduction in modern marine sediments, corresponding to an increase of $^{32}S/^{34}S$ in bacteriogenic sulfide between 15 and 45 ‰ relative to seawater sulfate. An alternative "open system" model allowing for diffusion of sulfate into interstitial waters has yielded maximum fractionation factors around 1.060 (33). Such fractionations come close to the extremes observed for bacteriogenic sulfide in semi-closed ("euxinic") sedimentary basins such as the Pacific basins off California and the Black Sea (see Fig. 10).

Impact of Dissimilatory Sulfate Reduction on the Isotope Geochemistry of Sulfur

Irrespective of the small number of bacterial genera engaged in dissimilatory sulfate reduction, the process as such is responsible for sizable turnover rates of sulfur in the exogenic exchange reservoir. The large fractionations inherent in this type of sulfur metabolism were consequently bound to work their way into the rock section of the sulfur cycle, the long-term effect being the conspicuous isotopic disproportionation of terrestrial sulfur into a "light" biogenic and a "heavy" inorganic reservoir (Fig. 1).

This disproportionation originates in the exogenic compartment where sulfate reducers continuously harvest light sulfur from the environmental sulfate pool. Due to the subsequent reaction of bacteriogenic sulfide with iron and other trace metals, substantial amounts of this ^{32}S-enriched sulfur are removed from the system as sulfide minerals, leaving the heavy isotopes to accumulate in the residual sulfate reservoir. This state of affairs is reflected by the isotope geochemistry of the present ocean (Fig. 10) where heavy sulfur is retained in dissolved sulfate ($\delta^{34}S \simeq +20$ ‰) and where bacteriogenic hydrogen sulfide and its derivatives are markedly enriched in the light isotope (average $\delta^{34}S$ between −15 and −25 ‰).

At steady state, any influx of sulfur into the exchange reservoir will be balanced by the two sinks that compete for its subsequent removal, namely, (i) bacterial sulfate reduction (with subsequent incorporation of bacteriogenic H_2S into sulfide minerals), and (ii) precipitation of sulfate evaporites.
The relative strengths of these sinks will be recorded by the isotopic compositions of both sulfur fluxes as a result of mass balance constraints. Assuming, for instance, an influx with the composition of the Earth's primordial sulfur ($\delta^{34}S_{prim} \simeq 0$ ‰) and an average bacteriogenic fractionation $\Delta\delta_S = \delta^{34}S_{sulfide} - \delta^{34}S_{sulfate} = -40$ ‰, marine sulfate values around +20 ‰ would

indicate a ratio $R = S_{sulfide}/(S_{sulfide} + S_{sulfate}) \simeq 0.5$ in compliance with the relationship

$$\delta^{34}S_{prim} = R\delta^{34}S_{sulfide} + (1-R)\delta^{34}S_{sulfate} \qquad (2),$$

(i.e., the total sulfur flux leaving the exogenic compartment would be equally divided [0.5 : 0.5] between sulfide and sulfate). In terms of Eq. (2), the $\delta^{34}S$ values of both sulfur species should, in principle, monitor the partitioning between biogenic and inorganic sulfur in the environment. However, because of the rapid mixing of dissolved SO_4^{2-} in the oceans, it is apparent that the sulfate mean integrates more faithfully over the exogenic reservoir as a whole. Thus, it provides the best index of the state of the system at any given time (in contrast to the scatter of the sulfide values each of which reflects a particular microenvironment). If, as at present, the marine sulfate reservoir is flushed once within 20 million years (40), it would not take long -- in terms of the geologic time scale -- for a stationary state to be established which is characterized by a specific $\delta^{34}S$ mean for sulfate.

Summing up, it is safe to state that microbial sulfate reducers are largely in control of the environmental geochemistry of sulfur. Any given state of the exogenic system is likely to be monitored by the $\delta^{34}S$ mean of marine sulfate which, in terms of Eq. (2), provides a quantitative measure of the involvement of bacterial sulfate reducers in the total turnover rate of sulfur in the exogenic cycle.

Biological Fractionation of Sulfur Isotopes Through Geologic Time

Since sulfide and sulfate are transferred from the exogenic compartment to sediments with little change in their isotopic composition, a record of the fractionation between reduced and oxidized sulfur is preserved in crustal rocks. Comparison of bacteriogenic H_2S from seawater and pore waters with sedimentary sulfide (pyrite) suggests that the $\delta^{34}S$ patterns of sulfide minerals preserve, albeit in a rather broad and general way, the values of the primary metabolic phase. Once diagenetically stabilized, sedimentary sulfides tend to retain their original isotopic composition even throughout advanced stages of metamorphism (76). Dissolved marine sulfate is precipitated with about 1 - 1.6 ‰ enrichment of ^{34}S in the solid phase (99) which is the equilibrium fractionation at surface temperatures, the $\delta^{34}S$ values of sedimentary (evaporite) sulfates thus closely reflecting the values of their parent brines.

Attempts to trace bacteriogenic fractionations back in time should, ideally, rely on an investigation of either coexisting or at least coeval sedimentary sulfide and sulfate. However,

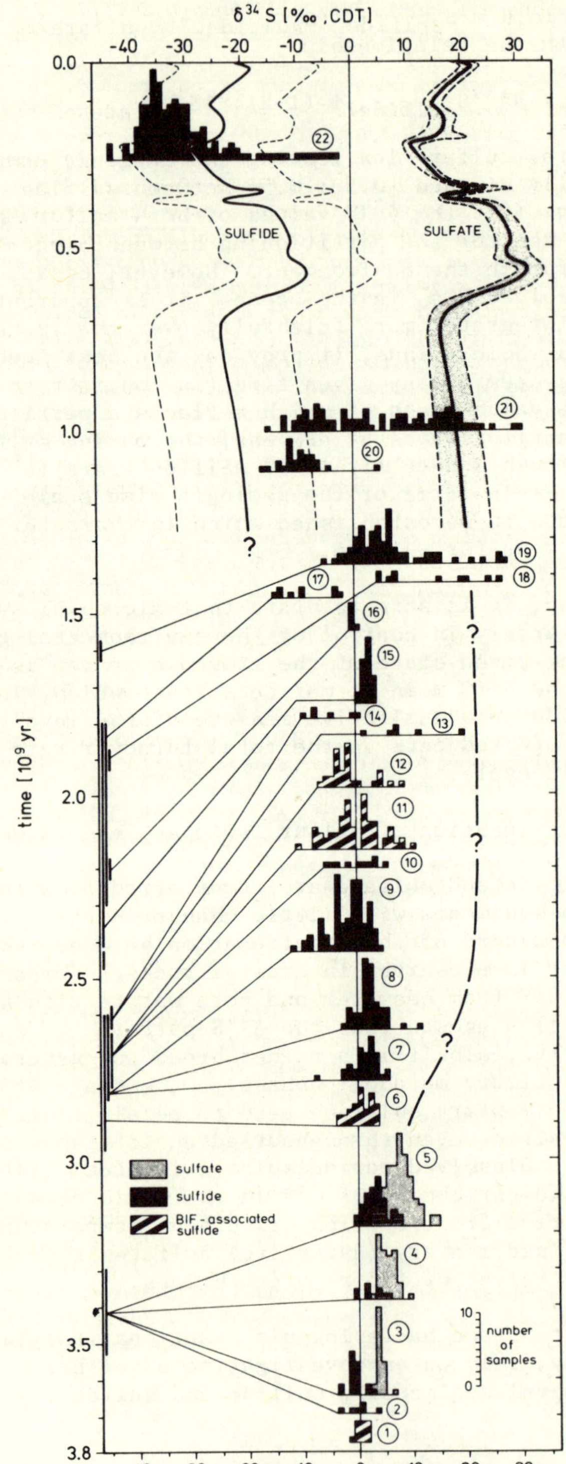
δ34 S [‰, CDT]
-40 -30 -20 -10 0 10 20 30
0.0
0.5
1.0
1.5
2.0
2.5
3.0
3.5
3.8
time [10^9 yr]
SULFIDE
SULFATE
?
1
2
3
4
5
6
7
8
9
10
11
12
13
14
15
16
17
18
19
20
21
22
sulfate
sulfide
BIF-associated sulfide
10
0
number of samples

Figure 11

Figure 11. Isotope age curves for sedimentary sulfide sulfide and sulfate. The sulfate curve for the Phanerozoic and latest Proterozoic ($t > 1.3 \times 10^9$ yr) is the best documented part of the record, the stippled band representing the estimated area of uncertainty (17). Prior to 1.3×10^9 yr ago, the record of sulfate evaporites is extremely scarce and restricted to sedimentary barite, notably from Early Archaean occurrences whose $\delta^{34}S$ values come close to zero permil. The age function for sulfide for the last 1.3×10^9 yr is a construct based on a fractionation of -40 ‰ between sulfide and sulfate which is consonant with the average observed in contemporary sedimentary environments (cf. Fig. 10). With the large spreads noted for the $\delta^{34}S$ values of bacteriogenic sulfides, actually observed values should fall into a range ± 12 ‰ of this average (dashed lines). With increasing age, sulfide data becomes more scanty, numbered occurrences 1 - 21 virtually representing the complete data base for the respective time interval (massive and detrital sulfides excluded). Note tendency of sulfide values to cluster around zero permil as from $t \geq 2.8 \times 10^9$ yr. -- (1) Banded iron-formation from Isua, West Greenland (56). (2) Swartkoppie Formation of Onverwacht Group, South Africa (84). (3) Warrawoona Group of Pilbara Block, Western Australia; sulfate occurring as bedded barite (47). (4) Fig Tree Group, South Africa; sulfate occurring as bedded barite (47, 65, 105). (5) Iengra Series of Aldan Shield, Siberia; sulfate occurring as barite and anhydrite (105). (6) Banded iron-formation from Rhodesian schist belts (mostly Sebakwian Group) (28). (7) Black shales from greenstone belts of Yilgarn Block, Western Australia (23). (8) Deer Lake Greenstone Sequence, Minnesota (74). (9) Syngenetic "barren" sulfide deposits of Birch-Uchi Greenstone Belt, Superior Province of Canadian Shield (88). (10) Fortescue Group, Hamersley Basin, Australia (84). (11) and (12) Michipicoten and Woman River banded iron-formation, Superior Province of Canadian Shield (34). (13) Steep Rock Lake Series, Canadian Shield (Veizer and Nielsen, unpublished results). (14) Ventersdorp Supergroup, South Africa (Veizer and Nielsen, Unpublished results). (15) Bedded sulfides, Cahill Formation of Pine Creek geosyncline, N.T., Australia (22). (16) Frood Series of Sudbury District, Canadian Shield (96). (17) Black shales from Outokumpu, Finland [(52) and Makela, pers. comm., 1978). (18) Onwatin Slate, Sudbury Basin, Canadian Shield (96). (19) Sediments of McArthur Basin, N.T. Australia (92, 93). (20) Adirondack sedimentary sulfides, Grenville Province, Canadian Shield (11). (21) Nonesuch Shale, Superior Province, Canadian Shield (12). (22) Permian Kupferschiefer, Central Europe (53).

this requirement is difficult to meet for the older record since evaporites are subject to preferential removal (intrastratal solution, etc.) from any existing pile of sediments. Therefore, they tend to disappear from the record with increasing age. In fact, sulfate occurrences older than $\approx 1.2 \times 10^9$ yr are mainly preserved as barite, $BaSO_4$, rather than in the common forms of soluble calcium sulfate like anhydrite and gypsum. Accordingly, an evaluation of the Precambrian record has to be based mainly on the isotope patterns of sedimentary sulfides.

Fig. 11 gives an overview of the principal age trends displayed by sulfide and sulfate through geologic time. Obviously, the data base for both curves fades with increasing age, being extremely scant for sulfide and almost nil for sulfate at times $t \gtrsim 1.3 \times 10^9$ yr.

Sedimentary Sulfate. In the case of sulfate evaporites, the best documented part of the isotope age curve is that for the Phanerozoic which is based on an impressive assembly of measurements contributed during the last decades (17, 40, 59, 99). As a result of these investigations, it is firmly established that $\delta^{34}S$ averages of marine sulfates have varied between a minimum of +11 ‰ and a maximum of +32 ‰ during the last 600 million years. Since $\delta^{34}S$ values of sulfates monitor the state of sulfur in the exchange reservoir (see "Impact of Dissimilatory Sulfate Reduction on the Isotope Geochemistry of Sulfur"), the observed variations would indicate secular imbalances in the operation of the sulfur cycle. As has been fully set out elsewhere (84, 86), such perturbations are likely to reflect, for the most part, changes in the relative intensities of the two sinks responsible for the removal of sulfur from the exogenic compartment, with $\delta^{34}S$ of marine sulfate increasing with increasing sulfide/sulfate ratios in the total sulfur flux leaving the exogenic system (cf. Eq. 2). This is in marked contrast to the carbon cycle where near-constant $\delta^{13}C_{carb}$ values through time indicate a correspondingly constant strength of the biological sink while, in the case of sulfur, the biological sink seems to have varied to a considerable extent.

The Phanerozoic record extends, albeit partly interrupted, for another 600 million years into the Late Precambrian where it fades between 1.2 and 1.3×10^9 yr ago (Fig. 11). Naturally, the confidence level of this part of the age curve cannot compare with the Phanerozoic section(cf. 17). For $t > 1.2 \times 10^9$ yr, a record of sulfate evaporites is virtually lacking, noteworthy exceptions being barites from several Archaean terranes. In the opinion of most investigators, these barites were either directly precipitated as chemical sediments or formed diagenetically, replacing an evaporitic calcium sulfate precursor without a significant change in bulk isotope geochemistry (3, 24, 37, 47, 50).

The widespread occurrence in Archaean sediments of barite and chert pseudomorphs after primary gypsum and anhydrite adds particular weight to the second alternative (while certainly not excluding the first one). The original existence of extended evaporite basins during the Early Archaean is also documented by abundant geological field evidence (3, 31).

Ample indirect evidence (e.g., halite casts, replacements of primary evaporite beds by chert and carbonate) attests to the original presence of salt-bearing strata also in the missing part of the sulfate curve between the Archaean barites and the onset of the quasi-continuous record some 1.2×10^9 yr ago. It seems likely that a better documented Precambrian record would reveal similar oscillations as noted in the Phanerozoic section.

Sedimentary Sulfide. Accepting an average fractionation of -40 ‰ between sulfate and bacteriogenic sulfide as observed in the present environment (Fig. 10), we may utilize the sulfate curve for constructing a corresponding age function for sulfide. Considering the larger spread of the sulfide values, the scatter band accompanying this curve would be much broader than in the case of sulfate, covering an envelope of perhaps ± 12 ‰ around the mean. Most values obtained for Phanerozoic and Late Precambrian sulfides do indeed fall into such envelope as does, for instance, the isotope pattern of the Permian Kupferschiefer (Fig. 11, No. 22) that reflects the "euxinic" facies of a semiclosed sedimentary basin. Since contemporary environments of this type have furnished maximum fractionations approaching -60‰, it is consistent with such observations that the bulk of the Kupferschiefer values exceed the average of -40 ‰ underlying the sulfide curve of Fig. 11 by about -6 ‰. Wholesale conversion into sulfide of the "heavy" sulfate content of completely barred basins may result in marked shifts of bacteriogenic sulfide patterns to the positive field as in the case of the McArthur Basin and the Nonesuch Shale (Fig. 11, Nos. 19 and 21).

Detailed investigations of Precambrian sedimentary sulfides are still limited in number which substantially detracts from the confidence level of the older record. Fig. 11 lists virtually the complete data base presently available for $t > 10^9$ yr, excluding only detrital and massive stratiform sulfides as well as those with obvious hydrothermal overprints. $\delta^{34}S$ patterns showing a marked encroachment on the negative field along with other distributional features suggestive of a bacteriogenic origin can be traced back to about 2.7×10^9 yr. The more ancient record -- and notably the oldest sedimentary sulfides from the Isua banded iron-formation -- definitely lack such characteristics.

IMPLICATIONS OF THE SEDIMENTARY CARBON AND SULFUR ISOTOPE RECORDS

With the presently available isotope age curves at hand, we may attempt to trace the fractionations inherent in the principal forms of carbon and sulfur metabolism back into the geologic past. The ultimate aim of such efforts would be the identification of the oldest signals in the geological record of biologically-mediated isotope effects, specifically those involved in (i) photoautotrophic carbon fixation and (ii) dissimilatory sulfate reduction.

The Carbon Record

In the case of carbon (Fig. 8), the implications of the age functions for both C_{carb} and C_{org} seem fairly straightforward. Since sedimentary $\delta^{13}C_{carb}$ values closely approximate those of the marine bicarbonate precursor, the basic message conveyed by the carbonate record is that of a remarkable constancy of the isotopic composition of marine bicarbonate through time. Accepting the well-documented Phanerozoic section as representative of the carbonate record as a whole, original bicarbonate values should have floated within a band about 5 ‰ wide, with excursions closely tethered to a mean between -1 and 0 ‰ (i.e., about 1 ‰ below the mean of +0.5 ‰ obtained for the $\delta^{13}C_{carb}$ function shown in Fig. 8). Since marine carbonates monitor the state of the exogenic carbon system (see "Impact of Biological Carbon Fixation on the Isotope Geochemistry of Carbon"), $\delta^{13}C_{carb}$ values around zero permil would imply values close to -25 ‰ for coevally produced organic matter if we assume

$$\Delta\delta_c = \delta^{13}C_{org} - \delta^{13}C_{carb} = -25 \pm 1 ‰$$

for the <u>average</u> fractionation associated with biological carbon fixation. This is confirmed by the actually observed mean for fossil organic matter (kerogen) which lies somewhere between -24 and -28 ‰, with the majority of measurements falling into the range -27 ± 6 ‰ and the total spread extending from about -10 to -52 ‰. This wide scatter seems due to both primary variations in the isotopic composition of organic substances (cf. Fig. 6) and to postdepositional processes that have imposed a secondary noise on the original biogenic patterns. Positive extremes in $\delta^{13}C_{org}$ usually result from metamorphism and concomitant ^{13}C exchange with carbonates while the conspicuous negative excursions suggest the involvement of methanotrophic pathways in the formation of the respective kerogen progenitors (cf. "Biological Carbon Isotope Fractionation Through Geologic Time").

We may reasonably assume that the prime responsibility for the characteristic enrichment in ^{12}C which we observe in ancient kerogens rests with the process that gave rise to the biological precursor materials, specifically with the enzymatic carboxylation of CO_2 in the Calvin cycle, the quantitatively most important pathway of carbon assimilation. Since the relatively constant average fractionation between C_{org} and C_{carb} in sediments largely conforms with respective fractionations in the contemporary environment, it is difficult to escape the conclusion that the sedimentary $\delta^{13}C_{org}$ record as a whole bears the isotopic signature of biological (viz., autotrophic) carbon fixation as from at least 3.5×10^9 yr ago. Furthermore, the uniformity of this signature would indicate an extreme degree of conservatism in the basic biochemical mechanisms of carbon assimilation which is consistent with the notion of a general conservatism of biochemical processes through time. In fact, the mainstream of the envelope covering the $\delta^{13}C_{org}$ values in Fig. 8 may be reasonably explained as the geochemical manifestation of the activity of one single enzyme, RuBP carboxylase. Such an interpretation is fully supported by the fossil record of apparently autotrophic life which includes the remnants of cellular microfossils as well as the "stromatolites" they built (that is, biosedimentary structures originating from the matting behavior of benthic prokaryotes). Stromatolites have been recently shown to extend to about 3.5×10^9 yr ago (25, 49, 109), while the potentially oldest cellular fossils hitherto reported (67) are objects of a current controversy (8).

For this interpretation of the carbon isotope record to be invalidated we would have to postulate an inorganic process of global relevance that had mimicked fractionations in biological carbon fixation with a remarkable degree of precision. Although we might think of two possible alternatives, both fail to meet the above requirement completely. Fractionations between oxidized and reduced carbon in Fischer-Tropsch processes range from -50 to -100 ‰ at 400 °K (48) and increase with falling temperature. Accordingly, they may be safely discounted as a potential source of the kerogenous materials stored in ancient sediments. Miller-Urey spark discharge syntheses are, on the other hand, associated with small isotope effects (usually > -10 ‰) though individual products like amino acids may approach -20 ‰ and thus overlap the biological range (16). In this case, the borderline between biological mediated organic and inorganic fractionations appears blurred. However, since the isotopically light fraction accounts for less than one percent of the total yield from such reactions, its quantitative importance is negligible. Maximum fractionations between -9 and -12 ‰ sometimes obtained for the gross reaction products at low reactant consumption encroach just marginally on the range attributed to biological fractionations and their fossil manifestations. In general, the relatively heavy organic matter comprising the bulk product of spark discharge reactions should lend itself to an easy detection in unmetamorphosed sediments. The virtual absence of such isotopic signature in the record (cf. Fig. 8) is likely to exclude any large-scale contribution from such sources to the kerogen constituents of rocks younger than 3.5×10^9 yr.

Altogether, the continuity of the sedimentary $\delta^{13}C_{carb}$ and $\delta^{13}C_{org}$ records seems to leave little doubt that biological (notably photoautotrophic) carbon fixation had been established as a geochemically relevant process as from at least 3.5×10^9 yr ago. Hence, the only part of the record which might appear negotiable is the oldest portion covered by the meta-sedimentary rocks from Isua, West Greenland ($\sim 3.8 \times 10^9$ yr) which display conspicuous shifts in both their $\delta^{13}C_{carb}$ and $\delta^{13}C_{org}$ values. However, as has been pointed out before ("Biological Carbon Isotope Fractionation Through Geologic Time") and is fully detailed elsewhere (82, 84), these shifts are consistent with an isotopic reequilibration between coexisting sedimentary carbonate and organic carbon in response to the amphibolite-grade reconstitution which the Isua series had undergone during its later history. Therefore, it would seem rather far-fetched to seriously entertain the possibility that the Isua organics were derived from abiogenically produced precursor materials (though their isotopic compositions as such might also be consistent with fractionations obtained in spark discharge reactions). If the interpretation of the Isua values in terms of a metamorphic overprint is correct, then the record of "normal" sedimentary $\delta^{13}C_{carb}$ and $\delta^{13}C_{org}$ values would have originally extended to 3.8×10^9 yr ago, indicating that biological control of the terrestrial carbon cycle had been established already at the time of formation of the oldest sediments hitherto known.

The Sulfur Record

The principal question arising with regard to the sulfur record is that of the antiquity of dissimilatory sulfate reduction. Among other things, the answer will depend on how far back isotope patterns indicative of bacterial sulfate reducers can be traced into the geologic past.

For the last 1.2×10^9 yr, there is ample evidence in sedimentary rocks of the characteristic bacteriogenic fractionation of -40 ± 12 ‰ between sulfide and sulfate, with sulfate values staying well within the positive field and the corresponding sulfide patterns occupying appropriate ranges on the negative side of the scale (Fig. 11). In terms of Eq. (2), the positive readings of the sulfate curve would, by themselves, attest to formidable turnover rates in the biological processing and biogeochemical cycling of sulfur, their oscillations reflecting pronounced secular variations in the global activity of sulfate reducers (cf. "Impact of Dissimilatory Sulfate Reduction on the

Isotope Geochemistry of Sulfur"). Minima of the sulfate curve would indicate times of extensive evaporite formation at the expense of sulfate reduction (e.g., Permian with $\delta^{34}S = +11$ ‰) while the extended maximum during the early Paleozoic (with a peak of +32 ‰ in the Cambrian) would signal a temporary boom in bacterial sulfate reduction with concomitant suppression of the formation of sulfate evaporites.

As from $t \gtrsim 1.2 \times 10^9$ yr, sulfate virtually disappears from the sedimentary column. Accordingly, attempts to define, or bracket, the time or emergence of dissimilatory sulfate reduction have to rely on the patchy record of sedimentary sulfide available for this time span, with efforts concentrating on the identification of bacteriogenic isotope distribution patterns.

As is illustrated in Fig. 10, bacteriogenic sulfide patterns are generally characterized by (i) marked shifts of the $\delta^{34}S$ values in negative direction relative to the parent sulfate, and (ii) extended spreads of these $\delta^{34}S$ values. These criteria hold for open systems where the sulfate pool is continuously replenished and its initial isotopic composition basically maintained. For systems closed to sulfate (where a limited SO_4^{2-} reservoir is almost completely converted into sulfide) the shift toward negative $\delta^{34}S$ values is much less pronounced, while the large spreads are retained or even augmented. The frequency distribution of bacteriogenic values from an "open" basin is exemplified by the histogram of the Permian Kupferschiefer shown in Fig. 11 (No. 22), whereas a corresponding graph typical of an environment closed to sulfate is furnished by the pattern of the McArthur Basin (Fig. 11, No. 19).

When trying to trace such bacteriogenic sulfide patterns back in time, the oldest analogs are supplied by the Deer Lake Greenstone Sequence and the Woman River Iron Formation (Fig. 11, Nos. 8 and 12). The $\delta^{34}S$ distributions displayed by sulfides from the Birch-Uchi Greenstone Belt and the Michipicoten Iron Formation (Nos. 9 and 11) have also come to be included among the oldest presumptive evidence of bacterial sulfate reduction (cf. 34, 46, 79, 84). This would imply that sulfate respiration as an energy-yielding metabolic process had arisen at or before $\sim 2.8 \times 10^9$ yr. During upper Precambrian times, bacteriogenic isotope patterns became more abundant, one of the best examples being furnished by the sulfide mineralization of the McArthur Basin (Fig. 11, No. 19).

All older sedimentary sulfides hitherto investigated lack convincing bacteriogenic features, showing instead narrow spreads around zero permil suggestive of an isotopically undifferentiated (magmatic) sulfur source. This is particularly true for a suite of sulfides from the Isua banded iron-formation whose mean of

112. Wong, W.W. and Sackett, W.M.: 1978, Geochim. Cosmochim. Acta 42, pp. 1809-1815.

113. Wong, W.W., Sackett, W.M. and Benedict, C.R.: 1975, Plant Physiol. 55, pp. 475-479.

114. Woodwell, G.M., Whittaker, R.H., Reiners, W.A., Likens, G.E., Delwiche, C.C. and Botkin, D.C.: 1978, Science 199, pp. 141-146.

ORGANIC MOLECULES AS CHEMICAL FOSSILS - THE MOLECULAR FOSSIL RECORD

Geoffrey Eglinton

University of Bristol, Organic Geochemistry Unit,
School of Chemistry, Bristol BS8 1TS, England.

ABSTRACT. The molecular record in the geosphere shows that carbon compounds can retain a high degree of structural integrity and stereochemical specificity over hundreds of millions of years, if they are not exposed to raised temperatures and are stored in the sedimentary column in such a way that they are not subject to oxidation or microbial attack. The signature of environments is written in sediments in the form of specific "biological marker" components, the presence of which, as "chemical fossils", can be interpreted in palaeoenvironmental terms. The thermal history of a sediment is also reflected in the epimerisation, hydrogenation, aromatisation and carbon-carbon bond cleavage reactions observed through the study of samples of differing burial histories. Of particular interest is the possibility of relating the lipid record to past microbial activity. Thus, the archaebacteria possess unusual membrane lipids which may well prove a useful clue for microbial activity as far back as the Precambrian. There is a need for the recovery and study of unmetamorphosed sediments of all ages, including the Precambrian, so that micropalaeontologists, inorganic and organic geochemists, geologists and especially geochronologists, can work together towards a progressive characterisation, through time, of the various types of record left in the rocks. Only in this way will deep understanding of Precambrian events become possible.

INTRODUCTION

The Earth's crust contains substantial quantities of carbon, mainly as carbonate. However, some 15% is estimated to be

C. Ponnamperuma (ed.), Cosmochemistry and the Origin of Life, 323–359.

organic carbon which represents fossilised organic matter derived ultimately from the remains of dead organisms. The estimate of organic carbon in the Earth's crust is ca. 6.4×10^{15} metric tons, which exceeds a thousandfold the carbon held in the present biosphere. Much of this organic carbon is held in sedimentary rocks such as claystones and shales and only minute proportions in the form of the recognised fossil fuel reserves, such as petroleum, coal and natural gas. Virtually all of this crustal carbon, including the carbonate rocks and graphitic sediments, has been processed by living organisms in the past and subsequently entombed in sediments deposited in the world oceans. The evidence of this biological activity lies partly in the observable fossil remains, partly in the molecular structure of the residual and other sedimentary organic matter, and partly in the isotopic composition of these materials. This article will be concerned solely with the molecular record.

Most sedimentary organic matter is insoluble in organic solvents and is given the generalised term of "kerogen". Kerogen is a highly heterogeneous and variable, amorphous, polymeric material, which represents the debris and condensation products of the more resistant organic matter deposited from the environment of palaeodeposition. Its nature and composition is, of course, highly dependent on the source of the material and on the thermal history of the sediment containing it.

In addition to the kerogen, most sediments contain a rather small fraction of organic material which is soluble in organic solvents. This material, the geolipid fraction, is usually associated with the kerogen and is found to be complex in composition. However, it can be fractionated by chromatographic and other means and individual compounds isolated and identified. The recognition of these individual molecular compounds and the study of their origin and conversion in the sediments forms the basis of modern, molecular, organic geochemistry.

The solvent-soluble geolipid fraction which can be extracted from sediments may vary from a few ppm to several percent of the rock. In general, the quantity of extract falls off rapidly with age from the high percentages found with some young, organic-rich sediments, to the minute traces encountered in many ancient sedimentary rocks, such as those from the Precambrian. However, where thermal alteration through deep burial and other events has affected the insoluble kerogen, then the proportion of soluble geolipids may increase again, possibly even leading to an accumulation of petroleum. It has become clear that the complex mixture of organic compounds represented in the geolipid fraction comprises many compounds which bear direct relationships to present-day, biologically-formed lipids (biolipids). It is assumed that these structural relationships between the geolipids and biolipids result from the preservation of the original carbon skeleton during the accumulation, diagenesis and maturation of the sediment. This preservation is a direct result of the strength of carbon-carbon covalent bonds. Hence, all or part of the carbon skeleton of the lipid molecules may be preserved over long periods of time, though diagenetic defunctionalisation results in loss of hydroxyl, carbonyl and other functional groupings. These compounds, whose carbon skeletons suggest unambiguous links with known biological products, have been termed "chemical (or molecular) fossils". An additional term "biological marker compound" is also used for such compounds but more especially where a relationship is proposed between the compounds and particular organisms.

In summary, chemical fossils, or molecular fossils, are compounds of carbon, hydrogen and possibly oxygen, nitrogen and other heteroatoms which are found in sediments and crude petroleums. Their origins may be inferred by relating the carbon skeletons, making up their molecular backbones, to those of the presumed biological precursor molecules (natural products) deposited in the sediment at the time of deposition. These compounds vary greatly in stability. Depending on structure and functionality, their effective half-lives can extend to 10^9 years and hence back into the Precambrian, provided that the sediment burial temperatures have remained low. Their history may be complex. For example, they may represent a free lipid molecule which has never been bound covalently into the sedimentary organic debris. On the other hand, they may have been components of the insoluble debris (kerogen) and released as related compounds by thermal alteration during deep burial. The compounds vary greatly in structural complexity: an important point here is that the more complex the structure of a molecule, the greater the information content. An additional source of information is the isotopic composition of a compound, i.e. $\delta^{13}C$ for the compound itself and for individual atoms in its structure. A final source of information is the relative abundances of individual compounds, members of homologous series and of classes of compounds. All of this information is, of course, related to the original source organism(s) of the compound. Organisms synthesise molecules which are of limited structural type and which display specific distributions of individual members of that type. The structures are often highly specific in their skeletal arrangements and in their stereochemical configurations. In some cases, a molecule, such as that of a biologically-synthesised alkane, e.g. pristane, may be completely unchanged through geological time and can be found as such in an ancient sediment. Other compounds are often encountered in partially altered form, as derived structures. Diagenetic and catagenetic processes may well result in very extensive alteration or even complete "mineralisation" to CO_2, etc., in which case all information, other than isotopic, is lost (Fig.1).

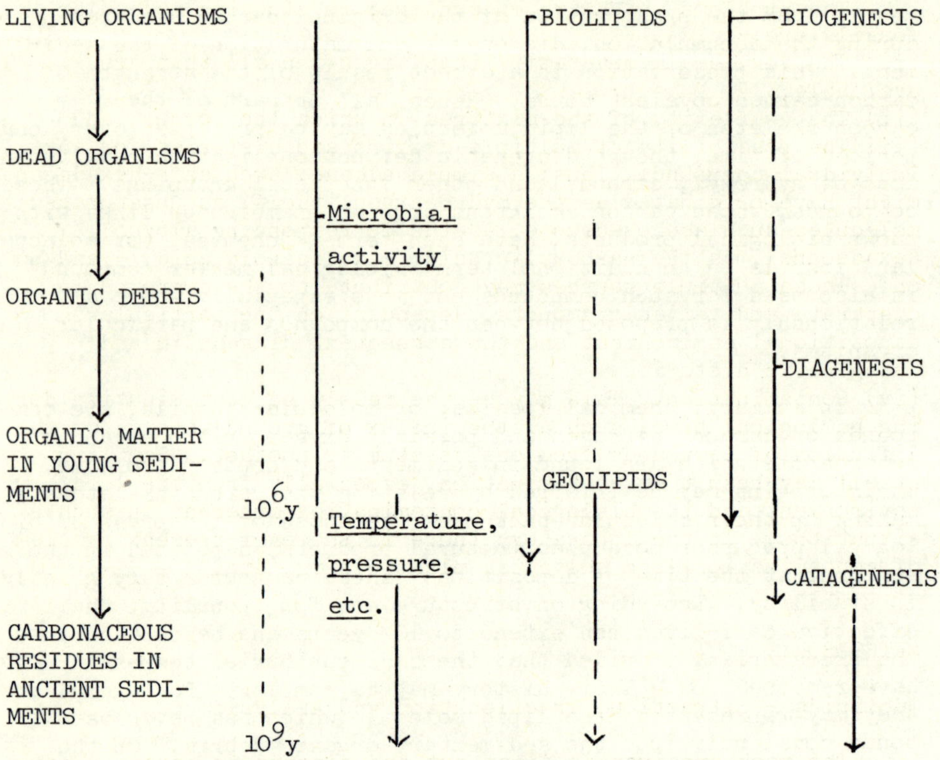

Figure 1. SCHEME FOR ORIGIN AND FATE OF GEOLIPIDS

The onset and cessation of processes or events, indicated by an incomplete arrow, vary from one sedimentary column to another.

The problems which are encountered in studying molecular fossils are:-

(i) The very small quantities that are often all that are available,

(ii) The complexity of the mixtures in which they occur, and

(iii) The possibility of multiple origins and histories for any individual compound. Thus, a simple alkane, such as n-hexadecane, might have originated as the hydrocarbon itself, as unsaturated molecules such as hexadecene, as the corresponding alcohol, hexadecanol, as the carboxylic acid, hexadecanoic acid, - and so on. These multiple sources may contribute to the compound in disparate and variable amounts, dependent on the particular depositional environment and the subsequent diagenetic and catagenetic histories.

(iv) Contamination. This may be the result of contamination during geological time, such as the influx of ground waters, or the diffusion of compounds from one stratum to another. Sampling itself may result in contamination, especially from the local environment and the biological contamination inherent in such situations. In the laboratory there is the ever-present problem of contamination from solvents, plasticisers, etc.

METHODS

Geolipids - Extraction, Fractionation and Analysis

The rock specimen is first cut and cleaned to obtain a sample free from external contamination, and then broken up into smaller pieces suitable for the next stage, which is pulverisation in a grinding mill or similar equipment. The finely powdered sediment is then extracted with organic solvents, such as toluene/methanol or methylene chloride, ultrasonically, or in a Soxhlet apparatus. The solvent extract can then proceed to the fractionation stage. Additional, prior demineralisation procedures are also used to release organic matter from the mineral matrix by treatment of the sediment with a mixture of purified hydrochloric and hydrofluoric acids. It is difficult to achieve complete demineralisation but the technique is important in the case of ancient, fully-lithified sediments, such as samples of Precambrian age. The soluble organic material (geolipid) in the solvent extracts is then fractionated by chromatography using liquid-solid chromatography carried out on columns or thin layers (LC or TLC), or a combination of these (Fig. 2). The fractions so obtained can be subjected to further LC or TLC prior to derivatisation (e.g. by conversion to the methyl esters and/or trimethylsilyl derivatives) and then taken on to the next stage, which is usually gas chromatography (GC). The distribution patterns of compounds, such as alkanes, aromatic hydrocarbons and so on, within a fraction is then revealed. Usually, the final step is to examine fractions by

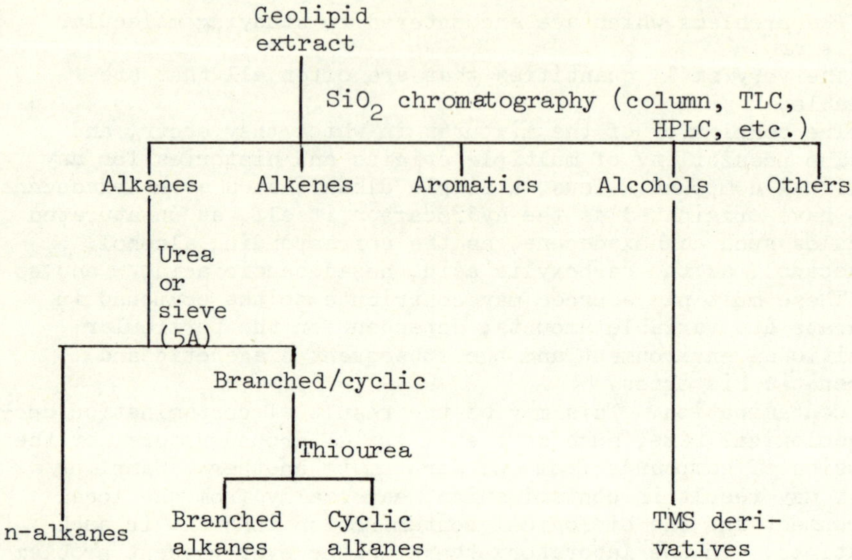

Figure 2. FRACTIONATION OF GEOLIPIDS

This scheme provides a basic procedure for obtaining geolipid fractions in the µg-mg range suitable for gas chromatography, followed by GC-MS analysis where appropriate.

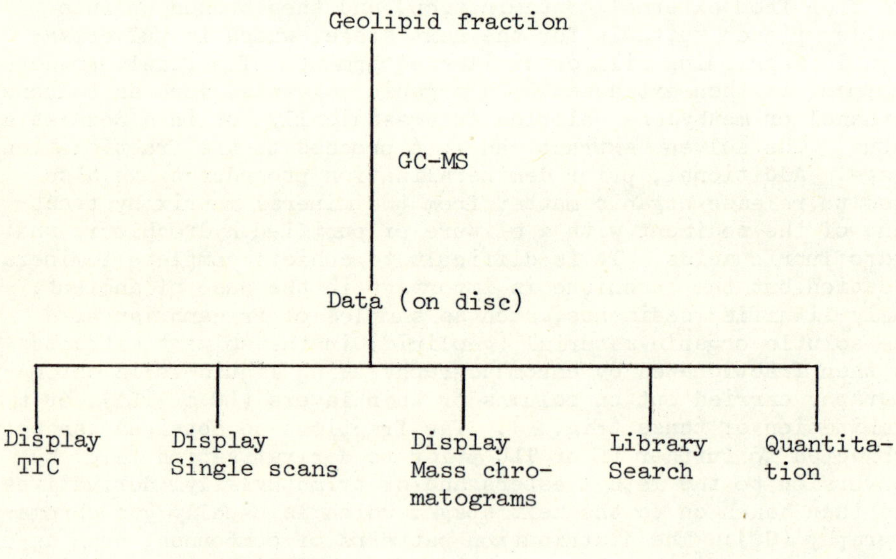

Figure 3. COMPUTERISED GC-MS ANALYSIS OF GEOLIPID FRACTIONS.

combined gas chromatography-mass spectrometry (GC-MS) or, less commonly, combined liquid chromatography-mass spectrometry (LC-MS). In either case, the aim is to obtain compound identification and quantitation. The development of computerised GC-MS with glass capillary columns or fused silica capillary columns has led to a substantial increase in the number of geolipids identified in sediment extracts. Indeed, the preliminary fractionation procedures may still result in fractions containing of the order of a thousand single compounds; hence, the data produced in a single GC-MS analysis require a computer both for acquisition and appraisal. Much use is made of computer-constructed, mass fragmentograms (plots of the intensity of an ion characteristic of a particular compound type against GC retention time). With commonly encountered compounds, the mass spectrum is taken as sufficient evidence for the presence of that compound, provided that the GC retention time is appropriate. However, with unusual samples or with previously unrecognised components, then it is appropriate to seek high resolution mass spectrometry for element composition of molecular ions and fragment ions, together with GC-MS coinjection studies with standard compounds. Additionally, it may prove necessary to isolate sufficient of the compound for more detailed structural studies by other techniques, such as nuclear magnetic resonance. It is usually difficult to obtain more than a few micrograms of a single pure compound from a geological extract and hence, high resolution capillary gas chromatography over phases of different polarity has been employed for detailed stereochemical characterisation of geolipids.

Computerised GC-MS of geolipid fractions usually involves capillary gas chromatographic separations lasting of the order of 1 hour, accompanied by fast, sensitive mass spectral data acquisition at the rate of ca. 1 scan sec^{-1} (Figs. 3 and 4). The scan data are acquired on magnetic disc, and may then be displayed as a reconstituted ion current (RIC), or total ion current (TIC), which should closely match the GC analyses previously obtained for the same sample using a similar column. These RIC traces give a general idea of the complexity of the mixture and the degree of separation obtained. The geochemist may then choose to display single scans of interest corresponding to peaks of significance. Additionally, the computer will provide the display of mass chromatograms (fragmentograms) - the intensity of each scan of a given mass. This type of display serves as a "fingerprint" for selected classes or types of compounds. It is also useful in revealing suites of analogous compounds, e.g. the C_{27}, C_{28} and C_{29} steranes. Another important facility available with computerised GC-MS systems is that of library search. If a given compound is present in a mixture under analysis, then provided it is in the main library of reference compounds available on the computer system, it may be identified and listed against the appropriate scan

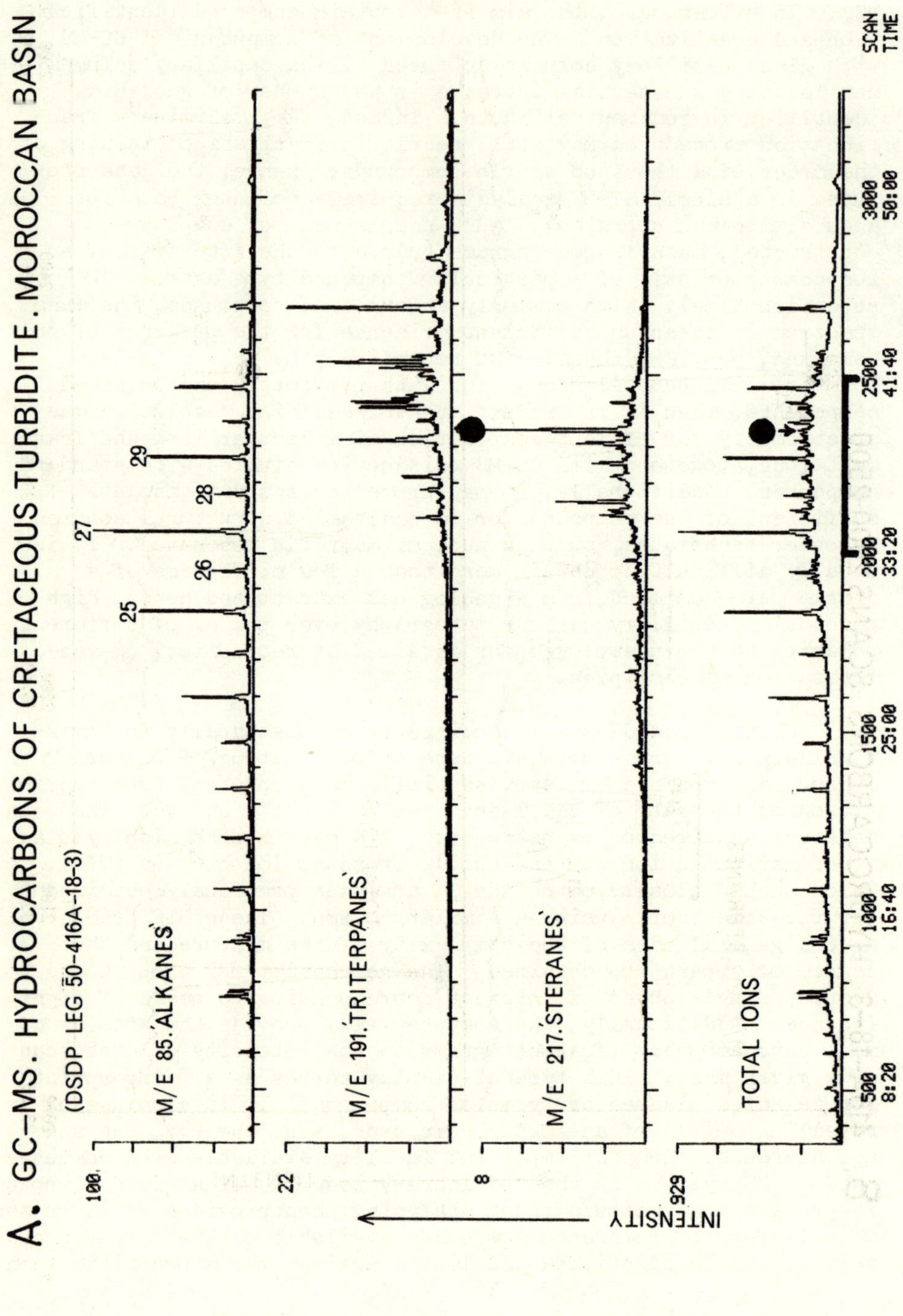
A. GC–MS. HYDROCARBONS OF CRETACEOUS TURBIDITE. MOROCCAN BASIN
(DSDP LEG 50-416A-18-3)
M/E 85. 'ALKANES'
25
26
27
28
29
M/E 191. 'TRITERPANES'
M/E 217. 'STERANES'
TOTAL IONS
100
22
8
929
INTENSITY
500
8:20
1000
16:40
1500
25:00
2000
33:20
2500
41:40
3000
50:00
SCAN
TIME

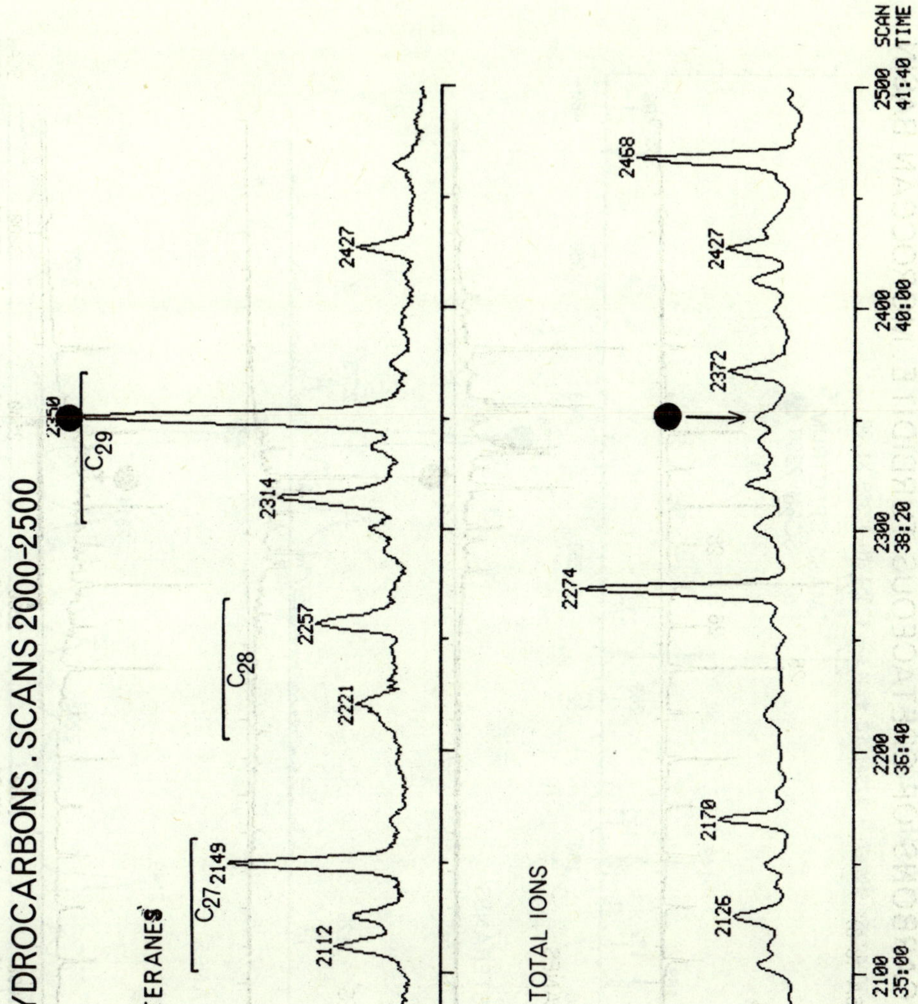
B.
416A–18–3 . HYDROCARBONS . SCANS 2000–2500
M/E 217. 'STERANES'
C27
C28
C29
2112
2149
2221
2257
2314
2350
2427
TOTAL IONS
2066
2126
2170
2274
2372
2427
2468
INTENSITY
2000 33:20
2100 35:00
2200 36:40
2300 38:20
2400 40:00
2500 41:40
SCAN
TIME

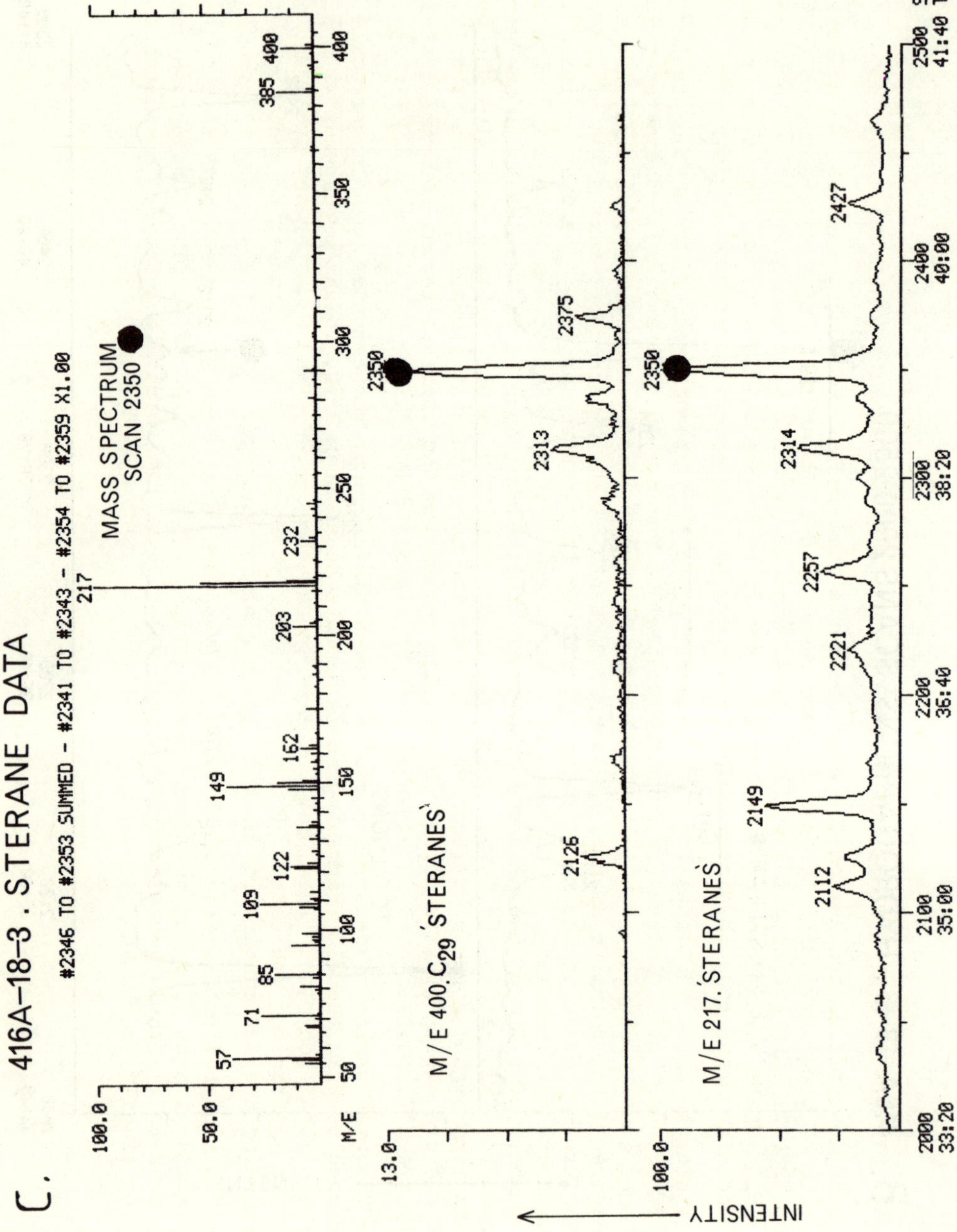
C.
416A-18-3 . STERANE DATA
#2346 TO #2353 SUMMED - #2341 TO #2343 - #2354 TO #2359 X1.00
MASS SPECTRUM
SCAN 2350
M/E 400, C29 'STERANES'
M/E 217. 'STERANES'
INTENSITY
SCAN
TIME

Figure 4. EXAMPLE OF COMPUTERISED GAS CHROMATOGRAPHY-MASS SPECTROMETRY (C-GC-MS).

This example typifies the search for characteristic families of compounds, such as the steranes. Recognition of the individual steranes is here based on C-GC-MS analysis of a sediment sample from the Deep Sea Drilling Project (Leg 50, Hole 416A, Core 18-3) which had been drilled in the Moroccan Basin. Full GC-MS conditions and other details are given by Brassell, Gowar and Eglinton, pp.421-426 in "Advances in Organic Geochemistry 1979" (eds. Douglas and Maxwell), Pergamon Press, 1980.

A. This diagram shows the Total Ion plot (Reconstructed Ion Chromatogram; RIC) for the greater part of the GC-MS run. Also plotted are mass fragmentograms for m/e 85, 191 and 217. These three ions are intense ions in the mass spectra of alkanes, triterpanes and steranes, respectively. A qualitative measure of the relative amounts of these components can be obtained by examination of the signal strength for the scan with the most intense signal for that particular ion. These intensities indicate the sequence: n-alkanes >> triterpanes > steranes. The m/e 85 plot has been marked with the carbon numbers of a few of the n-alkane homologues but the dominance of the n-alkanes in the mixture of hydrocarbons is seen in the Total Ion plot. An arrow on the Total Ion trace indicates the position of the maximum signal for m/e 217, corresponding to the most abundant sterane component, but it is not easy to discern a peak at this point. Hence, the region for the steranes (scans 2000-2500, as indicated by m/e 217) is presented in expanded form in Figures B and C.

B. The "Sterane region" (scans 2000-2500) has been displayed by the computer on a larger scale. The most intense peak in the 217 fragmentogram is seen to be at scan 2350, which corresponds with a weak peak in the Total Ion plot. From previous experience, the C_{27}, C_{28} and C_{29} steranes are found on the 217 plot in the regions indicated. For more details, it is necessary to proceed to additional fragmentogram displays and examination of full spectra as in Figure C.

C. The "Sterane region" is seen here for m/e 217 and m/e 400. Scan 2350 corresponds to a C_{29} sterane (molecular ion at 400) and the full spectrum (scan 2350) confirms this. The further assignment of this peak to 24-ethyl-5α-cholestane requires relative retention data or, better, coinjection data using standard compounds.

number. Such searches can be made automatic and their reliability and effectiveness much improved by incorporation of relative retention time data into the search.

Kerogen - Analytical Procedures

Usually more than 95% of the total organic matter in sediments is comprised of material insoluble in water and organic solvents. In very young sediments, this material is comprised of, at least in part, fulvic and humic acids, and hence can be extracted by aqueous base and subjected to fractionation and isolation procedures, followed by detailed spectrometric analysis. However, with age and deeper burial, this fraction is rapidly converted into completely insoluble material and becomes part of the heterogeneous polymeric material known as kerogen. In sediments which have not been subjected to deep burial and raised temperatures, the kerogen may be regarded as a composite cross-linked, heteropolycondensate, intermixed with and bound to fragments of biological debris. With increased temperatures and the passage of time, functional groups are lost and the content of hydrogen, oxygen and nitrogen falls as the composition tends towards amorphous carbon/graphite.

Several approaches to kerogen analysis have been used with varying success:-
(i) Investigation of overall composition by elemental analysis and by non-destructive methods of direct examination such as optical techniques and by infrared, nuclear magnetic resonance and electron spin resonance spectrometry, etc.
(ii) Use of pyrolysis techniques, either flash or step-wise, to break up the kerogen polycondensate into pyrolysate mixtures of smaller molecules amenable to GC or GC-MS, particularly as combined techniques, i.e. PY-GC, PY-MS, and PY-GC-MS.
(iii) Use of chemical degradation methods to attack selectively functional groups and hence yield mixtures of degradation products for analysis by conventional GC and GC-MS methodology (e.g. degradations effected by O_3, BBr_3, or $LiAlH_4$). The composition, as indicated by the compounds released, does vary from one type of depositional environment to another: thus, some kerogens are high in aliphatic components, which include straight, branched- and cyclic hydrocarbon moieties, while others are high in aromatic rings. Where the trend to graphite/amorphous carbon has not proceeded far along the diagenetic-catagenetic pathway, then aliphatic versus aromatic content is found to relate fairly closely to the morphological information gleaned by optical and electron microscope studies:- i.e. the highly aliphatic kerogens are associated with organic matter of an algal (phytoplankton) origin, whereas aromatic kerogens appear to be associated with land-derived, woody (higher plant) debris and humic acids and similar phenolic materials. Increasing attention is being given to the

1. Skeletal Type	Principal Biochemical Pathways for Carbon Skeletons
Linear chain	Acetate - malonate
Acyclic-branched	Acetate-malonate and methyl malonate
Acyclic-branched (isoprenoid)	Mevalonate
Alicyclic (isoprenoid)	Mevalonate
Aromatic ring	Acetate-malonate, shikimate.

2. Compound Classes

Alkanes*, alkanols, etc. ...
Arenes*, PNAs*, Furans*
Steroids, Triterpenoids
Porphyrins and metalloporphyrins* (chlorins)
Amino acids*, polypeptides, proteins
Sugars, polysaccharides

(* = reported for Precambrian sediments)

TABLE 1. CHEMICAL FOSSILS.

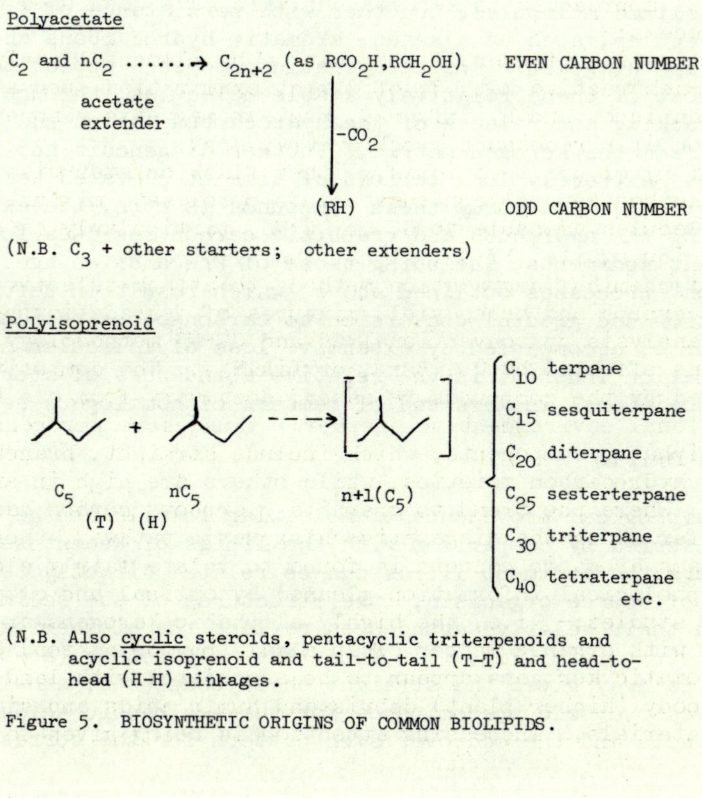

Figure 5. BIOSYNTHETIC ORIGINS OF COMMON BIOLIPIDS.

study of kerogen, for it is hoped that the wealth of knowledge concerning the extractable geolipids will be directly transferable to the interpretation of components released from kerogen by improved degradative techniques.

THE MOLECULAR RECORD

At this point it would seem appropriate to survey the occurrence of molecular fossils in the geological record. Most analytical effort has been expended on young bottom sediments and sediments of the Tertiary, with special emphasis on sediments associated with the major oil fields. Sediments of Precambrian age have received much less attention. There is, unfortunately, a substantial gap in time between the Precambrian and the Tertiary. The young bottom sediments contain a wealth of compounds of direct biological origin, together with some geolipids corresponding to early diagenetic products. Hence, these sediments contain a wide range of carbon compounds, including sensitive biolipids, such as carotenoid and chlorophyll pigments. Studies made of older samples of a single depositional type, either down the geological column or for a single depositional horizon buried to different depths, show progressive destruction and modification of biolipid molecules, together with persistance of fairly stable molecules, such as alkanes, aromatic hydrocarbons and porphyrins. The diagenetic and catagenetic processes appear to generate most of these relatively stable molecules, through thermal cracking and release of the hydrocarbon chains and cyclic molecules from the kerogen matrix. Further diagenesis and catagenesis over extremely long periods of time at elevated temperatures progressively destroy these compounds in turn, the end result being the amorphous and graphitic carbon residues found in many ancient sediments, including those of Precambrian age. The "maturation" processes outlined above, which result in defunctionalisation and gradual conversion to carbonaceous residues, are of course, accompanied by extensive loss of molecular order and information inherent in the relative abundances of stereoisomers, positional isomers and of members of homologous series.

Biological Origin

The biological provenance of molecular fossils can be most readily inferred by comparison with the lipids of Recent sediments. Here the sediment lipids can be related directly to the biolipids of source organisms; the structures of the sediment lipids and their relative abundances reflect the biochemical pathways of their origin (Fig. 5). Thus, the polyacetate (acetate/malonate) pathway gives the straight chain compounds, showing the typical even over odd predominance for the n-alkanoic acids and the n-alkanols and the odd over even pattern for the corresponding n-alkanes. With the polyisoprenoid compounds then the nC_5 pattern of carbon number (C_{10}, C_{15}, C_{20} etc.) is prominant. Additionally, carbon number maxima corresponding to the enzymatic "peaking" typical of biosynthesis show unimodal, bimodal and higher distributions, depending on the input from the different types of organism. The classic example of this is provided by the n-alkanes, where the carbon numbers around C_{17} are prominant in phytoplankton while those around 29 or 31 are dominant in the leaf surface waxes of higher plants. Many aquatic bottom sediments show bimodal distributions of n-alkanes centred around C_{17} and C_{29} or C_{31}, corresponding to input from both phytoplankton and terrigenous higher plant debris. Studies under way are aimed at surveying the whole range of lipids in Recent sediments and in the organisms expected to contribute to and modify them (Table 1 Fig.6). The great diversity of molecular parameters afforded by the several thousand discrete carbon compounds which would appear to be present in sediments offers a flexible and powerful method by which organisms, environments and sediments could be inter-related. Here, the true interdisciplinarity of the subject becomes apparent, for it will be necessary to consider inter alia, chemotaxonomic aspects, the underlying biochemistry and the food webs in the ecosystems. There are a substantial number of environmental situations which can be expected to give rise to sedimentary facies which differ at the molecular level. In young bottom sediments the correlations may possibly include those between the lipid patterns, biomass measurement and the level of biochemical activity. Microbiologically-mediated processes are of crucial importance here. Such processes not only make substantial contributions of microbial lipids but also modify pre-existing lipids. Hence, we have first to determine the molecular structures, including stereochemistry, and relative abundances of biolipids and geolipids found in contemporary aquatic environments and in Recent bottom sediments. From there, we need to deduce what these data reveal concerning the direct input of biosynthesised compounds and the operation of the various processes taking place in the environments and in the bottom sediments. This large task can be expected to occupy many years but while identifying this need, it is equally important to examine the lipids of older sediments. The problems of diagenesis and catagenesis have to be dealt with, even though the full picture for contemporary bottom sediments remains uncertain.

The contemporary situation in the study of molecular fossils may be compared with the early "natuial history" period of palaeontology, when the description and documentation of fossils were essential preliminaries to the development of the major unifying theory of evolution and of key, powerful tools such as biostratigraphy. We are now in a position, through C-GC-MS, to document a significant portion of the molecular fossil record and then to exploit it. However, the analytical and documentation problems

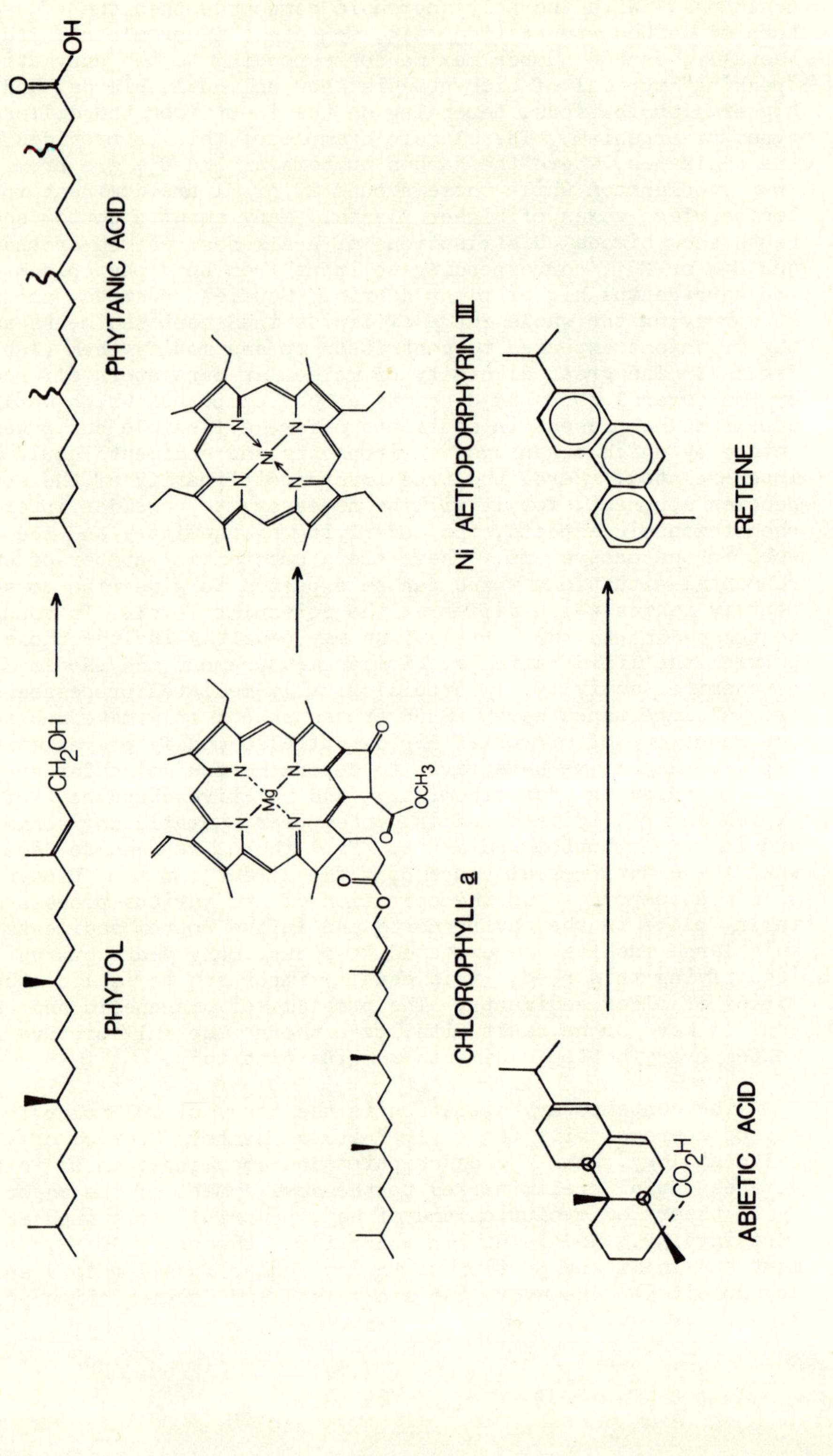
PHYTOL
CH2OH
PHYTANIC ACID
O
OH
CHLOROPHYLL a
Mg
OCH3
Ni AETIOPORPHYRIN III
Ni
ABIETIC ACID
CO2H
RETENE

Figure 6. MOLECULAR FOSSILS: PROPOSED PRECURSOR-PRODUCT RELATIONSHIPS.

The figure gives examples of proposed precursor/product relationships. The precursors are common biolipids found in organisms contributing to aquatic systems. The molecular fossils shown as products are not known to be biosynthesised directly but all are found in sediments. (Adapted from Mackenzie, Brassell, Eglinton and Maxwell, Science, in press).

The phytanic acid is found in immature bottom sediments and is believed to be derived from phytol. Stereochemical data indicate that this is so, but in more mature sediments, the mixtures of stereoisomers indicate that diagenetic changes have operated.

Chlorophylls are vital photoreceptor molecules, present in all green plants. The most abundant chlorophyll, chlorophyll a,is composed of a macrocycle and a phytyl side chain. In ancient sediments both of these component parts can survive separately in slightly altered forms,as the tetrapyrrolic porphyrins and the isoprenoid hydrocarbons, respectively. It is the skeletal relationships between these products and the precursor chlorophyll that attest their biological origin. The nickel aetioporphyrin-III is a C_{32} porphyrin which has been identified in the bitumen Gilsonite, isolated from rocks of Eocene age from the Green River formation, U.S.A. The precursor is believed to be chlorophyll a, an assumption which would require a loss of the phytol side chain, reduction of the vinyl and carbonyl groups, loss of the methyl ester function and finally, cleavage of the isocyclic ring, aromatisation of the macrocyclic ring and sequestration of the nickel ion. The phytyl side chain follows separate diagenetic pathways, giving rise to inter alia, phytanic acid (above) and pristane and phytane.

The diagenetic formation of retene from abietic acid is commonly believed to take place in young sediments. Related compounds such as dehydroabietic acid, dehydroabietane, dehydroabietin, simonellite and 1,2,3,4-tetrahydroretene are also known in sediments.

we face are of a different scale: where a single sample of sediment may yield up to 10^2 different microfossils, the same sample will likely contain some 10^2-10^5 different molecular fossils awaiting recognition and documentation.

Exploitation of molecular fossils is already under way in the earth sciences. For example, as follows:-

1. Amino acid dating and stratigraphy.
2. Use of sterane and triterpane patterns in the correlation of oil and source rocks.
3. Use of changes (epimerisation, aromatisation and carbon-carbon bond fission) in steroidal and triterpenoidal hydrocarbons for the assessment of the maturity of sediments and hence of geothermal gradients.
4. Determination of the abundances of straight and branched-chain alkanes in the assessment of the extent of biodegradation of oils.

In the following two sections, by way of exemplification, attention is given to the molecular assessment of first, deep sea sediments and second, ancient sediments and crude petroleums.

Molecular Assessment of Deep Sea Sediments

Solvent extracts of immature marine sediments contain a bewilderingly complex array of lipids. The ability to recognise these compounds has been vastly increased in recent years by the use of fast-scanning C-GC-MS systems equipped with glass capillary columns. For example, recent C-GC-MS studies of the sterol composition of marine sediments have illustrated that their distributions are markedly more complex than previously realised. Many of the minor components now recognised appear to be indicators of specific inputs to the sediments, e.g. of dinoflagellates or of sponges, so that the range of biota reflected in sterol distributions is highly diverse and includes both pelagic and benthic organisms. Additionally, series of long-chain unsaturated ketones, hydrocarbons and acids have been identified in numerous marine sediments, ranging from Miocene to Quaternary age, and in the marine coccolithophore, *Emiliania huxleyi*. These compounds appear to be biological markers for coccolithophore inputs and are present in sediments from the Japan Trench that contain no skeletal remains of these organisms in that they were deposited below the carbonate compensation depth (CCD). Hence, these organic geochemical markers can assess inputs that palaeontological data cannot. In addition, consideration of the distributions of the di- and triunsaturated ketones found in the various sediments suggests a relationship with the water column temperature (i.e. warm or cold water for coccolithophore growths). This possibility is under investigation.

Specific acyclic isoprenoids may be valuable indicators of inputs from methanogenic bacteria. They have been found in marine sediments from a variety of sites (e.g. Walvis Bay, Gulf of California and the Cariaco Trench), thus opening up possibilities for the evaluation of past microbial activity in such geological materials. Similarly, the presumed precursors of extended hopanoids - polyhydroxy-bacteriohopanes, have recently been recovered intact from lacustrine sediments and offer new potential in the evaluation of aerobic bacterial inputs.

The investigation of sediments from the Japan Trench has shown that in areas of low geothermal gradient, labile biolipids can be remarkably well preserved, hence providing important information concerning the palaeoenvironment of deposition. However, studies of this type require parallel information from contemporary environments of deposition. Hypotheses based on the expected contributions of biolipids from organisms need to be tested with the lipid abundance patterns found in sediments beneath different water column regimes, where the input of debris is at least partially understood from direct observation and microscopic studies. At a simple, preliminary level, the compound types that are being manipulated are characterised in terms of land-derived and water-derived contributions (Table 2). In some cases, the two overlap, e.g. chlorophyll _a_ is present in land plants as well as in phytoplankton; however, chlorophyll is believed to have been destroyed by the time that most land-derived debris reaches deep sea situations. Similar considerations apply to the sensitive carotenoid pigments. By contrast, cutin, the component polyester of higher plant cuticles, is not present in phytoplankton, but it is relatively resistant to decay. Numerous other possible marker compounds and biopolymers exist and are under scrutiny. An interesting point here is that ancient sediments exposed on land and eroded into the marine situation may contribute ancient, mature organic matter and possibly, free mature geolipids. These geolipids would include hydrocarbons such as steranes and hopanes, which are highly resistant to microbial attack and could thereby add an ancient "fingerprint" of mature hydrocarbons to an otherwise young and immature lipid pattern. Indeed, muds in estuaries have been shown to contain hydrocarbons of this type eroded from nearby organic-rich shales.

Molecular Assessment of Ancient Sediments and Crude Petroleums

Analyses to date have concentrated on the alkane and aromatic hydrocarbon fractions of extracts derived from ancient sediments, although the alkyl metalloporphyrins and monocarboxylic acids have received attention. A single chromatography step is normally sufficient to provide alkane and aromatic hydrocarbon fractions. Ratios of single components can then be measured by

	LAND (transported debris)	WATER (phytoplankton)
n-Alkanes	..C_{29}, C_{31}, C_{33}.. (ex. plant waxes)	C_{17} (ex. algal lipids)
n-Alkanoic acids	..C_{14}, C_{16}, C_{18}.. ..C_{26}, C_{28}, C_{30} (ex. plant waxes)	..C_{14}, C_{16}.. (ex. algal lipids)
Chlorophylls	+	++
Carotenoids	+	++
Cutin acids	+++	-

+, ++, +++ = relative contributions (-, no contribution)

TABLE 2. LAND AND AQUATIC CONTRIBUTIONS - AN ABBREVIATED SYNOPSIS OF LIPID INPUTS.

FACTOR	COMMENT
Temperature	Geothermal gradient, thermal conductivity of sediment.
Time	Time-temperature history.
Mineral matrix	Clay minerals, hydration, adsorbed metal ions.
Organic matrix	Phenolic and acidic groups, H-transfer groups.
Pore water chemistry	Porosity and permeability of sediment, pH and Eh of pore waters.
Pressure	Dissolved gases.

TABLE 3. BURIAL HISTORY OF SEDIMENTARY ORGANIC MATTER - FACTORS AFFECTING MOLECULAR DIAGENESIS/CATAGENESIS.

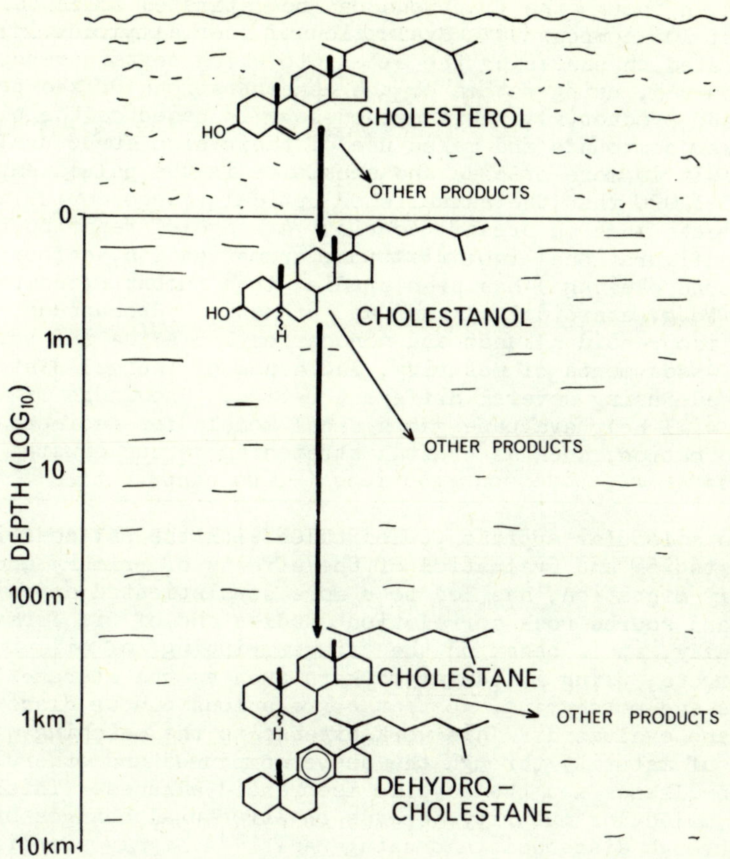

Figure 7. SEDIMENTARY FATE OF BIOLIPIDS: PRESERVATION OF STEROID SKELETON.

Steroids, such as the sterol cholesterol, are important constituents of aquatic animals and plants which contribute to bottom sediments. They are relatively resistant biolipids and can be found both in young sediments and in ancient sediments which have suffered little diagenesis. The C_{27} tetracyclic skeleton of cholesterol persists in deeply buried sediments, even though functional groups are lost. The trend to increased diagenesis is marked by the successive dominance of the reduced steroid, cholestanol, the cycloalkane, cholestane, and the partially aromatised hydrocarbon, dehydrocholestane. Petroleums frequently contain steranes and dehydrosteranes.
(The scheme shown is greatly simplified for ease of presentation; a detailed account is given by Mackenzie, Brassell, Eglinton and Maxwell in Science, 1982, in press).

GC-MS, even though the fractions can be estimated to contain in excess of 10^3 components. Evaluation of thermal maturation has concentrated on assessing the extent to which certain reactions have occurred, using ratios of the concentrations of the precursor(s) and product(s). This methodology is based on the biological marker compounds and makes use of their high structural specificity. It is more precise and sensitive in the palaeotemperature range 30-120°C than the established optical (microscopic) maturity measurements made on organic debris. A range of reaction types (e.g. configurational isomerisation, aromatisation, carbon-carbon single bond cleavage) has been studied in a number of compound classes (e.g. steroidal alkanes and aromatic hydrocarbons, acyclic isoprenoid alkanes and monocarboxylic acids and porphyrins). Assessments of maturity, and hence of thermal history, can be made using several different ratios. Knowledge of thermal history will help evaluate geophysical models for sedimentary basin formation, such as crustal stretching versus crustal flexure.

The molecular approach, when taken with the palaeoenvironmental studies and evaluation of the effects of primary and secondary migration, has led to a more sophisticated appraisal of oil and source rock correlation studies and of oil formation. Essentially, it is based on the "fingerprinting" of oils and source rocks, using molecular markers such as the steranes, triterpanes and porphyrins, so that correlations can be discovered, tested and evaluated. This work extends to the matching of degrees of maturity through the above compounds and others, such as the n-alkanes and the acyclic isoprenoid alkanes. This concept of molecular maturity depends on structural changes brought about through diagenesis and catagenesis. A highly simplified scheme (Fig. 7) illustrates how the biolipid sterols may be related diagenetically to the steranes and dehydro derivatives.

THE MOLECULAR RECORD: ALTERATIONS DUE TO DIAGENESIS AND CATAGENESIS

In a hypothetical situation where organic debris is being incorporated into a bottom sediment and buried by successive deposition of more sediment, then after the passage of millions of years, the sediment accumulation reaches, first hundreds of metres and then several kilometres. The temperature of the original layer of organic debris will rise steadily and may eventually reach the range of early metamorphism, say 200-300°C. The earlier, low-temperature changes in the sediment which take place as it forms, is compressed and becomes lithified, are part of diagenesis: changes brought about during deeper burial involve temperatures in excess of 40°C, and are in the realm of catagenesis. At the very deepest levels, the sediment temperatures may reach those (over 200°C) resulting in the formation of the green schist facies, which is termed the zone of metagenesis. No discussion will be attempted here of metagenesis, wherein organic matter will be destroyed with the evolution of methane and CO_2 and graphitisation or of the later stages of catagenesis, where these processes are well under way (Table 3 and Fig.8).

In early diagenesis, the main changes occur in the first few centimetres of sediment and are believed to be largely mediated by microorganisms and are therefore biochemical in nature. Additionally, non-biochemically mediated reactions may take place, especially as the chemistry (for example, Eh or pH) of the pore waters changes. Many other factors may play a part but much of the original organic matter will be recycled here and lost as CO_2, methane, etc. Biolipids may suffer modification and appear as corresponding geolipids, with structures reflecting their origin in the precursor biolipids. As burial continues and the organic compounds experience continued slow diagenesis, functional groups are progressively lost and the proportion of recognisable biolipids falls as the proportion of geolipids (molecular fossils) rises. The lipids of the Toarcian shales, when this horizon is followed down into the Paris Basin syncline, clearly show these gradual trends. They are also seen in the samples taken from the long cores of sediment obtained by the Deep Sea Drilling Program from sites such as the Japan Trench. A variety of reactions have now been recognised as proceeding under diagenesis and catagenesis. These include dehydration, hydrogenation, dehydrogenation, aromatisation, epimerisation, isomerisation, rearrangement, polymerisation and depolymerisation. The reactions and processes are, of course, accompanied by the loss of hydrogen, oxygen, nitrogen, sulphur and other atoms. There is a sequence of reactions but different compounds are, of course, different in their susceptibilities to various types of reaction. Hence, the reactions are not localised to brief intervals of time but some generalisations are emerging. Thus, during early diagenesis, polymerisation may occur, resulting in the incorporation of biolipids into the kerogen, whereas in the later stages of diagenesis and during catagenesis the kerogen will be broken down by depolymerisation, so giving rise to newly-evolved geolipids. Always the progress will be towards more stable molecules. Hence, if examination of a sediment reveals the presence of lipid components which are not of a maturity appropriate to that particular rock, then geological "contamination" must be inferred. Most of this information concerning diagenetic and catagenetic stages is presently empirical, as it is based on geolipid data obtained directly from sediment samples taken in the field or from cores. However, laboratory heating experiments do provide a way to determine the reaction pathways and rates at which they will occur in a variety of matrices and under different conditions of temperature and pressure, etc. There is much work

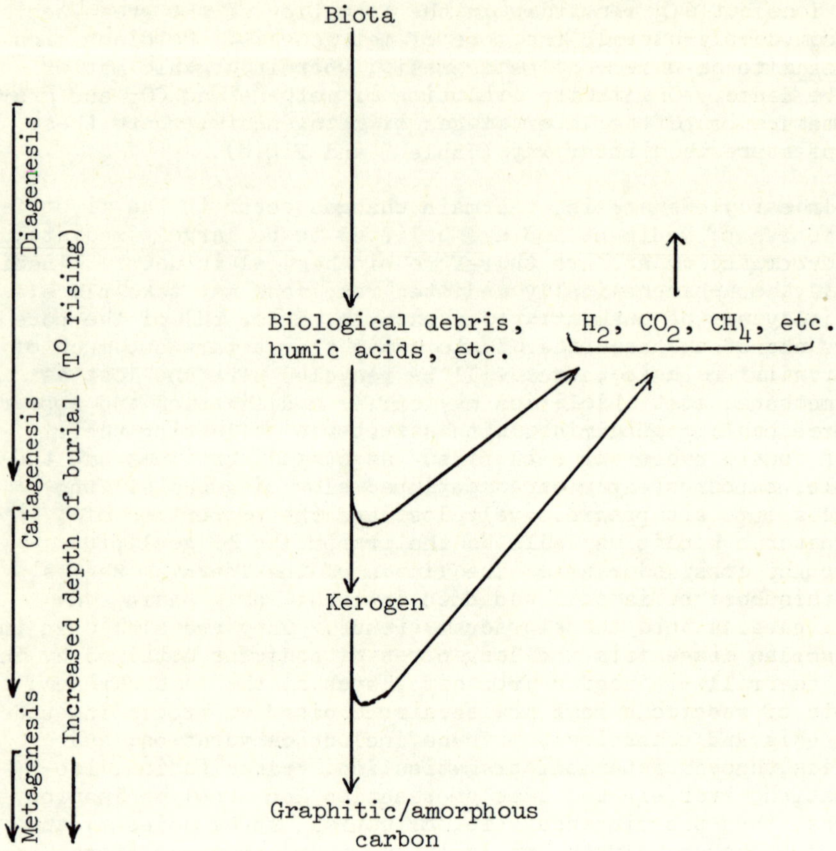

Figure 8. ULTIMATE FATE OF DEEPLY BURIED ORGANIC MATTER.

to be done but a general pattern is emerging. The generation of oil from deeply-buried, kerogen-containing source rocks is beginning to be understood and the knowledge applied in a predictive sense. One rather delicate measure of the effects of the slow maturation of organic matter (diagenesis/catagenesis) is that of stereochemical change.

Stereochemical Change

Enzymatic control of the biosynthesis of lipids results not only in the appropriate carbon skeletons being assembled in the right sequence of carbon-carbon bonds, but also in the strict observance of the required stereochemistry. The question of functionalities aside, these stereochemical configurations are often not the most stable and diagenesis results in conversion to more stable configurations. Hence, stereochemical studies are of considerable value in following diagenetic changes in, for example, suites of hydrocarbons, such as the acyclic isoprenoids, steranes and triterpanes. Of course, selective attack on particular stereoisomers may well take place in the bottom sediments during the early biochemical diagenesis but examination of a depth sequence of samples should reveal such phenomena. Much more significant is the slow epimerisation of carbon centres during late diagenesis. These epimerisations are believed to be slow, thermally-induced reactions. However, the most well known example of stereochemical diagenesis is provided by the α-amino acids (Fig.9). This low temperature reaction occurs readily and proceeds through intermediates which involve the functional groups. However, in the case of the chiral centres of hydrocarbons, then the diagenetic epimerisation may be mediated by clays or other catalysts but is little understood. Differently substituted carbon atoms isomerise at different rates but it has proved possible to discern an order in the rates which is followed rather generally by a wide variety of molecules possessing these particular types of centre. Hence, the presence in a sediment of a number of compounds bearing chiral centres of differing ease of epimerisation, provides a range of potential geochronometers and geothermometers. For example, the typical epimerisation half lives for burial in the sedimentary columns would be in the range of 10^2-10^7 years for α-amino acids and 10^7-10^9 years for acyclic isoprenoids, steranes and triterpanes.

MICROBIAL LIPIDS

Two important structural classes of molecular fossil found in sediments and petroleums are acyclic isoprenoids of presumed archaebacterial origin and hopanoids ascribed to procaryote microbial sources. The acyclic isoprenoids are of a special type and represent a relatively recent discovery (Fig.10). Hopanoid

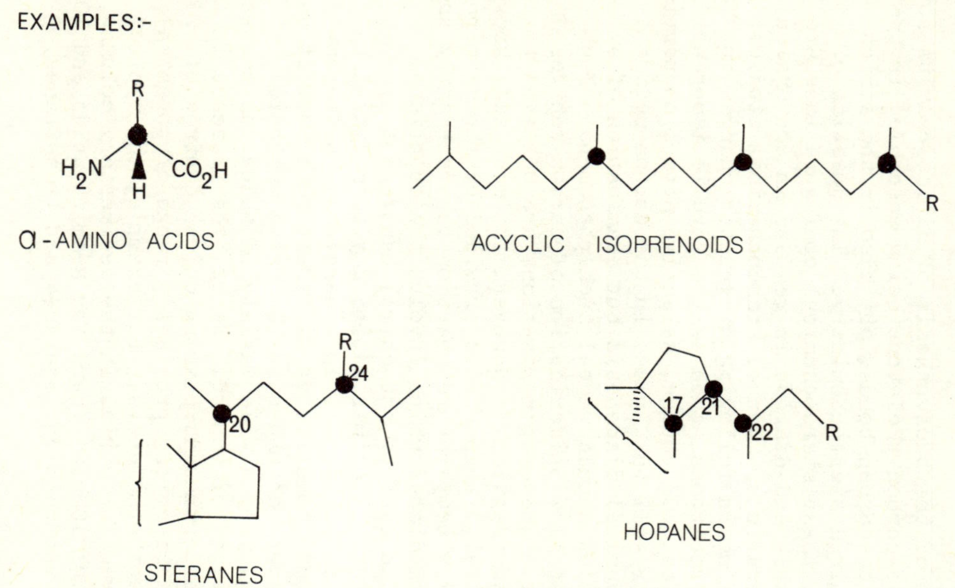

Figure 9. EXAMPLES OF MOLECULAR FOSSILS WHEREIN CHIRAL CARBON ATOMS PROVIDE THE OPPORTUNITY FOR STEREOCHEMICAL CHANGES WHICH CAN BE BROUGHT ABOUT DURING DIAGENESIS.

In most α-amino acids the only chiral centre is that of the α carbon, which bears the amino and carboxyl groups and for which racemisation is usually fairly rapid. Several families of acyclic isoprenoid are known, e.g. with phytane there are three chiral centres in the molecule which may be epimerised, giving rise to a mixture of eight possible stereoisomers. In the steranes, where R = methyl (C_{28}) or ethyl (C_{29}), carbons 20 and 24 epimerise slowly during diagenesis. For the hopanes, there are three centres which can be followed in their epimerisation; carbons 17, 21 and 22 (22 only where R = methyl). The example of homohopane is discussed in more detail in Figures 11-13.

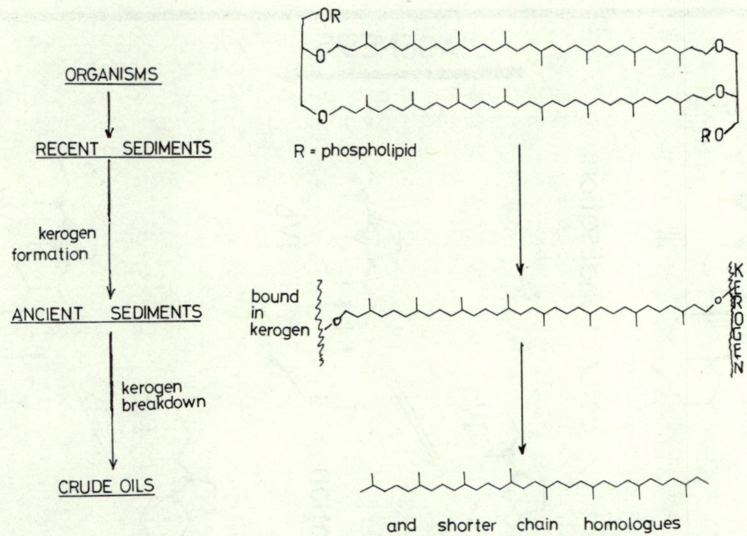

Figure 10. POSSIBLE FATE IN THE SEDIMENTARY COLUMN OF GLYCERYL ETHER PHOSPHOLIPIDS PRESENT IN ARCHAEBACTERIA.

Archaebacteria have been shown to contain unusual phospholipids as membrane constituents. These biolipids consist of acyclic isoprenoid chains bound at each end by ether links to glycerol moieties bearing a phospholipid side chain. Such dibiphytanyl glyceryl ether molecules are relatively stable and persist in Recent and older sediments. Indeed, in sediments as old as the Cretaceous they may survive, bound into the kerogen, from which they can be released in the laboratory through the action of ether cleavage reagents, such as BBr_3. For example, work at Strasbourg with kerogen isolated from the Messel shale (Eocene) has provided proof of the presence of ether-bound C_{40} acyclic isoprenoid skeletons with the appropriate head-to-head linkage of C_{20} units, presumably reflecting archaebacterial activity in Eocene times. Furthermore, diagenetic/catagenetic reactions over geological time must also cleave these kerogen-bound ethers, for the corresponding acyclic isoprenoid alkanes have been detected in many petroleums.

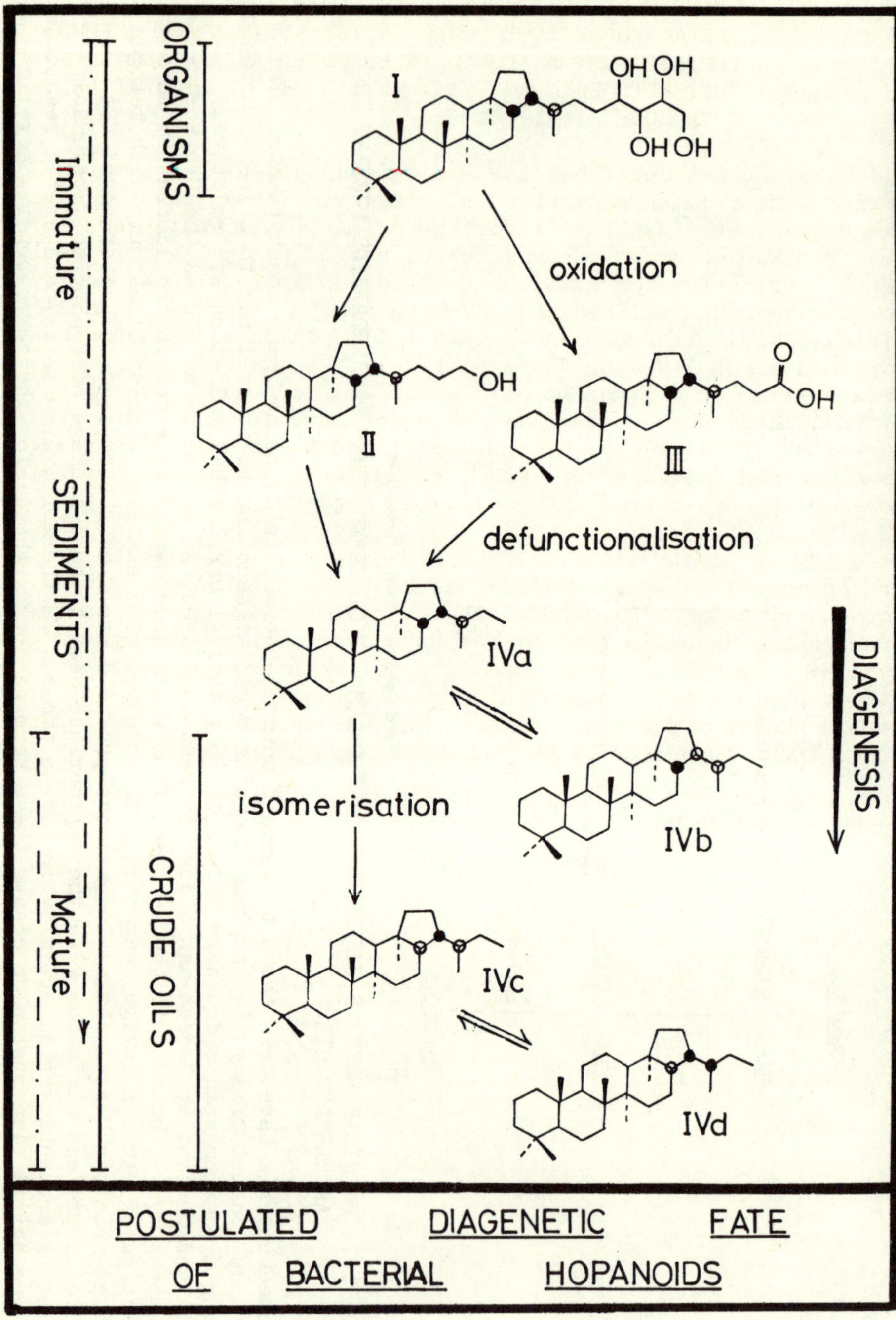

POSTULATED DIAGENETIC FATE OF BACTERIAL HOPANOIDS

Figure 11. EXAMPLE OF POSTULATED DIAGENETIC FATE. FATE OF BACTERIAL HOPANOID TETROLS IN THE SEDIMENTARY COLUMN. (DETAILS MAY BE FOUND IN BRASSELL AND EGLINTON, IN "HETEROTROPHIC ACTIVITY IN THE SEA", NATO ARI (ED. JOHN HOBBIE), 1982.)

The scheme describes the geological fate of tetrahydroxybacteriohopane (I), as understood at present. The postulated early-stage diagenetic pathways involving defunctionalisation are inferred from the observed occurrences of hopanoids in immature sediments. Oxidative degradation of tetrahydroxybacteriohopane (I) would give rise to 17β(H),21β(H)-bishomohopan-32-ol (II) and 17β(H),21β(H)-bishomohopanoic acid (III), which are the dominant hopanoid alcohol and acid, respectively, in immature sediments. Decarboxylation of the C_{32} acid would be expected to generate 17β(H),21β(H)-homohop-29(31)-ene, which may, in turn, be a precursor of 17β(H),21β(H)-homohopane (IVa), the major triterpane of immature sediments. The 17β(H),21β(H) configuration in hopanoids, however, is not thermally stable; it appears, initially, to undergo diagenetic isomerisation to 17β(H),21α(H) (IVb), and at higher maturity levels to 17α(H),21β(H) (IVc and d). These geochemical processes reflect the relative thermodynamic stability of the three series (αβ > βα > ββ). The overall scheme represents one prominent diagenetic pathway for hopanoids; there are others that lead, for example, to the formation of aromatic components. Despite the diagenetic changes discussed above, much of the hopanoid carbon skeleton remains unaltered and can therefore be recognised as an input from microbes to oils and mature sediments.

hydrocarbons had been known for some years as ubiquitous constituents of sediments and crude oils. The limited occurrence of suitable biolipid precursors presented somewhat of a puzzle until the recognition of polyhydroxybacteriohopanes as components of blue-green algae (cyanobacteria) and bacteria. Hopanoids (Figs. 11-13) have now been shown to be widely distributed among bacteria, probably functioning as membrane rigidifiers for these procaryotes in much the same way as sterols act in eucaryotes. These hopanoids include C_{35} compounds that possess several hydroxyl groups (typically 4 or 5) in their "extended" side chain. However, extended hopenoids were known from the evidence of chemical fossils prior to their recognition in extant living biota. Over one hundred derivatives, including alkanes, acids, ketones and alcohols, of bacteriohopanes have so far been isolated either from sediments or from microorganisms. This is a convincing argument in favour of the microbiological origin of at least part of the organic matter of sediments.

The widespread occurrence of both hopanes and isoprenoid alkanes in petroleums testifies to the importance of the role of microbial organic matter as an initial ingredient of sediments that later give rise to crude oils. This record seems to extend back well into the Precambrian and is remarkably general over both space and time, presumably reflecting the successful exploitation of the aquatic sedimentary environment by microorganisms.

The interpretation of hopanoids as molecular markers of microbial contributions to sediments stemmed initially from consideration of their geological occurrence. A better appreciation and understanding of microbial lipid compositions should lead to the search for further key compounds in geological materials. Indeed, there is much to be done in the study of microbial lipids. Many microorganisms have yet to be examined for their lipid composition and this is especially true of microorganisms which inhabit niches in the sedimentary column. We need, if possible, to be able to distinguish between the lipids of living microorganisms and those of dead or fossil populations. Thus, there is considerable controversy at the present time as to the depth in the sedimentary column to which the microbial biosphere extends. There are suggestions that slow diagenesis through anaerobic fermentation of residual organic matter, such as kerogen and resistant lipids, continues to great depths under the action of small populations of appropriate fermentative microorganisms. Others contend that these changes are brought about through non-biochemical reactions mediated by the minerals and the pore waters. Microbial debris is not easy to discern within the sediment matrix and the best geological record of microbial activity may well lie in the molecular residues. Following up this theme, if diagenetic changes can be documented properly, then it is conceivable that the molecular fossil record could

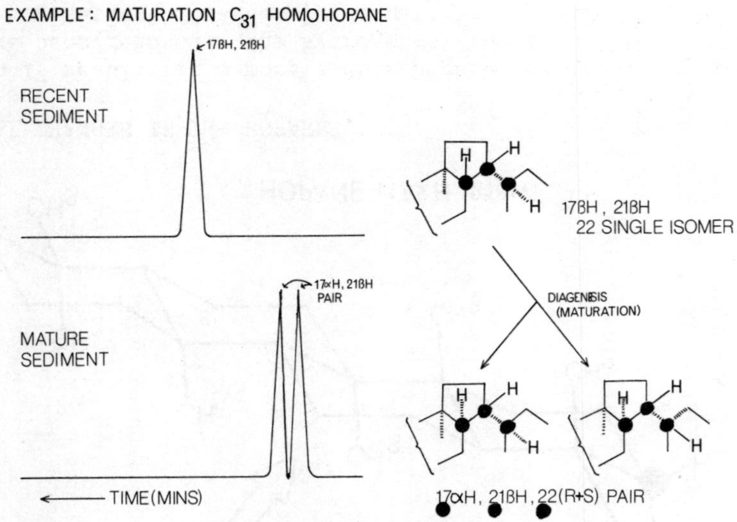

Figure 12. DIAGENETIC FATE OF THE C_{31} TRITERPANE, HOMOHOPANE.

The C_{31} triterpane hydrocarbon, homohopane, is found in many young bottom sediments. It is believed to be an early diagenetic product of higher functionalised hopanoids, such as tetrahydroxybacteriohopane (Fig. 11). It is found to have the biosynthetically derived 17β(H),21β(H) configuration and is hence seen as a single peak in capillary GC-MS analysis. It is not thermally stable and in ancient mature sediments, the C_{31} homohopane is found as a GC-MS doublet shifted to shorter retention time than the original 17β(H),21β(H) isomer. This doublet has been found to be the 17α(H), 21β(H) stereoisomer, present as approximately equal amounts of the 22<u>R</u> and 22<u>S</u> diastereoisomers. With this particular molecular fossil, the progress of diagenesis can be followed by the disappearance of the 17β(H),21β(H) single peak and the appearance of the 22<u>R</u> and <u>S</u> doublet of the 17α(H),21β(H) isomer. The extent of this conversion can be used as a measure of the burial history of the sediment.

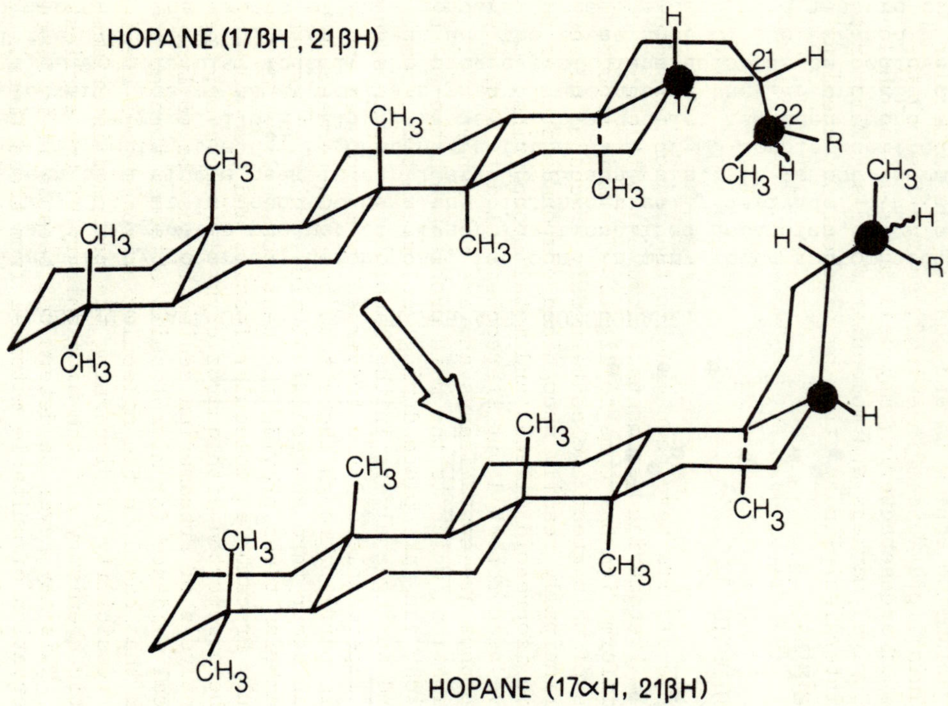

Figure 13. STEREOCHEMICAL CHANGES IN THE HOPANES.

Epimerisation at carbon 17 results in a more stable hopane skeleton, in which certain stereochemical interactions have been reduced. The hydrogen at the 17-position moved from the β face to the α face, while the epimerisation of carbon 22 results in approximately equal amounts of the two diastereoisomers, the mass spectra of which are virtually identical.

provide some indication of the evolution of microorganisms, at least in terms of their lipid components. A converse approach which can be expected to operate over the relatively short time span of the Quaternary and the Tertiary is to view the patterns of fossil lipids as environmental signatures characteristic of particular depositional environments. A full description of a sedimentary environment must include the microorganisms inhabiting it and, by inference, their lipids. These in turn represent the obligate biomolecular machinery and structural lipids evolved by the microorganisms to cope with and control the particular environment. This lipid "bioengineering", as typified by the membrane lipids of archaebacteria, may well underlay the whole concept of biological markers and molecular fossils. This area needs further interdisciplinary study by biochemists, chemists and microbiologists.

PRECAMBRIAN SEDIMENTS

Considerable efforts have been made in the search for molecular fossils in Precambrian sediments. However, little non-controversial information has come to light. The main problem is that most rocks of this great age contain only traces of molecular fossils, principally readily recognised compounds such as n-alkanes, acyclic isoprenoid alkanes and metalloporphyrins. α-Amino acids have been reported but stereochemical data indicate that these compounds are contaminants and are not syngenetic. Indeed, contamination, which is the main problem with the soluble geolipid fraction, can take place during geological time, sampling in the field and during analysis in the laboratory. Rocks such as cherts are not impermeable and are traversed by micropores and cracks and it is partly through these that the slow permeation of contaminants can occur through geological time. Mineral crystals formed in Precambrian times provide the best possibility for the analysis of truly Precambrian molecular fossils and some work of this type has been reported with quartz crystals. However, the main aim for analysis of Precambrian sediments must be to link the morphological evidence for microfossils obtained through the microscope with that based on the molecular approach. Of special concern is the possibility of discerning the presumed interface between prebiotic chemistry and the biochemistry of the earliest forms of life. This may never be determined because it is dependent upon preservation of suitable material from those times. If sediments have survived without being heated much above ambient temperatures, then there is the possibility of retention of molecular information, as the strength of the carbon-carbon bond is such that hydrocarbons have an estimated half life in the region of 10^9 years. Of the greatest importance for the origin of life studies must be increased attention to the lipids of microorganisms since the principal processes taking place in Precambrian times would have been microbial photosynthesis, heterotrophy and chemoautotrophy.

The best hope for successful preservation of molecular information from the early Precambrian lies in the insoluble organic debris entombed in the rocks. However, much of this organic debris, some of it associated with the morphological remains of microorganisms, has progressed far along the diagenetic/catagenetic pathway and is substantially carbonised or graphitised and therefore lacking in molecular information. Suitable, well-preserved samples may yet be secured but improved chemical techniques for the degradation of such ancient kerogens will be needed to release molecular fragments for detailed analysis.

Undoubtedly, what is needed is a determined, interdisciplinary attack on the molecular record for the Precambrian, which should include samples selected over the full time span from the present back to the Precambrian. These samples should be such that the burial history is understood with the main variable being time rather than diagenetic/catagenetic influences. Such a time sequence of samples which have undergone progressively greater diagenesis/catagenesis should be selected for a single sedimentary facies, such as the lithified cyanobacterial mats known as stromatolites. It will be important to validate the microbial nature of the sedimentary structure by detailed microscopic examination. This work should be paralleled by laboratory experiments designed to distinguish the effects of thermal alteration. Other important developments would include the ability to determine the molecular fossils released directly from a microscopic site (e.g. a few μm^3) in a specimen, e.g. through bombardment techniques that would release molecules or molecular fragments direct into a mass spectrometer. Such techniques have shown promise but extensive control experiments are required for this work to be substantiated and extended.

There can be no guarantee that we can eventually solve such questions as:- firstly, the validity of the molecular approach for microbial ecosystems of Precambrian age, or, secondly, the related question as to whether or not syngenetic and non-syngenetic molecules can be reliably distinguished. Certainly, the delineation of the pre-biogenic/biogenic interface in the Earth's history is an outstanding and worthy aim for molecular organic geochemists. Work of this type has additional and parallel significance in relation to the general problem of life detection at the surface of a planet. To establish a firm chronology for the molecular imprint of life on Earth would be to erect an important milestone in our understanding of the Universe.

CONCLUSIONS

The study of molecular fossils - molecular organic geochemistry - has made substantial strides in recent years. The geological fate of some biolipids is now partially understood. The process of diagenesis and catagenesis requires much more study but it is now possible to explore the use of molecular fossils as tools for the study of past environments. The approach is based on two main concepts: first, that of using single compounds as markers of particular biological inputs and second, that of using summation data for selected groups of compounds, which can then be ratioed to provide a different input assessment. Uses include the assessment of inputs, the extent of reworking, palaeogeography, water column productivity, bottom depositional conditions and, especially, the characterisation of events. "Events" refer to oxic/anoxic transitions, climatic changes and marine transgressions.

Other applications for molecular fossils include the assessment of basin history and prospects for oil exploration. The latter are again based on the presence or absence of specific compounds and compound classes but also on the use of the molecular approach to determine diagenetic and catagenetic alteration and hence the burial history of the sediments. Applications here are in the history of the geothermal gradient in particular areas, including faulting and subsidence and the correlation of oils with each other and with source rocks. The maturation aspects include the basin evaluation approach and the twin questions of migration of hydrocarbons and biodegradation of hydrocarbons by populations of microorganisms.

As molecular tools, geolipids are somewhat variable in their applications as is indicated in Table 4. This table is an oversimplification but gives some idea of the present situation.

In the future, molecular organic geochemistry must look first to carefully selected suites of sediment samples which carry detailed information from other disciplines, e.g. in relation to lithology, facies, age, thermal history and depth of burial. The wealth of organic geochemical data now emerging will need to be entered in computerised data bases, along with sedimentological, palaeontological and other appropriate information. Further processing will involve data manipulation by statistical programs as well as by the more direct subjective techniques for establishing correlations. All of this will need to rely on the continued development of the fundamental, molecular organic geochemistry, through improved analytical methods for the molecular fossils themselves and through the study of the reactions which are taking place in the sediments. The synthetical organic chemist and the physical organic chemist have much to offer in

TYPE	CARBON NO. RANGE	DEPOSITIONAL ENVIRONMENT	BURIAL HISTORY	BIODEGRADATION
STRAIGHT CHAIN	12-40	+	++	+++
BRANCHED CHAIN	14-33	+	+	++
ACYCLIC ISOPRENOID	14-40	++	+	+++
STEROID	19-30	+++	+++	+
TRITERPENOID	27-35	++	+++	++
PORPHYRIN	26-33	++	+++	?

The symbols (+, ++, +++) give an indication of the values of each lipid type in the assessment of depositional environment, burial history and the extent of petroleum biodegradation.

TABLE 4. GEOCHEMICAL SIGNIFICANCE OF LIPID TYPES.

this essentially interdisciplinary field of scientific endeavour.

ACKNOWLEDGEMENTS

I thank the Natural Environment Research Council for their continuing support of the C-GC-MS facilities and of the research projects at the University of Bristol, and also the National Aeronautics and Space Administration for their sub-contract (NGL-05-003-003) through the University of California, Berkeley. My thanks go to Dr. S.C. Brassell, Mrs. Ann Gowar and other colleagues in the Organic Geochemistry Unit at Bristol for assistance and helpful discussions, and also to Sue Trott and Pam Eglinton for the onerous task of "the typing" and Tim Eglinton for some of the Figures.

BIBLIOGRAPHY

1. Calvin, M. "Chemical Evolution": 1969, Oxford Univ.Press, London.
2. "Advances in Organic Geochemistry 1979" (eds. A.J. Douglas and J.R. Maxwell): 1980; Pergamon Press, Oxford, and other volumes in the "Advances in Organic Geochemistry" series.
3. "Organic Geochemistry: Methods and Results" (eds. G. Eglinton and M.T.J. Murphy): 1969, Springer Verlag, Berlin.

4. Maxwell, J.R., Pillinger, C.T. and Eglinton, G. "Organic Geochemistry": 1971, Quarterly Reviews 25, pp.571-628.
5. McKirdy, D: 1974, Precambrian Research 1, pp.75-137.
6. "Environmental Chemistry" Vol.1 (Senior Reporter, G. Eglinton): 1974, Specialist Periodical Reports, The Chem.Soc., London.
7. Philp, R.P., Maxwell, J.R. and Eglinton, G. "Environmental Organic Geochemistry of Aquatic Sediments": 1976, Science Progress (Oxford) 63, pp.521-545.
8. Papers from the Symposium on "Concepts in Marine Organic Chemistry", Edinburgh, Sept.1976: 1977, Mar.Chem. 5.
9. "Petroleum Formation and Occurrence" (eds. B.P. Tissot and D.H. Welte): 1978, Springer Verlag, Berlin.
10. Hunt, J.M. "Petroleum Geochemistry and Geology": 1979, Freeman Co., San Francisco.
11. Ourisson, G., Albrecht, P., Rohmer, L.: 1979, Pure and Applied Chemistry 51, pp.709-729.
12. Chappe, B., Michaelis, W. and Albrecht, P.: 1980, in "Advances in Organic Geochemistry 1979", Pergamon Press, Oxford.
13. Edmunds, K.L.H., Brassell, S.C. and Eglinton, G. "The organic geochemistry of benthic microbial ecosystems": 1982, in "Mineral Deposits and the Evolution of the Biosphere" (eds. H.D. Holland and M. Schidlowski), pp.31-50, Springer Verlag, Berlin.
14. Mackenzie, A.S., Brassell, S.C., Eglinton, G. and Maxwell, J.R. "Chemical Fossils - the geological fate of steroids": 1982, Science, in press.

This bibliography provides a brief list of articles and texts which should enable the reader to follow up some of the material outlined in the article.

APPENDIX

Cosmochemistry and the origin of life

Geoffrey Eglinton, Ann Henderson-Sellers and
Stephen Moorbath

At a recent NATO conference* on 'Cosmochemistry and the origin of life,' Ponnamperuma (University of Maryland) summed up the overall picture by showing that there is a continuity stretching from space through to the present in terms of the formation of organic molecules and those that are found in present-day living organisms, and held that this is the natural course of events. Viola (Indiana University) started with the 'big bang' and then discussed nucleosynthesis and the abundance of the elements, particularly those essential to life. He also discussed the life cycle of stars and models of the Universe - salutary topics which should be compulsory course work for all scientists at some stage in their careers. Greenberg (University of Leiden) and Irvine (Chalmers University, Sweden) continued with details of the nature, composition and distribution of the interstellar dust and gas medium, and organic molecules and most basic raw materials of life known to exist in interstellar space. Conference participants thought it unnecessary, at this state of our knowledge, to invoke Panspermia - the seeding of our planet, at any stage in its history, by life from extraterrestrial sources. In this much publicized context the conference generated neither science fiction nor science friction.

NASA's spaceship Enterprise, ably piloted by de Vincenzi (NASA Headquarters), took us on a tour of the Solar System, culminating in an account of the Mars Lander's biological experiments. Initial optimism and excitement soon gave way to the more sober realization that surface oxygen release could be better accounted for by presence of labile, active inorganic oxides in martian soil than by biological activity. Prospects for contemporary life on Mars are bleak, but perhaps in its orginally airer and wetter manifestation it might have temporarily harboured the seeds of early life. The rest of the Solar System is decidedly no place for protoplasm!

The conference then rapidly came down to Earth. However improbable it seems, life and mind flourish, and some 4 billion years separate the earliest replicating organic molecules from the deliberations of the present conference. After presenting an inventory of terrestrial surface carbon, Henderson-Sellers (University of Liverpool) noted that, without its hydrosphere and biosphere, the Earth would be very similar to Venus, where carbon dioxide is not tied up in carbonate rocks. She, as well as Ponnamperuma, summarized much of the evidence that the atmosphere of the early Earth was dominated by carbon dioxide, water, nitrogen and possibly some carbon monoxide, and not by ammonia, methane and hydrogen as was once widely believed. The bulk of the hydrosphere was undoubtedly present from very early prebiotic times. Ample energy sources, including electrical discharges, ultraviolet radiation, radioactivity and volcanic/geothermal heat, were available on the early Earth to foster complex, prebiotic organic syntheses from relatively simple mixtures of volatiles and gases, with clay minerals as a likely base for the earliest biotic processes.

Moorbath (University of Oxford) showed that isotopic dates on early biological or molecular fossils could frequently be obtained from stratigraphically closely associated igneous rocks, and more rarely from the sediments themselves. Algal stromatolites and other simple cellular structures were known from sediments probably as old as 3.5 billion years, although in some cases there is a worrying stratigraphic gap between the sediment and the dated igneous rock. Whether the Isua metasediments of West Greenland, dated at about 3.8 billion years, contain biogenic markers is still a matter of controversy. Schidlowski (Max-Planck-Institut, Mainz) is convinced from detailed carbon isotope studies that Isua carbonaceous sediments represent a 3.8 billion year-old record of biological activity. There is a continuity of the C record from the present back to Isua tiems, and yet the Isua rocks also exhibit a marked shift in C which may result from their strong metamorphic overprint. Indeed, some workers claim that the C record of metamorphism alone. Walters (University of Maryland), after painstakingly pyrolizing 27 Isua graphite samples, observed no trace of biogenicity, with one possible exception. The metamorphism was certainly intense enough to erase any biogenic evidence.

Ochiai (University of Maryland) pointed out that bioinorganic chemistry was essential for understanding chemical evolution. Many trace metals, such as Fe, Cu, Mo, Mn, Co, Zn, Ni, Cr, V, form complexes which catalyse organic reactions considered essential for life processes. He also discussed the problem of the major Precambrian Banded-iron Formations, ranging in age from about 3.8 to 2.2 billion years, and regarded by many as biogenic markers in the sense that their precipitation may have required oxygen released from marine organisms, in the virtual absence of atmospheric oxygen. This "iron age" was followed at about 1.5 billion years onwards by copper sulphide deposits - the "copper age." Before this, prevailing redox conditions would not permit copper to appear in a dissolved, mobile state. A tentative correlation might be made between the sudden appearance of the copper deposits and the distribution of copper enzymes in present-day organisms. In this view, the copper superoxide dismutase is a later and more efficient development than the manganese and iron superoxide dismutases.

From the organic record in more recent sediments, Eglinton (University of Bristol) produced abundant evidence that molecules can retain a high degree of structural and stereochemical specificity over hundreds of millions of years if they are not exposed to raised temperatures or subject to oxidation or microbial attack. Specific biological marker compounds, or 'chemical fossils,' can be interpreted in terms of the palaeoenvironment. The thermal history of a sediment can also be understood in terms of epimerization, aromatization and carbon-carbon bond cleavage reactions observed in samples with different burial histories. Of particular interest is the possibility of relating the molecular lipid record to display unusual membrane lipids which may provide a clue to their activity during the Precambrian.

All participants agreed that there is an urgent need to continue the search for ancient, unmetamorphosed sediments from the earliest accessible Precambrian, so that micropalaeontologists, inorganic and organic geochemists, clay mineralogists, geologists and geochronologists can work closely together towards a progressive characterization, through geological time, of the biological record in the rocks. Recent geological work clearly shows that the earliest known (about 3.8-3.5 billion years) sedimentary and volcanic rocks characterize a depositional environment which is totally compatible with conditions now widely regarded as favourable for the earliest life on Earth. Of course, one big question was forcibly put to us by the students - "when did life begin?" The best bet was 4.0 (plus or minus 0.1) billion years ago. The origin of life could be even older, but most people agreed that it was unlikely to be younger.

Altogether, it was a creative and original Study Institute, which looked as much into the future as into the past. Is it just conceivable that a conference on the same topic is at this moment being held somewhere else up there in the night sky?

PARTICIPANTS

NATO/Advanced Study Institute
"Cosmochemistry and the Origin of Life"
Maratea, Italy
June 1 - 12, 1981

Blair S. Allen
Science Department
Harcum Junior College
Bryn Mawr, Pennsylvania 19010

Ulker Bakir
41 Sok 5/14
Bahcelievler-Ankara
Turkey

Patrizia Barone
Department of Chemistry
Georgetown University
Washington, D.C. 20057

Barbara Blair
Department of Chemistry
Sweet Briar College
Sweet Briar, Virginia 24595

V. M. Canuto
2800 Broadway
NASA/Goddard Institute for
Space Studies
New York, New York 10025

Donald DeVincenzi
Life Sciences Division
NASA Headquarters
400 Maryland Avenue, S.W.
Washington, D.C. 20546

Geoffrey Eglinton
University of Bristol
School of Chemistry
Cantock's Close
Woodland Road
Bristol BS8 ITS
United Kingdom

Gozen Ertem
University of Istanbul
Faculty of Science
Vezneciler, Istanbul

Maurizio Fedi
V. Luco-Giordano 13
80127 Napoli, Italy

Paolo Gasparini
Instituto di Geologia e Geofisica
Univsita di Napoli
Osservatorio Vesuviano
Ercolano, Napoli, Italy

Casimir Gnanadickam S. J.
Curia Praepositi Generalis S. J.
C. P. 6139, 00195 - Rome
Italy

J. Mayo Greenberg
University of Leiden
Leiden
The Netherlands

Pat Groff
Laboratory of Chemical Evolution
University of Maryland
College Park, Maryland 20742

Juergens Hahn
Max-Planck-Institut fur Chemie
(Otto-Hahn-Institut)
Postfach 3060
D-6500 Mainz, F.R.G.

A. Henderson-Sellers
Department of Geography
University of Liverpool
P.O. Box 147
Liverpool, United Kingdom

William M. Irvine
Astronomy
GRC Tower B
University of Massachusetts
Amherst, Massachusetts 01003

Antonio Lazcano-Araujo R.
Instituto de Astronomia, UNAM
Apdo. Postal 70-264
Cd. Universitaria
Mexico 20, D.F., Mexico

Susan Leschine
Department of Microbiology
University of Massachusetts
Amherst, Massachusetts 01003

Richard Monier
c/o Dr. Francois Raulin
Lab. Physico Chimie de l'Environnement
Universite Paris Val de Marne
Av. General de Gaulle
94010 Creteil Cedex
Paris, France

Stephen Moorbath
Department of Geology and Mineralogy
University of Oxford
Parks Road
Oxford OX1 3PR, England

William E. Moore
Department of Chemistry
Norfolk State University
2401 Corprew Avenue
Norfolk, Virginia 23504

Sharon Norris
Department of Earth Sciences
University of Cambridge
Downing Street
Cambridge CB2 3EQ, England

Ei-Ichiro Ochiai
Department of Chemistry
Juniata College
Huntingdon, Pennsylvania 16652

Massimo Olivucci
Viale Dell'Appennino 52
47100 Forli, Italy

Cyril Ponnamperuma
Laboratory of Chemical Evolution
Department of Chemistry
University of Maryland
College Park, Maryland 20742

Joaquim Moura Ramos
Centro de Quimica-Fisica Molecular
Complexo Interdisciplinar
Instituto Superior Tecnico
Av. Rovisco Pais
1000 Lisboa, Portugal

David Randall
Department of Chemistry
Open University
Walton Hall
Milton Keynes
Bucks, England

Sherman Randerson
Biology Department
University of Wisconsin
Oshkosh, Wisconsin 54901

Manfred Schidlowski
Max-Planck-Institut fur Chemie
(Otto-Hahn-Institut)
Postfach 3060
Saarstrasse 23
D-6500 Mainz, F.R.G.

Yoshihiro Shigemasa
Department of Industrial Chemistry
Faculty of Engineering
Tottori University
Tottori, Japan 680

Raffaele Stella
V. Kerbaker 86
80129 Napoli, Italy

R. Suresh
Department of Geology
Central College
Bangalore 560 001, India

Vic Viola
Department of Chemistry
Chemistry Building
Indiana University
Bloomington, Indiana 47405

Clifford Walters
Laboratory for Chemical Evolution
Department of Chemistry
University of Maryland
College Park, Maryland 20742

John Wright
Department of Physical Sciences
Southeastern Oklahoma State University
Durant, Oklahoma 74701

Betul Yeter
Asagi Ayranci
Menevis sok. 41-10
Ankara, Turkey

Index